Primary Mathematics
Capitalising on ICT for today and tomorrow

Second edition

Now in its second edition, *Primary Mathematics* continues to provide a comprehensive introduction to teaching and learning mathematics in today's classrooms. With links to both the Australian and New Zealand curriculums, this book covers the core learning areas of measurement, space and geometry, early number concepts, data and statistics, chance and probability, and patterns and algebra.

Primary Mathematics encourages the integration of technology into a pedagogically sound learning sequence for primary mathematics and provides teachers with detailed activities and examples to cater for the mathematical needs of all learners. The book comes with free 12-month access to Cambridge HOTmaths, a popular, award-winning online tool with engaging multimedia to help students and teachers learn and teach mathematics concepts.

In this edition, the authors emphasise the Proficiencies and General Capabilities of the Australian Curriculum, and provide comparable references to the New Zealand Curriculum – including cultural competencies, teaching as inquiry, and numeracy project curriculum links, which are all firmly embedded in teaching mathematics in New Zealand.

Each chapter features:

- definitions of key terms
- classroom activities, including HOTmaths activities
- 'pause and reflect' points
- classroom vignettes
- suggested further reading and websites for exploration.

Grounded in evidence and linked to practical classroom activities, *Primary Mathematics* is an indispensable resource for pre- and in-service teachers who wish to integrate contemporary technology into teaching and learning key mathematical concepts.

Penelope Serow is Associate Professor of Mathematics Education at the University of New England. In 2009 she received the Vice Chancellor's Teaching Excellence award. She is actively involved in research in the areas of primary, secondary and pre-service mathematics education.

Rosemary Callingham is Associate Professor in Mathematics Education at the University of Tasmania. She has worked on projects in Hong Kong and North Korea, as well as in many parts of Australia, and has been a Chief Investigator on several large research grants.

Tracey Muir is Senior Lecturer in Mathematics in the Faculty of Education at the University of Tasmania. She is also a member of the Mathematics Association of Tasmania, Australian Association of Mathematics Teachers, National Council of Teachers of Mathematics and the Mathematics Education Research Group of Australasia.

Primary Mathematics

Capitalising on ICT for today and tomorrow

Second edition

PENELOPE SEROW,
ROSEMARY CALLINGHAM
AND TRACEY MUIR

CAMBRIDGE
UNIVERSITY PRESS

477 Williamstown Road, Port Melbourne, VIC 3207, Australia

Cambridge University Press is part of the University of Cambridge.

It furthers the University's mission by disseminating knowledge in the pursuit of education, learning and research at the highest international levels of excellence.

www.cambridge.org
Information on this title: www.cambridge.org/9781316600757

© Cambridge University Press 2016

This publication is copyright. Subject to statutory exception and to the provisions of relevant collective licensing agreements, no reproduction of any part may take place without the written permission of Cambridge University Press.

First edition 2014
Reprinted 2015
Second edition 2016

Cover designed by Leigh Ashforth, watershed art + design
Typeset by Newgen Publishing and Data Services
Printed in Singapore by C.O.S. Printers Pte Ltd

A catalogue record for this publication is available from the British Library

A Cataloguing-in-Publication entry is available from the catalogue of the National Library of Australia at www.nla.gov.au

ISBN 978-1-316-60075-7 Paperback

Reproduction and communication for educational purposes
The Australian *Copyright Act 1968* (the Act) allows a maximum of one chapter or 10% of the pages of this work, whichever is the greater, to be reproduced and/or communicated by any educational institution for its educational purposes provided that the educational institution (or the body that administers it) has given a remuneration notice to Copyright Agency Limited (CAL) under the Act.

For details of the CAL licence for educational institutions contact:

Copyright Agency Limited
Level 15, 233 Castlereagh Street
Sydney NSW 2000
Telephone: (02) 9394 7600
Facsimile: (02) 9394 7601
E-mail: info@copyright.com.au

Cambridge University Press has no responsibility for the persistence or accuracy of URLs for external or third-party internet websites referred to in this publication and does not guarantee that any content on such websites is, or will remain, accurate or appropriate.

© Australian Curriculum, Assessment and Reporting Authority (**ACARA**) 2009 to present, unless otherwise indicated. This material was downloaded from the ACARA website (www.acara.edu.au) (**Website**) (accessed as indicated) and was not modified. The material is licensed under CC BY 4.0 (https://creativecommons.org/licenses/by/4.0/). ACARA does not endorse any product that uses ACARA material or make any representations as to the quality of such products. Any product that uses material published on this website should not be taken to be affiliated with ACARA or have the sponsorship or approval of ACARA. It is up to each person to make their own assessment of the product.

Contents

How to use HOTmaths with this book · page ix

Chapter 1 Teaching mathematics today with tomorrow in mind
    in the primary setting · 1
    The TPACK framework   4
    Summary of chapters   9
    How to use this book   11
    Websites for exploration   12

Chapter 2 Exploring early number concepts   13
    Early number concepts   15
    Number frameworks   16
    Linking with curriculum documents   17
    Early number activities and strategies   21
    Extending counting beyond 10   23
    Linking numbers with quantities   26
    Subitising   26
    Early operations with number   33
    Mental computation strategies   35
    Drill and practice   37
    Alternatives to traditional drill and practice   39
    Further reading/websites for exploration   44

Chapter 3 Exploring measurement   46
    Learning sequence for measurement   47
    Measurement topics   51
    Measuring with non-standard units   65
    Measurement misconceptions and difficulties   67
    Extending measurement concepts into older years   74
    Reflection   81
    Websites for exploration   81

Chapter 4 Exploring geometry — 82
   Geometrical concepts — 83
   Taking geometry outside the classroom — 85
   Theoretical framework — 86
   Geometry in the primary classroom — 92
   The van Hiele teaching phases — 112
   Reflection — 119
   Websites for exploration — 120

Chapter 5 Exploring whole number computation — 121
   Operations with whole numbers — 123
   A word about calculators — 133
   Drill and practice — 135
   Reflection — 137
   Further reading/websites for exploration — 137

Chapter 6 Part-whole numbers and proportional reasoning — 140
   Some background — 142
   Proportional reasoning — 153
   Reflection — 157
   Further reading — 157

Chapter 7 Exploring patterns and algebra — 159
   Linking with curriculum — 160
   Pattern and structure — 161
   Developing an understanding of relationships — 168
   Equals and equivalence — 170
   Generalisation in upper primary — 178
   Reflection — 187
   Websites for exploration — 188

Chapter 8 Exploring data and statistics — 189
   The importance of understanding statistics — 190
   Asking questions (problem) — 193
   Collecting and recording data (plan, data) — 194
   Analysing and representing data (analyse) — 200
   Telling a story from the data (conclusions) — 209
   Using data in social contexts — 212
   Reflection — 213
   Websites for exploration — 214

Chapter 9 Exploring chance and probability — 215
   Why is probability important? — 216
   Understanding probability — 217
   Early primary years — 220

Middle primary years	222
Upper primary years	226
Representing probability	234
Reflection	237
Websites for exploration	237

Chapter 10 Capitalising on assessment for, of and as learning — 239

Assessment	240
Assessment for learning	241
Quality of student responses	246
National testing	254
Online assessment item banks	262
Assessment as learning	265
Reflection	265
Websites for exploration	266

Chapter 11 Capitalising on ICT in the mathematics classroom — 267

Supporting students' learning in technology	270
Considering teacher knowledge	271
Developing an understanding of place value	271
Other suggestions for capitalising on ICT in the classroom	276
HOTmaths as a one-stop shop	278
Capitalising on ICT through an integrated unit	283
Pedagogical commentary	291
Reflection	293
Websites for exploration	294

Chapter 12 Diversity in the primary mathematics classroom — 295

Teachers' beliefs	298
Children with physical disabilities	299
Children with intellectual disabilities	305
Children with social, emotional and behavioural disabilities	309
Multicultural classrooms	311
Children who are gifted and talented	313
Reflection	314
Further reading/websites for exploration	315

Chapter 13 Surviving as an 'out of field' teacher of mathematics — 316

Shortage of secondary mathematics teachers and associated issues	317
Secondary lesson structures	322
Issues to think about in the secondary context within each strand	328
Sample secondary sequence	331
Reflection	336
Websites for exploration	337

Chapter 14 Teaching mathematics beyond the urban areas — 338
  Rural and remote areas — 339
  Being flexible — 340
  Considering classroom structures — 346
  National testing — 350
  Resourcing — 351
  Reflection — 357
  Websites for exploration — 358

Chapter 15 Planning and sustainability in the mathematics classroom — 359
  The TPACK framework — 360
  TPACK in the mathematics classroom — 362
  Planning for teaching mathematics with technology — 367
  Leading the teaching of ICT within your school — 369
  Professional reading and viewing — 374
  Website evaluation — 376
  Collating resources — 377
  Involving the community — 378
  Leadership, ICT and change — 380
  Reflection — 381
  Websites for exploration — 381

Glossary — 382
References — 388
Index — 401

How to use HOTmaths with this book

Once you have registered your HOTmaths access code, found on the inside front cover of this book, for subsequent visits the below navigation instructions provide a general overview of the main HOTmaths features used within this textbook.

Log in to your account via at www.hotmaths.com.au.

Upon logging in you will automatically arrive at your Dashboard. This screen offers you access to **FUNdamentals** (colourful maths games and activities for Foundation to Year 2 students), **Games** and the HOTmaths **Dictionary**. The Dashboard can also be accessed via the icon on the right-hand side of the toolbar at the top of any HOTmaths lesson page.

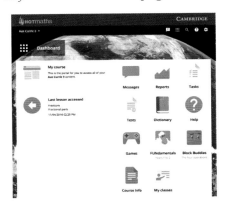

Different HOTmaths streams can be accessed via the Course list dropdown. You can change the **Course list** and **Course** (year level) using the dropdown on the left-hand side of the toolbar.

You can then select a **Topic**, and finally a **Lesson**.

Most lessons contain a number of interactive and printable activities, which can be accessed via the links on the right-hand side of the orange toolbar. These include: **Resources**, **Walkthroughs**, **Scorcher** and **Questions**.

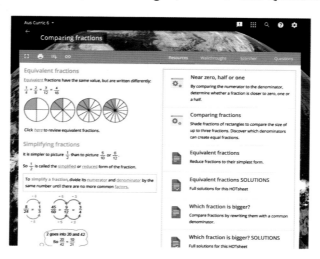

The Resources tab within lessons contain **widgets** (animations) and **HOTsheets** (activities). By clicking on the 'Comparing fractions', you will access the widget below. Clicking the 'Equivalent fractions' link will give you access to the HOTsheet below.

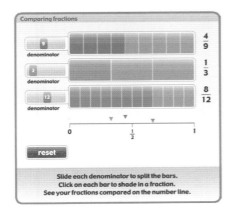
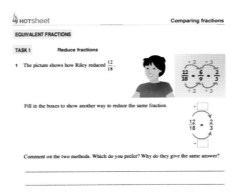

The toolbar at the top of each lesson page is also the location for the search function, where you can enter the name of any widget or HOTsheet for quick access. The results page will automatically display widgets based on the keywords searched – as denoted by the 'widget' tab being highlighted in blue. If you are looking for a HOTsheet, simply click onto the 'Hotsheets' tab and these results will appear. Using the above HOTsheet as an example, searching 'equivalent fractions' and clicking on the 'Hotsheets' results tab will provide you a link to the Equivalent Fractions HOTsheet.

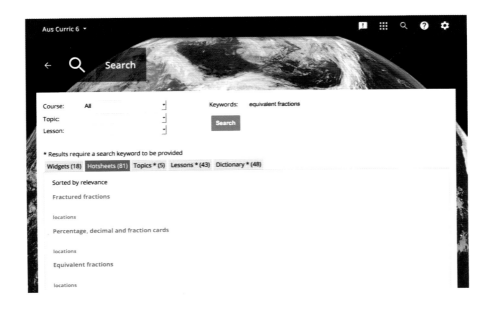

Throughout this textbook you will find numerous references to lessons and resources from three of the HOTmaths Course lists: **Cambridge Primary Maths**, **Australian Curriculum** and **HOTmaths Global**. Whenever these references occur, a margin icon will be present to denote which Course list you should be logged-in under in order to access the item. A plain flame icon denotes Cambridge Primary Maths, a flame with AC below it denotes Australian Curriculum and a flame with Global below it denotes HOTmaths Global.

Please note that given its nature HOTmaths is constantly being updated. All pathways and references are correct as of March 2016 and every effort has been made to provide you with an accurate picture of the functions within HOTmaths.

CHAPTER 1

Teaching mathematics today with tomorrow in mind in the primary setting

Mr Brookes held his grand-daughter's hand and proudly walked into her Year 4 classroom. It was Grand-Friends Day, and he was looking forward to seeing what Bonnie was achieving – particularly in mathematics, his favourite subject while at school. He remembered the feeling of achievement when his page was filled with a column of ticks.

Mr Brookes looked around the room, and already he was feeling a little uneasy. On one wall he saw some contextual addition problems with different techniques such as a jump method, split method and compensation method. He thought to himself, 'What is wrong with lining up the two numbers, beginning with the units column, and borrowing and pay-back?' Next to the strategies that were already disturbing him, he saw the formal strategy with which he was familiar; however, it was labelled with words such as 'trade'.

On closer inspection, he observed that the students had previously made their own metre rulers, were tracing around feet on grid paper and using string to find the perimeter, and were building as many different rectangular prisms as they could with 24 small cubes. There were group results on the wall of throwing dice and tabulating the results, and strips of paper used to find the average of the class by breaking the strips into pieces until they were approximately the same length. In another display, he saw patterns being explored on the 100 chart. He couldn't believe his eyes when he saw times tables presented next to rows and columns of dots.

Mr Brookes was about to ask Bonnie whether he could look in her exercise book when he noticed an electronic board of some kind on the wall where he expected to see the blackboard. On this board – which Bonnie called the interactive whiteboard – a group of children were predicting the shape that would be made when they cut an object at particular positions. After choosing the shape, they cut the object with a virtual knife and checked their solution. On another table, two children were sharing an iPad, creating as many different four-sided figures as they could, and were exploring everything else they could find out about the shapes. On the far side of the classroom, three children sat at computers and were entering the data about the area and perimeter of the feet in the class into a special program. When he turned to see what three other children were doing on computers at the back of the room, he noticed that they were completing a series of review questions against the clock.

The changes that had occurred in the classroom since Mr Brookes had visited his own children's classroom were undeniable. He could see that the students had been doing mathematics, but he was astounded by the variety of the concepts covered, how the tasks were accessible to all students, the level of engagement in the class and the use of concrete materials with computers and mobile technology. Bonnie proudly shared her achievements, but they weren't about the number of ticks in her maths book. Instead, she showed her mathematics discoveries and the interesting findings she had recorded electronically in her maths journal, stored on the school network.

Teachers are searching for a resource that will address issues surrounding the teaching of mathematics within the broad and ever-changing context of their particular classroom setting. To meet this need, this book is grounded in empirically evidenced developmental models and linked closely to practical classroom practice. While many classrooms have been resourced with equipment such as computers, interactive whiteboard (IWB) technology and mobile devices, extensive professional development is required to enable these pieces of technological equipment to be transformed into teaching tools. The difficulty faced by the teaching profession is in integrating a wide range of hands-on concrete materials with information and communication technology (ICT) to weave a pedagogically sound learning sequence. Technological change has occurred at an extremely fast pace, which means that many educators across the field, without sufficient development of skills in this area, have been forced into adopting teacher-centred techniques in an effort to use the provided technologies. This book provides the opportunity to meet this challenge and provide mathematics teachers with detailed teaching activities that are designed with developmental models as their basis. It does not pretend to deliver everything you need to know about teaching primary mathematics. Its focus is on the sensible and achievable integration of technology with research-based approaches to mathematical development that provide for the mathematical needs of all learners. As such, it is intended for primary pre-service teachers, as well as those teachers already working in the classroom who want to use a range of technologies in meaningful and educationally sound ways to improve their mathematics teaching.

In writing the book, we were guided by a moral imperative: we wanted students to be using engaging technologies in purposeful ways with teachers who were willing and able to use technology's power to enhance students' learning. As part of a YouTube clip that shows a vision of 'K–12' students today (<http://www.youtube.com/watch?v=_A-ZVCjfWf8>), a student laments that, 'At least once a week 14 per cent of my teachers let me create something new with technology, but 63 per cent never do.' It is hoped that this book will encourage you to teach school children to become successful learners who are prepared for and responsive to the dynamic demands of an ICT-rich future.

As the expectation of accreditation increases, teachers around the world are being asked to consider issues such as student diversity, behaviour management and assessment for learning techniques within the key learning areas. Another contentious – but very real – issue is the impact of external assessment 'of' learning, in the form of national testing. Mathematics teachers are asking, 'How can we use this information to enhance the mathematics learning in our classroom?'

In many countries, we have reached a crisis point in relation to staffing schools with mathematics teachers in rural and remote areas. The problem is not being addressed, and 'out of field' teachers of mathematics – such as primary-trained teachers working in the secondary setting – are in need of a tool kit to assist them to design student-centred mathematics activities. As in any teaching area, if the appropriate knowledge base is not strong, teachers resort to teaching in the manner in which they were taught in an effort to survive the situation. Australia and the wider Pacific region are not in a position to fill these vacancies with qualified teachers of mathematics. We need to accept this situation, and to consider ways of supporting 'out of field' teachers while the workforce in this area grows.

Currently, there is an extensive market for online mathematics resources. These span 'the good, the bad and the ugly'. With time constraints, teachers and pre-service teachers are often not in a position to make a critical assessment of the merits of those available. While a wide range of classroom resources will be explored in the following chapters, this book features the benefits for enhancing mathematical content knowledge and pedagogical content knowledge of the online mathematics resource HOTmaths (<http://www.hotmaths.com.au>) alongside other available online tools. Due to the generosity of the HOTmaths team and Cambridge University Press, university students and lecturers in teacher education programs across Australia have been provided with free access to the HOTmaths program. This text aims to take this a step further through the linking of HOTmaths tools to other tools in the mathematics teacher's tool kit, and presenting these to pre-service and practising teachers. To broaden pre-service and in-service teachers' access to HOTmaths, the purchase of this text includes six months' online access to HOTmaths, which supplements the pedagogical approaches explored in each chapter.

Throughout the remaining 14 chapters of this book, the contents of each chapter are informed by the TPACK framework described below.

The TPACK framework

Technological Pedagogical Content Knowledge (TPACK) is a framework that builds on Shulman's (1987) formulation of Pedagogical Content Knowledge (PCK); it describes how teachers' understanding of technology and pedagogical content knowledge interact with one another to produce effective teaching with technology (Koehler & Mishra 2008a, 2008b). As shown in Figure 1.1, there are seven components to the framework, which can be summarised as follows:

Technological knowledge (TK)

This refers to knowledge about various technologies, so is continually in a state of flux as technology constantly changes. A person with good technological knowledge would be able to:

- apply it broadly in their everyday life and at work
- recognise when it could be used to assist in the achievement of a goal
- continually adapt to changes in new technology.

Content knowledge (CK)

Content knowledge is knowledge about the actual subject-matter that is to be taught or learnt, and includes knowledge of concepts, theories and ideas,

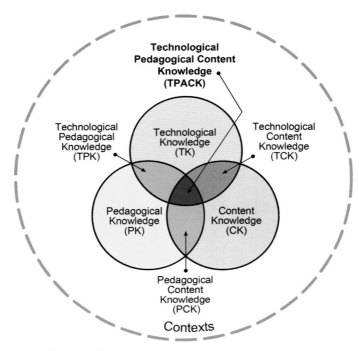

Figure 1.1 TPACK framework
Source: <http://www.tpack.org>

evidence and proof. In mathematics education, concerns are often raised about the level of pre-service and primary teachers' knowledge of mathematics, and there is an ongoing debate about what level of mathematics is required to teach effectively in these areas. The American Council on Education (ACE) states that 'a thorough grounding in college-level subject-matter and professional competence in professional practice are necessary for good teaching … students learn more mathematics when their teachers report having taken more mathematics' (cited in Mewborn 2001, p. 28). However, it appears that simply having studied mathematics at a higher or even advanced level before undertaking teacher training does not necessarily equate with having effective content knowledge. In their wide-scale study into effective teaching of numeracy, Askew and colleagues (1997) found that there was a lack of evidence to support a positive association between formal mathematical qualifications and pupil gains, and that even teachers with high-level mathematics qualifications displayed knowledge that was compartmentalised and framed in terms of standard procedures, without underpinning conceptual links. Content knowledge is important, as 'you cannot teach what you do not know' (Rowland et al. 2010, p. 22), but considering other forms of teacher knowledge is also relevant.

Pedagogical knowledge (PK)

PK is knowledge about the processes and methods of teaching, and is a generic form of knowledge that applies to student learning, classroom management, lesson plan development and implementation, and student evaluation. PK requires an understanding of cognitive, social and developmental theories of learning and how they apply to students in the classroom (Koehler & Mishra 2008b).

Pedagogical content knowledge (PCK)

Shulman (1987, p. 8) defines PCK as 'an understanding of how particular topics, problems, or issues are organised, represented, and adapted to the diverse interests and abilities of learners, and presented for instruction'. A teacher with good PCK in mathematics would be aware of likely student misconceptions, the importance of recognising and catering for students' prior knowledge and how to make connections between different topics within mathematics.

Technological content knowledge (TCK)

TCK can be defined as an understanding of how technology and content influence and constrain one another. Teachers need to have a good understanding of which specific technologies can be used to create new representations and how the content can dictate or even change the ways in which the technologies are used. Technology can also be used to make the content accessible to students – perhaps in more ways than traditional methods allowed. TinkerPlots software (Konold & Miller 2005), for example, allows younger students to access many graphing concepts that traditionally were introduced in secondary school. Similarly, graphics calculators have allowed students to access higher-level mathematical concepts that traditionally were the domain of upper secondary studies.

Technological pedagogical knowledge (TPK)

TPK represents an understanding of the way in which teaching and learning change when particular technologies are used. An important part of TPK is developing creative flexibility with available tools in order to use them for specific pedagogical purposes. Chapter 2 shows how a teacher used the IWB to effectively scaffold students' learning about bridging 10 using 10-frames. Tools such as the IWB can be used to change how teachers teach, and TPK requires an understanding of the use of technology – not for its own sake, but for that of advancing student learning and understanding.

Technological pedagogical content knowledge (TPACK)

TPACK is an understanding that emerges from an interaction of technology, content and pedagogy knowledge, and is different from knowing all three concepts individually. According to Koehler and Mishra (2008b, pp. 17–18), TPACK represents:

- an understanding of the representation of concepts using technologies
- pedagogical techniques that use technologies in constructive ways to teach content
- knowledge of what makes concepts difficult or easy to learn and how technology can redress some of the problems that students face
- knowledge of students' prior knowledge and theories of epistemology
- knowledge of how technologies can be used to build on existing knowledge and to develop new epistemologies or strengthen old ones.

In this book, the reader will find references to some of the online HOTmaths tools. The online tools that are referred to are:

- *Lesson notes covering a wealth of topics from Foundation level to Year 10.* These lesson notes are written for students in Years 3 to 10. There are teaching notes included for Foundation to Year 4, which include creative teaching ideas for integrating ICT into the mathematics classroom. The course list includes a drop-down menu, which enables the user to structure the topics according to a global mathematics content structure that is generic, to the Australian or New Zealand content structure, and to various Cambridge resources to provide further support to teachers and students. Once the course is selected, the level is selected; at this point, the list of topics with lesson notes appears. From these lesson notes, links can be found to the tools described below, such as HOTsheets, widgets, walk-throughs, scorchers and topic quizzes. Each of these tools is described below, and examples of each are explored throughout this book.
- *Approximately 1000 PDF worksheets, named HOTsheets*. These are appropriately named, as the HOTsheets prompt student action, investigation and higher-order thinking, rather than the completion of a routine worksheet.
- *Over 650 interactive investigations and animations known as 'widgets'.*

- *Walkthroughs*, where students learn the procedures for a concept, and are given detailed and instant feedback on their errors.
- *Banks of timed questions called Scorchers*, for introduction, review and promoting automatic recall of known facts. Achievement is recorded in reports and on a series of leader boards.
- *Banks of questions*, in every lesson, at four levels of difficulty.
- *Diagnostic topic quizzes*, where students can diagnose their strengths and weaknesses across a topic. They will then be directed to HOTmaths content they need to review.
- *A test generator*, where questions from a range of topics, lessons and levels of difficulty can be selected manually or automatically to create a customised online or printable test. These tests can be set as assessments that need to be completed in one sitting, or as revision activities that students can complete over a number of sittings.
- *Interactive games*.
- *An illustrated mathematics dictionary* that links definitions to lessons throughout each course.
- *A complete reporting system* that tracks individual student and whole-class progress and results.

HOTmaths also includes:

- *A task manager*, where teachers can set activities for students to complete as class work or homework.
- *A search tool* that will find HOTmaths resources based on keywords.

While this book is relevant to the teaching of mathematics in many nations, particular reference is made to the Australian and New Zealand curriculum documents. However, the content can be related to most primary mathematics curriculum structures. The New Zealand Curriculum, known as NZmaths, is divided into three strands: Number and Algebra; Geometry and Measurement; and Statistics. A fourth strand, Problem Solving, overlays each of the three main content strands. Each of the strands contains units that target various levels of understanding within each topic – usually from Level 1 to Level 5. The units adhere to a student-centred activity design, described as either Exploration design or Station design, where the students rotate among various activities over a one-week period. Similarly, the Australian curriculum has three content strands: Number and Algebra; Measurement and Geometry; and Statistics and Probability. The syllabus includes stage outcomes, stage statements, content outcomes, content, background information and language sections. Both curriculum documents include support materials such as assessment and teaching mathematics for children with special needs. Readers are encouraged to relate the ideas explored in this book to their own context and curriculum structure.

The next section provides the reader with an advance organiser that presents a summary of the key objectives of the remaining 14 chapters in the book.

Summary of chapters

Chapter 2: Exploring early number concepts examines the key ideas associated with learning early number concepts. Readers will become familiar with counting principles and how they underpin number understanding. The capacity of learning frameworks and curriculum documents to guide teaching early number concepts is explored. The effective use of technology to develop an understanding of early number concepts is examined through active engagement in a number of different activities designed to help you think about and evaluate the role of technology as a teaching tool.

Chapter 3: Exploring measurement presents the measurement sequence used to introduce and develop an understanding of each measurement attribute – that is, length, area, volume and capacity; angle, mass, time, temperature; and money and value. The principles of measuring with units will be explored and the importance of developing estimation skills and meaningful benchmarks emphasised. Strategies for utilising technology to support, develop and extend students' measurement experiences are addressed.

In *Chapter 4: Exploring geometry*, we consider the breadth of concepts included in the geometry section of curriculum documents generally, and in the Australian Curriculum: Mathematics in particular. The chapter presents a theoretical framework, known as the van Hiele theory, as a lens through which to view students' geometrical thinking and a pedagogical framework that is useful for designing sequential student tasks to assist students to grow in their understandings of geometrical concepts. The important role of language and maintaining 'student ownership' of the geometrical ideas is explored, as is the use of technological tools to enhance our teaching of geometrical concepts for the e-generation.

The difference between additive and multiplicative thinking and the appropriate use of drill and practice activities are both explained in *Chapter 5: Exploring whole number computation*. This chapter explores various representations that teachers can use to illustrate different ways of thinking and the effective use of the range of technological tools available to explore whole number computation.

The importance and representation of part-whole numbers such as fractions, decimals and percentages are investigated in *Chapter 6: Part-whole numbers and proportional reasoning*. The importance of understanding proportional reasoning concepts in daily life is explored, together with strategies for using technology effectively in this domain.

Underpinning *Chapter 7: Exploring patterns and algebra* is an understanding of the central importance of patterns in early childhood and primary school mathematics, and the importance of mathematical structure and its relevance to children's learning of mathematics. The chapter explores the process of using sequences effectively to find and justify rules, and to explain phenomena. Strategies for representing and resolving number sentences, equivalence and equations are presented. Effective ways to use technology to explore algebraic situations where students are encouraged to describe relationships between variables are investigated.

Chapter 8: Exploring data and statistics examines suitable statistics questions for investigation by children of different ages, using a cycle of problem, plan, data, analysis and conclusion (PPDAC) (Wild & Pfannkuch 1999). The importance of variation in data and different types of variables, and the difference between a population and a sample, are investigated. Readers will explore different ways of displaying data to 'tell a story'. The importance of drawing inferences from data, and the uncertainty associated with these inferences, are discussed. Readers will engage in activities that use technology to support the development of statistical understanding.

Chapter 9: Exploring chance and probability begins with a consideration of the difference between objective and subjective views of probability. A range of tools for investigating probability are explored, and applications of probability in daily life are provided. Strategies for using technology effectively to develop ideas about uncertainty in the primary classroom are presented.

Chapter 10: Capitalising on assessment for, of and as learning focuses on the notion of assessment 'for', 'of' and 'as' learning, and how these forms of assessment work together in the mathematics classroom. A developmental framework to assist in designing an assessment item and assessing the quality of a student's response – the Structure of the Observed Learning Outcome (SOLO) model (Biggs & Collis 1982) – is presented. Issues surrounding national testing data are raised in the light of positive ways to support growth in mathematical understanding. Readers will engage with various ICT tools that can assist in creating valid assessment items.

Chapter 11: Capitalising on ICT in the mathematics classroom considers the role played by digital technologies in today's classrooms, and how ICT impacts upon mathematics teaching. The chapter explores the ways in which ICT is described and integrated in curriculum documents and strategies to support students in their use of ICT. Activities in this chapter demonstrate the way technology can be incorporated into classroom routines to enhance learning experiences for students.

Chapter 12: Diversity in the primary mathematics classroom raises issues concerning the complexity of primary mathematics classrooms and a range of potential barriers to learning mathematics experienced by many children. Strategies for planning for diversity in the mathematics classroom are explored; these utilise the potential of technology to meet all learners' mathematical needs.

In *Chapter 13: Surviving as an 'out of field' teacher of mathematics*, we identify issues pertinent to 'out of field' teachers of mathematics. Strategies are offered to assist 'out of field' teachers of mathematics while considering the invaluable knowledge a primary-trained teacher can bring to the secondary mathematics classroom. Examples are presented of teachers' critical pedagogical content knowledge that has a high impact on students' growth and development across the main strands of the curriculum.

Chapter 14: Teaching mathematics beyond the urban areas delves into the challenges and rewards of working in remote areas of countries such as Australia and New Zealand, and small Pacific nations. Teaching strategies are presented to assist in maintaining a positive learning environment in remote and small Pacific nation classrooms. The importance of the relationships among and between parents, students, teachers and other community members is explored, along with practical suggestions for making the most of the available resources. The chapter demonstrates that, even with limited or no internet access, ICT can still be utilised as a tool to enhance and extend upon more traditional classroom materials; it is about making the most of the tools and resources that are available.

Finally, *Chapter 15: Planning and sustainability in the mathematics classroom* reacquaints readers with the TPACK framework and the different types of teacher knowledge required for effective teaching for numeracy. It provides opportunities to develop strategies for evaluating and reflecting upon teachers' own use of technology. The notion of a community of practice is explored, as are ways to establish one within your school. Strategies are presented for evaluating the use of online resources and insights are provided into the change process and how school-wide approaches to teaching mathematics with ICT can be sustained.

How to use this book

Each of the chapters includes common key components that provide different ways to consider the material.

- *Learning outcomes* describe the learning that is expected as a result of engaging with the different components of the chapter.
- *Key terms* identify and define the common terms that are used throughout the chapter.

- *Classroom snapshots* are short vignettes of real classroom situations that exemplify some pedagogical principles, including technology use.
- *Pause and reflect* sections are points where the reader is asked to think deeply about some specific issue or dilemma.
- *Activities* present a task or short investigation to enable the reader or students to become familiar with a tool or way of presenting mathematics in a technological environment.
- *HOTmaths icons* indicate an activity or example that links to HOTmaths.
- *Further reading/websites for exploration* at the end of each chapter provide some of the research background, as well as places to find other ideas and material.

As with all texts, the more engagement there is with the ideas and activities presented, the better the learning is likely to be. Technology is changing so fast that this book can present only what is available at a single point in time. Teachers starting their careers will find that what is new and cutting edge when they first enter the classroom is obsolete before they are ready to retire. Bonnie's grandfather probably used a slate and chalk when he started school. Bonnie's parents used a ballpoint pen and saw the introduction of whiteboards instead of chalkboards. In three generations, technology has changed teaching from literally 'chalk and talk' to a dynamic and creative endeavour, in which teacher and students are co-learners and partners. It is difficult to predict what skills and knowledge school students will need in an ICT-rich future, but it is hoped that this book will provide teachers with ways in which they can capitalise on ICT in order to engage and inspire mathematical development.

Websites for exploration

HOTmaths: <http://www.hotmaths.com.au>

Learning and Teaching: <http://www.learningandteaching.info/learning/references.htm>

CHAPTER 2

Exploring early number concepts

LEARNING OUTCOMES

By the end of this chapter, you will:

- have gained an understanding of the key ideas associated with learning early number concepts
- have become familiar with counting principles and how they underpin number understanding
- understand how learning frameworks and curriculum documents can be used to guide teaching early number concepts
- be able to use technology effectively to develop an understanding of early number concepts
- be able to engage in a number of different activities designed to help you think about and evaluate the role of technology as a teaching tool.

KEY TERMS

- **Array:** An arrangement of rows and columns
- **Constructivism:** A theory of learning whereby new knowledge is constructed on the basis of past experiences
- **Counting principles:** The principles that govern and define counting: the one-to-one principle, the stable order principle, the cardinal principle, the abstraction principle and the order irrelevance principle
- **Drill and practice:** Repetition of a skill or procedure
- **Hundreds square:** An array of 10 columns and rows depicting the numbers 1 to 100

- **Mental computation:** Performance of mathematical computations in the mind without the aid of pen and paper or calculator
- **Part-part-whole:** The relationship between the parts of something and its whole
- **Subitise:** Recognise how many are in a set without counting
- **10-frames:** An array of two rows of five used to provide a visual representation of the base-10 number system
- **Virtual manipulative:** A virtual representation of a physical manipulative

PAUSE AND REFLECT

A recently viewed YouTube clip showed a 1-year-old girl becoming increasingly frustrated with a magazine that did not 'work'. A caption that accompanied the clip stated, 'My 1-year-old thinks that a magazine is an iPad that doesn't work.' Further on in the clip, the magazine is replaced with an iPad and the child happily clicks away and manipulates the images on the screen. Another clip shows a 9-month-old baby using an iPad to view pictures with captions underneath, while yet another shows a toddler using his mother's mobile phone to play *Angry Birds*.

If you type in 'Baby works iPad perfectly. Amazing must watch!' you can view footage of a toddler who shows dexterity, word recognition and even understanding through operating an iPad. His father explains the context and is positive about the benefits of technology:

> My son Bridger just turned 2 last week and I bought him an iPad, mostly an excuse for me to get one and he actually can use it perfectly! His speech, understanding, word recognition and even hand–eye coordination have improved within just a short while! I am so amazed and thankful for this amazing learning tool that my son has! My son can read tons of words now, he knows every animal and dinosaur and he just turned 2 years old! If you have a child around 2, don't rob him/her of knowledge, go buy him/her an iPad!

How will this child's affinity with technology be supported when he enters formal schooling? While we know the importance of recognising prior learning before children enter school, how many early childhood teachers are really aware of, and in a position to capitalise on, the technical knowledge that many children will bring to school today? We may expect them to be able to count, and even write their name, but the above vignette shows the potential experiences in which some children may have been immersed. Are they likely to be satisfied with traditional teaching tools and practices, which are often static and do not provide for immediate feedback or interaction? On the other hand, as teachers, can we be confident that the tools and software actually enhance learning, or are they simply replacing practices that might more appropriately be carried out by teachers?

Early childhood research has recognised that early number development begins before children enter formal schooling, and that it involves much more than just being able to count. This chapter looks at some of the fundamental early number concepts, and at how technology can be used to supplement, support and extend more traditional teaching approaches in order to develop students' understanding in this area.

Early number concepts

Children come to school with many ideas about number, on which the teacher can build to develop understanding through making connections in the mind of the learner. The concepts of number and number operations can be understood through building a network of cognitive connections between concrete experiences, symbols, language and pictures. This assists children with *constructing their own knowledge*, which is the basic tenet of **constructivism** (von Glasersfeld 1996). The general principles of constructivism are based primarily on Piaget's notions of assimilation and accommodation. Assimilation refers to the use of an existing schema to give meaning to new ideas. Sometimes this requires restructuring of that network in order to *accommodate* the new experiences (Haylock & Cockburn 2008). Researchers have also recognised the importance of both the individual's construction of meaning and the social context within which this occurs (e.g. Simon 2000), leading to the realisation that children are actively contributing to their mathematical development.

It could be argued that counting and understanding number are at the heart of mathematics, and while many children will begin school knowing how to count, it is likely that they will vary considerably in what they know and how it can be demonstrated in practice. It is also important to recognise that learning to count requires a considerable background of experience, including the ability to be able to sort and categorise, the notion of 'more or less' and even the ability to distinguish between small numbers – such as 1, 2, 3 (Haylock & Cockburn 2008). While the recitation of the number sequence up to 10 or even 20 is one indicator of a child's counting ability, there are a number of other understandings that are required. Siemon (2007, p. 2) identifies that early numeration involves:

- one-to-one correspondence
- recognising that '3' means a collection of three, whatever it looks like

- recognising that the last number counted represents the total number in the collection
- recognising small collections to five without counting (**subitising**)
- matching words and/or numerals to collections less than 10 (knowing the number naming sequence)
- being able to name numbers in terms of their parts (**part-part-whole**)
- identifying one more, one less, what comes after and what comes before a given number.

Other authors have also agreed that there are a number of principles that can help us to understand how children have mastered this knowledge or assist us in supporting those children who haven't. While there are variations, the following principles, originally developed by Gelman and Gallistel (1978), have been identified as underpinning counting:

- *The one-to-one principle.* A child who understands the one-to-one principle knows that we count each item once.
- *The stable order principle.* A child who understands the stable order principle knows that the order of number names always stays the same. We always count by saying 1, 2, 3, 4, 5 … in that order.
- *The cardinal principle.* A child who understands the cardinal principle knows that the number they attach to the last object they count gives the answer to the question, 'How many …?'
- *The abstraction principle.* A child who understands the abstraction principle knows that we can count anything – they do not all need to be the same type of object. So we can count apples, we can count oranges, or we could count them all together and count fruit.
- *The order irrelevance principle.* A child who understands the order irrelevance principle knows that we can count a group of objects in any order and in any arrangement and we will still get the same number (Cotton 2010, p. 55).

Number frameworks

The ways in which students demonstrate an understanding of the **counting principles** are commonly described using number or learning frameworks.

Learning frameworks – also known as progress maps or learning trajectories – provide a description of skills, understandings and knowledge in the sequence in which they typically occur, giving a virtual picture of what it means to progress through an area of learning. Thus they provide a pathway or map for monitoring individual development over time. A student's location on a framework can be utilised as a guide to determining the types of learning experiences that will be most useful for meeting the student's individual needs at that particular stage in their learning (Bobis 2009).

For example, the First Steps Framework (Department of Education & Training WA 2004) includes a diagnostic map in which terms such as 'emergent', 'matching' and 'quantifying' are used to describe the phases through which students typically pass in developing a sense of number and operating with numbers.

The Count Me in Too (CMIT) Learning Framework in Number was initially developed by Wright (1994), and outlines how students move from using naïve strategies to adopting increasingly sophisticated strategies in order to solve number problems. The CMIT website (<http://www.curriculumsupport.education.nsw.gov.au/countmein>) provides information about the framework, resources for teachers, children and parents, and a number of interactive activities that can be used to develop children's understanding.

ACTIVITY 2.1

Visit the Count Me in Too website (<http://www.curriculumsupport.education.nsw.gov.au/countmein>) and access the Learning Framework in Number. View the video excerpts showing emergent counting, perceptual counting, figurative counting and counting on and back.

- How do the excerpts illustrate the differences in the early phases of counting?
- What would you do as a teacher to move children from the emergent phase to the perceptual phase?

The New Zealand Number Framework

The New Zealand Number Framework (Ministry of Education 2008), was developed from the Numeracy Development Projects, and assists with understanding the requirements of the Number and Algebra strand of the Mathematics and Statistics curriculum learning area. Distinction is made between strategy and knowledge, with early number strategies including emergent, one-to-one counting, counting from one on materials, counting from one by imaging and advanced counting. The framework booklets can be accessed at <http://nzmaths.co.nz/new-zealand-number-framework>.

Linking with curriculum documents

National curriculum frameworks in Australia and New Zealand cover the early childhood years from ages 0–8. *Belonging, Being & Becoming: The Early Years Learning*

Framework for Australia (EYLF) (DEEWR 2009) provides a holistic approach to learning and development, and covers children from ages 0–5. The Australian Curriculum: Mathematics (ACARA 2011), which is directed more at school-aged children (ages 5–8), is focused on content and proficiencies, with more specific learning outcomes identified. In New Zealand, Te Whāriki is the Early Childhood Curriculum (Ministry of Education 1996), with the New Zealand Curriculum (Ministry of Education 2007) beginning at Level 1, which encompasses the first two years of formal schooling.

The Early Years Learning Framework

The EYLF outcomes and key components that are particularly relevant to mathematics are not explicit, as specific learning areas are avoided (Perry, Dockett & Harley 2012), but the following two outcomes and their components can be linked with mathematical development:

Outcome 4: Children are confident and involved learners

Components: Children develop dispositions for learning such as curiosity, cooperation, confidence, creativity, commitment, enthusiasm, persistence, imagination and reflexivity.

Children develop a range of skills and processes, such as problem-solving, inquiry, experimentation, hypothesising, researching and investigating.

Outcome 5: Children are effective communicators

Component: Children begin to understand how symbols and pattern systems work.

Technology is also mentioned in Outcomes 4 and 5, in the following components:

- Children resource their own learning through connecting with people, places, technologies and natural and processed materials.
- Children use ICT to access information, investigate ideas and represent their thinking.

The links between mathematics and technology are documented as being evident when children 'use information and communication technologies (ICT) to investigate and problem solve' (DEEWR 2009, p. 37).

The implications for educators are that children need to be provided with access to a range of technology; technology needs to be incorporated into children's play experiences and projects; and skills and techniques need to be taught in order for children to use technologies to explore new information and represent their ideas (DEEWR 2009).

Te Whāriki

This document is a bicultural curriculum statement developed in New Zealand that covers the education and care of children from birth to school entry age (Ministry of Education 1996). Similar to the EYLF, this document provides broad principles and goals for early childhood, rather than stipulating particular mathematical understandings. The principles are Empowerment (*Whakamana*), Holistic Development (*Kotahitanga*), Family and Community (*Whānau Tangata*) and Relationships (*Ngā Hononga*). The five strands that arise from the four principles are:

- Well-being – *Mana Atua*
- Belonging – *Mana Whenua*
- Contribution – *Mana Tangata*
- Communication – *Mana Reo*
- Exploration – *Mana Aotūroa*.

References to developing mathematical abilities and interests are included in Contribution, Communication and Exploration. For example, Goal 3 of Communication includes the following learning outcomes:

- familiarity with numbers and their uses by exploring and observing the use of numbers in activities that have meaning and purpose for children
- skill in using the counting system and mathematical symbols and concepts, such as numbers, length, weight, volume, shape and pattern, for meaningful and increasingly complex purposes
- the expectation that numbers can amuse, delight, illuminate, inform and excite
- experience with some of the technology and resources for mathematics, reading, and writing.

Te Whāriki also links each strand with the New Zealand Curriculum Framework. For example, in relation to belonging, 'children learn to use numbers in relation to family members, children in a group, and ordering the environment in patterns and relationships' (p. 95).

The Australian Curriculum: Mathematics

The stable order principle, which refers to the key understanding that the number names are said in sequence and must be said in a particular order, is one of the first connections children make when counting. The ability to rote count and then extend this to rational counting is fundamental to early number development. While many children will come to school able to rote count to 5, 10 or beyond, they are expected to further develop this in the Foundation Year of the Australian Curriculum: Mathematics. The descriptors related to counting for the Foundation Year include:

- Establish understanding of the language and processes of counting by naming numbers in sequences, initially to and from 20, moving from any starting point (ACMNA001).
- Connect number names, numerals and quantities, including zero, initially up to 10 and then beyond (ACMNA002).

The elaboration statements include a reference to the understanding that numbers are said in a particular order and that there are patterns in the way we say them.

The New Zealand Mathematics Curriculum

The New Zealand Mathematics Curriculum begins at Level 1, which is related to Year 1 and Year 2 of the National Standards. During these years, students are expected to develop counting skills and to:

- use a range of counting, grouping, and equal-sharing strategies with whole numbers and fractions
- know the forward and backward counting sequences of whole numbers to 100
- know groupings with five, within ten, and with ten
- communicate and explain counting, grouping and equal-sharing strategies using words, numbers and pictures
- generalise that the next counting number gives the result of adding one object to a set and that counting the number of objects in a set tells how many.

There are a number of activities and strategies that can be used to facilitate the development of counting, and many of these are interactive and accessible via the internet. The following section looks at some established routines and practices, and provides some alternative approaches that encompass ICT.

Early number activities and strategies

Rhymes and stories

Nursery rhymes, number poems and counting books offer an engaging way to expose young children to numbers and how they are used. The stable order principle can be introduced and reinforced through rhymes and songs, and through counting story books such as *One is a Snail, Ten is a Crab* (Sayre & Sayre 2003) and *Ten Out of Bed* (Dale 2010). Rhymes and songs can be accompanied by actions, which also help to reinforce the principle of one-to-one correspondence, such as in the following songs:

> One, two, three, four, five
> Once I caught a fish alive.
> Six, seven, eight, nine, ten
> Then I let it go again
> Why did you let it go?
> Because it bit my finger so.
> Which finger did it bite?
> This little finger on my right.

and

> One little, two little, three little, four little,
> Five little birds so small;
> One little, two little, three little, four little,
> Five little birds on the wall.
> Kitty Cat came from a nearby bush ... MEOW!
> Gave the garden gate a push ... SQUEEEEAK!
> And one little, two little, three little, four little,
> Five little birds went WHOOSH!

As you sing, you can role-play using children or stuffed animals and, in New Zealand, use Te Reo Māori for numbers and words.

Many examples of rhymes and songs can be sourced on the internet – for example, on the Nursery Rhymes website, <http://www.nurseryrhymes4u.com/NURSERY_RHYMES/counting.html>), with a number of sites providing interactive elements, such as audio-recordings and video clips – for example, the animation of 'Five Little Monkeys Jumping on the Bed' at Flash Apps, <http://www.eflashapps.com>. YouTube has literally thousands of examples of animated rhyming songs for children. Simply enter 'counting songs for children' into the search field to begin exploring.

Many counting activities and games are also available as apps, and can be downloaded for free or cheaply from the iTunes Store or its Android equivalent. Counting Bear (see Figure 2.1) is an example of an application that is compatible with iPhones and iPads, and focuses on consolidating the correct sequencing of counting numbers to 10.

Figure 2.1 Counting Bear app

Sequencing activities

One commonly used strategy that is popular in many early childhood classrooms involves the use of a clothes line where numbers are 'pegged' to designate their correct sequencing. This activity is extremely versatile, as the end numbers can be varied to represent, for example, 0 and 10, 0–20, 10–20 and even 0–1 for older children. Once the numbers have been placed in order, further teaching opportunities can occur, such as looking at which number comes before and after, one more, one less, what number comes between and so on. Versions of this activity can also be found online, and can be used either as substitutes for the real materials in the classroom or as follow-up activities to provide both reinforcement and individual consolidation. The NZmaths website, <http://nzmaths.co.nz/nzmaths-learning-objects-0>, has a number of interactive digital learning objects, including 'Number Line' where users can individually construct and label their own number lines.

ACTIVITY 2.2

Access the interactive washing line activity on the Count Me in Too website, <http://www.curriculumsupport.education.nsw.gov.au/countmein/children_washing_line.html>.

Figure 2.2 Count Me in Too website

- Play around with placing the numbers on the line. What happens when you place a card incorrectly?
- Do you think this activity would assist with developing an understanding of the correct ordering of numbers?
- What would be the advantages and disadvantages of doing this activity online as compared with using real materials in the classroom?

Extending counting beyond 10

As children develop increased facility with counting, they will learn to associate the name of the number with the numeral (for example, one – 1, two – 2) and extend their counting beyond 10 to 20 and so on. The **hundreds square** (see Figure 2.3) provides a useful visual representation of the numbers 1–100, and can assist with counting numbers up to 100. The spatial arrangement of numbers emphasises the ordinal aspect of number and the relationships associated with place value. Numbers are defined by their position in the chart – for

1	2	3	4	5	6	7	8	9	10
11	12	13	14	15	16	17	18	19	20
21	22	23	24	25	26	27	28	29	30
31	32	33	34	35	36	37	38	39	40
41	42	43	44	45	46	47	48	49	50
51	52	53	54	55	56	57	58	59	60
61	62	63	64	65	66	67	68	69	70
71	72	73	74	75	76	77	78	79	80
81	82	83	84	85	86	87	88	89	90
91	92	93	94	95	96	97	98	99	100

Figure 2.3 Hundreds square

example, 47 is in a square that is in a row between 46 and 48 and in a column between 37 and 57. Along with being an excellent tool for identifying patterns in the number system and demonstrating place value concepts, the hundreds square is often used in early childhood settings as an opportunity to engage in group counting and class discussions about the counting of numbers up to 100.

The Count Me in Too website has an interactive hundreds square (<http://www.curriculumsupport.education.nsw.gov.au/countmein/children_hundred_chart.html>), which has a number of different squares with a range of missing numbers that have to be filled in. While completing such activities would help to reinforce the structure of the chart, one of the limitations of this particular interactive activity is that incorrect responses are not immediately noted, thereby providing little scaffolding to assist children and even potentially leading to the reinforcement of incorrect sequencing. HOTmaths has a version of this called 'Hidden Numbers' (see Figure 2.4), in which incorrect answers are noted and there are options to try again. You can access this resource by logging into HOTmaths and clicking on the FUNdamentals icon on the Dashboard. Next, click on 'Numbers', then 'Numbers to 100', and finally 'Hidden numbers'.

The interactive hundreds square on the Apples for the Teacher website, <http://www.apples4theteacher.com/math/games/100-number-chart-one.html#interactive100chart>, is most suitable for use by the teacher

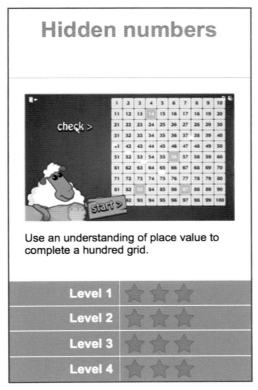

Figure 2.4 HOTmaths Hidden Numbers activity for Foundation to Year 2 students

in whole-class or small-group situations. It contains a number of different-coloured pens that can be used to colour the squares in the chart, providing a nice visual pattern; it is particularly effective for demonstrating skip counting.

ACTIVITY 2.3

Investigate the range of interactive number charts that are available on the internet. Each of the examples below has different features and functions, and requires different levels of interaction.

- Ambleweb: <http://www.amblesideprimary.com/ambleweb/mentalmaths/interactivenumbergrids.html>; <http://www.amblesideprimary.com/ambleweb/mentalmaths/countersquare.html>
- Crickweb: <http://www.crickweb.co.uk>
- ABCya.com: <http://www.abcya.com/one_hundred_number_chart_game.htm>

1 Consider how you would use the charts within the context of a lesson.
2 Would you incorporate the IWB and engage the whole class in modelling and demonstrating skip counting?
3 Would you encourage students to access the charts individually, in pairs or in small groups?
4 What aspects of TPACK (see Chapter 1) would you need to have in order to effectively incorporate such tools and capitalise on their affordances for learning about the place value of numbers?

Linking numbers with quantities

Another important connection for children to make is recognising the relationship between the number name and the quantity it represents. Children need to understand that the numeral 5, which can also be represented as 'five', can be used to show there are five objects. They also need to understand that a set of objects can be counted in any order or arrangement, and the result will be the same each time. Activities that reinforce this concept typically involve 'matching' experiences or drawing tasks that focus on the number words, symbols and visual representations. The Illuminations website, <http://illuminations.nctm.org/ActivityDetail.aspx?ID=73>, provides an example of such an activity, and requires the user to match the numeral and name with representations showing various quantities. Similarly, 'Counting birds' (see Figure 2.5) in HOTmaths shows representations of these concepts and includes support materials in the form of HOTsheets (Collection Cards and Using Numbers to 10). You can access this resource by logging into HOTmaths and clicking on the FUNdamentals icon on the Dashboard. Next, click on 'Numbers', then 'Numbers to 10', and finally 'Counting birds'. The support materials can be found via the 'Resources' button on the 'Numbers to 10' page.

Drawing programs such as Kid Pix could be utilised for children to design their own representations, providing an alternative to the often labour-intensive task of drawing and writing. If real materials were utilised, digital photographs could be used to record what was done physically, and these could be accompanied by audio-recorded verbal explanations.

Subitising

Subitising refers to the ability to recognise how many objects are in a group without counting. It is a fundamental skill in the development of number,

Figure 2.5 HOTmaths Counting Birds activity for Foundation to Year 2 students

and one that needs to be developed and practised through experiences with patterned sets and dot patterns. Subitising is a useful skill, as it saves time, reduces cognitive load, develops number sense – particularly in terms of understanding the relative sizes of numbers – and accelerates addition and subtraction skills. Clements (1999) distinguishes between perceptual subitising, which is the immediate recognition of numbers up to 4, and conceptual subitising, where groups can be recognised in terms of their subitised parts. The part-part-whole relationship that is developed through conceptual subitising is an important forerunner to both number and computational sense.

Appropriate materials to use in pattern-recognition activities include dot plates and 10-frames. Both can be made cheaply by using paper plates, cardboard and round sticky dots. Alternatively, PowerPoint can be used effectively to produce a series of patterns that can be flashed up for a short, predetermined period of time (see Activity 2.4). When designing the aids, include variations of arrangements and gradually increase the number of dots shown (see Figure 2.6). When using the activity with children, ask questions such as:

- How many do you see?
- How do you know?

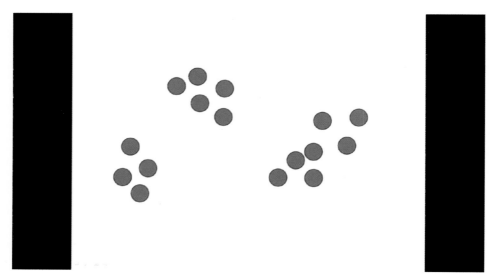

Figure 2.6 Dot patterns on a PowerPoint slide

- Who saw something different?
- How many more do you need to make ...?

Interactive websites that support and reinforce subitising can also be used to complement and extend whole-class activities.

ACTIVITY 2.4

To create subitising flash screens using PowerPoint:

- Open PowerPoint.
- Begin with a blank slide.
- Click on the 'Select shapes' icon and select circle.
- Copy the circle and paste in desired formation (e.g. see Figure 2.6).
- Click on 'Insert new slide' and repeat as desired.
- Begin with one formation, then extend into different variations and combinations.
- To transition, select all dots and put a box around them.
- Select 'cut' or 'fade' and 'advance slide' after one to three seconds (or select 'on mouse click' if you want children to explain what they see).

In her book, *Number Sense Routines*, Shumway (2011) discusses how 'quick images' can be organised in such a way as to enhance and extend children's subitising skills. She recommends using a variety of configurations and models to encourage children to think flexibly, and to consider selecting combinations that help to consolidate and reinforce mental strategies. For example, if you were encouraging children to use double facts (use of doubling, such as 8 + 8 = 16) to assist them with adding number combinations such as 8 + 7, this could be built by using dot cards or slides to visually display 7 + 8 as being two lots of 7 and one additional dot (see Figure 2.7). Sequences to encourage this strategy can be devised using similar facts, such as 4 + 5, 6 + 7 and so on.

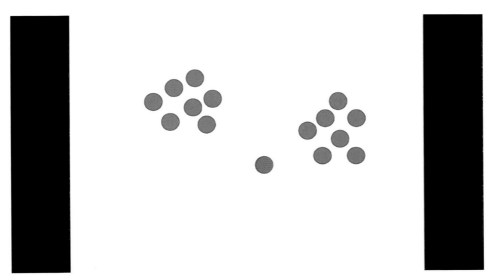

Figure 2.7 Dot formations to demonstrate doubling strategy

HOTmaths FUNdamentals has an interactive activity where cupcakes are placed in trays to visually represent the double and near double facts (see Figure 2.8). You can access this resource by logging into HOTmaths and clicking on the FUNdamentals icon on the Dashboard. Next, click on 'Addition and subtraction', then 'Addition skills', and finally 'Doubles and near doubles'.

When introducing combinations of numbers up to 20, begin by asking for estimates. For example, you may ask students to indicate with a 'thumbs up' if they think there are more than 10.

10-frames

10-frames are also useful tools for encouraging children to subitise and to build up an understanding of the part-part-whole relationship. Activities and routines with 10-frames encourage students to think in groupings and foster their development in using the five- and 10-structures of our number system. The structure

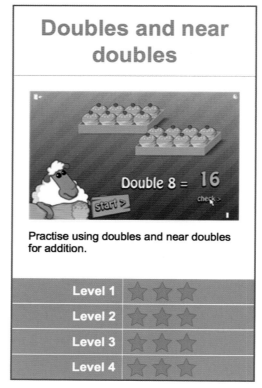

Figure 2.8 HOTmaths Doubles and Near Doubles activity for Foundation to Year 2 students

of the frame always shows a quantity's relationship to 10, and assists students with composing and decomposing 10, and hence the addition/subtraction combinations of 10 (Shumway 2011). Classroom snapshot 2.1, taken from a transcript of a lesson conducted with a Year 1 class, demonstrates how the teacher used 10-frames and an IWB to teach children how to use the strategy of bridging 10.

CLASSROOM SNAPSHOT 2.1

TEACHER: We're going to talk about a different strategy that you can use today. So you could use knowing your doubles facts, you could use counting in ones and you could use this other strategy as well. And that's what we are going to learn today. Just to give you another way of working out maths problems. How does that sound?

STUDENTS: Good.

[Displays the following on screen]

Teacher: So do you know what these are called?
Jake: Um, counters.
Teacher: Well we do use them with counters, but we call them 10-frames. So the first step in using bridging 10, which is the strategy we're finding out about today, is to imagine these 10-frames in your head. So imagine an eight in your head and imagine a seven in your head. Do you think you can do that? Close your eyes and see whether you can see an eight in the frame in your head, or a 10-frame with an eight in your head. How many dots is it missing?
Jake: Um, two.
Teacher: Great. Now see if you can imagine a 10-frame with seven counters in your head for me. Cover up, your eyes are meant to be closed … How many are missing?
Students: Three.
Teacher: Fantastic! Okay. This is what we do when we bridge 10. We make one of the 10-frames up into 10 by moving the dots. Which one would be the sensible one to move the dots in up here? [referring to the two 10-frames showing eight and seven]
Amit: What do you mean by sensible?
Teacher: So that you don't have to move too many dots.
Sarah: From the yellow frame [showing seven] to the purple frame [showing eight].
Teacher: Yes. Would you like to do that, Jack? Fill up the 10-frame showing eight from the 10-frame which shows seven.
[Jack uses his finger to move two counters from the frame showing seven, and fills up the other frame, leaving five counters in the right-hand frame.]
Teacher: So what have we got?
Sarah: Five and ten, which makes 15.
Teacher: Let's move them back and do it again. Sarah, would you like to do that – move the two dots back to the yellow frame?
[Sarah goes to the IWB and moves two dots from the first frame to the right-hand frame, leaving the original eight.]
Teacher: Thank you. Roy, would you move those two dots over to the frame showing eight? So we're bridging 10 by moving two from the seven.
[Roy moves the two dots into the frame showing eight, filling it up.]
That's it. So we bridged 10 by moving two over on to the eight. What did we turn the eight into now?

Students: Ten.
Teacher: Yes, so we've got 10 and …?
Sarah: Five.
Teacher: Makes …?
Jade: Fifteen.
Teacher: Beautiful. Roy, could you put us onto the next slide?

PAUSE AND REFLECT

After reading the above lesson transcript, reflect on the following:

- How did the teacher encourage student interaction with the IWB?
- In what ways did the teacher capitalise on the features of the IWB to demonstrate and model the bridging 10 strategy?
- Do you think the lesson would have been as effective if each child had their own individual 10-frame in front of them and placed counters on it according to instructions? Why/why not?

There are also a number of websites that use 10-frames to reinforce the part-part-whole relationship, and their use can serve to reinforce the concepts discussed or modelled in class. Again, the Count Me in Too website provides an example of this, through an interactive game involving placing butterflies on a 10-frame (see Figure 2.9): see <http://www.curriculumsupport.education.nsw.gov.au/countmein/teachers_teaching_ideas_-_butterfly_ten_frame.html>.

Figure 2.9 Butterfly 10-frame from the Count Me in Too website

Early operations with number

According to Siemon (2007), the idea that the numbers up to 10 can be conceptualised as units in their own right is fundamental to establishing 10 as a countable unit. As children learn about number in terms of the part-part-whole relationship, these ideas can be related to addition and subtraction, with the numbers being seen as *composite units*. Similarly, multiplication and division require students to think about numbers as units (Van de Walle, Karp & Bay-Williams 2013): in 4 × 3, each of the four threes is counted as a unit. The Australian Curriculum: Mathematics expects that by Year 1 students will be able to count to and from 100, locate numbers on a number line, carry out simple addition and subtraction problems using counting strategies and partition numbers using place value.

Addition and subtraction

Addition and subtraction are related: addition names the whole in terms of the parts and subtraction names a missing part (Van de Walle, Karp & Bay-Williams 2013). Researchers have separated addition and subtraction problems into categories based on the types of action or relationship described in the problems (Carpenter et al. 1999). Cognitive guided instruction (CGI) describes four basic classes of problems that can be identified for addition and subtraction: join, separate, part-part-whole and compare. Within these structures, there are different quantities involved and various actions that are undertaken to solve the problem. For example, the following problem involves a situation where the two quantities are *joined* and the result is unknown:

> John had 3 toy cars. Michael gave him 6 more toy cars. How many toy cars does John have now?

Consider how this structure is different in the following problem:

> John has 3 toy cars. Michael gave him some more. Now John has 9 toy cars. How many did Michael give him?

In the above problem, the result and the initial amount are known, but the change is unknown.

Similarly, separate problems also involve different actions, whereby the result, the change or the initial amount can be unknown. Can you identify the different structures in the following problems?

Oscar had 9 toy cars. He gave 3 toy cars to Dong. How many toy cars does Oscar have now?

Oscar had 9 toy cars. He gave some to Dong. Now he has 7 toy cars. How many did he give to Dong?

Oscar had some toy cars. He gave 3 to Dong. Now Oscar has 6 toy cars left. How many toy cars did he have to begin with?

This focus on different problem structures, the analysis of story problems and the importance of teachers' interpretation of student responses are key components of the CGI program, which is highly regarded in the United States. Developed by Carpenter and colleagues (1999), the program provides a framework for teachers to consider students' mathematical thinking, and workshops demonstrate how to put this framework into practice in the classroom. A number of resources – including examples of video footage of children carrying out activities – are available to help teachers with developing their own pedagogical content knowledge in this area.

ACTIVITY 2.5

View the footage of children solving addition problems on the following website: <http://www.curriculumsupport.education.nsw.gov.au/countmein/learning-framework_in_number_21.html>.

- What type of problem structures did you observe (result unknown, change unknown, initial unknown)?
- What did the children's approaches to solving the problem tell you about their thinking?
- What prior experiences do you think the children would have had before being asked to solve this problem?
- What other problems could you have given the children that would illustrate other problem structures?

PAUSE AND REFLECT

Think back to the types of problems you were asked to solve in class. How many of them were typically of the 'result unknown' type? What do you see as the limitations in teachers providing students only with these types of problems? What are the benefits of exposing students to a variety of problem types?

Mental computation strategies

As children are exposed to a variety of situations that require them to use addition and subtraction, they should be encouraged to make use of **mental computation** strategies to assist them to solve these types of problem. As demonstrated earlier in this chapter, bridging 10 was an appropriate strategy to use to solve 8 + 7. McIntosh and Dole (2005) endorse this strategy, as well as others, as suitable ways to solve basic addition and subtraction facts:

- *Commutativity (addition).* Reversing the order of the numbers so that the larger comes first (e.g. to solve 2 + 9, it is easier to begin with 9 and add 2).
- *Counting on and back in ones.* Efficient for adding 1, 2 or 3, but not for adding larger numbers.
- *Doubles/near doubles.* Many children find it relatively easy to double numbers; e.g. 'to add 5 and 6, I know 5 + 5, so 5 + 6 is one more'.
- *Bridging 10.* Knowledge of pairs of numbers whose sum is 10 is very valuable and can be used to derive other facts (e.g. 8 + 5; 8 + 2 = 10, + 3 = 13).

These strategies are also effective when adding and subtracting larger numbers. Sometimes the addition of other tools, such as the empty number line, can be used to informally record the thinking process. Figure 2.10 shows a

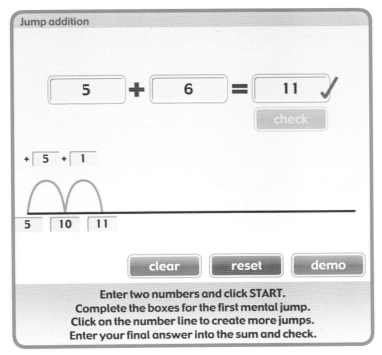

Figure 2.10 Screenshot of HOTmaths Jump Addition widget

screenshot from the HOTmaths Jump Addition widget, where the empty number line is used to add 5 and 6. You can access this widget by logging into HOTmaths and entering its name in the search field. As the widget provides for the entry of different numbers, it can be used to add and subtract single-digit numbers through to multiple-digit numbers. The notes on the left include an explanation of the thinking that occurred when solving 78–43; children should be encouraged to share their strategies and articulate their thinking on a regular basis.

Multiplication and division

Like addition and subtraction, there are problem structures that represent different multiplication and division situations. When the number and size of the groups are known, multiplication is the process to use, whereas division is used when either the number of the sets or the size of the sets is unknown. Multiplication and division are first directly referred to in the Australian Curriculum: Mathematics at Year 2 level, with the following descriptors:

- Recognise and represent multiplication as repeated addition, groups and **arrays** (ACMNA031).
- Recognise and represent division as grouping into equal sets and solve simple problems using these representations (ACMNA032).

Because children need to be familiar with the addition process, it seems logical to base early multiplication experiences around situations that require 'repeated addition'. For example, 5 x 3 can be seen as 'five groups of three' and the answer can be found by adding three five times (3 + 3 + 3 + 3 + 3 = 15). Links can also be made with tools that were useful in learning about place value, addition and subtraction, such as the hundreds chart. The interactive hundreds charts referred to previously are a great source for demonstrating skip counting and investigating patterns. The calculator can also be used to link multiplication and addition and to demonstrate skip counting, as Activity 2.6 demonstrates.

ACTIVITY 2.6

Calculator activities

- Choose a number to count by (e.g. 4).
- Key this into the calculator then press + +. Now press the = sign. With each subsequent press of the = sign, the calculator will 'count' in 4s (4, 8, 12, etc.).

- Have children predict what the next number in the sequence will be before pressing the = sign.
- Try it with different numbers and try predicting what numbers will be in the sequence (e.g. Will we get to count 64? To 83?).

Broken multiplication key

Ask children to use the calculator to find the answer to various multiplication problems without using the multiplication key. For example, 5 × 3 can be found by using the constant function feature (as above) or by keying in 3 + 3 + 3 + 3 + 3 = . Variations on the broken key can include a broken division key (e.g. find 15 ÷ 5 without using the ÷ sign) or a broken number key (e.g. find 6 × 4 without using the 4 button).

Drill and practice

Unfortunately, many children's early experiences with multiplication and division in particular tend to focus on **drill and practice**. There are many examples in the literature of children being unable to make real-life connections with these operations, despite them apparently knowing their 'tables' and being confident about recalling facts such as 7 × 3. Van de Walle, Karp and Bay-Williams (2013), as well as others, recommend that children be given contextual problems that are closely related to their lives, rather than school mathematics. Despite these recommendations, in practice children are often required to complete worksheets, pages of 'sums', and drill and practice routines that are not contextual and do not cater for a range of learning styles and abilities. As drill and practice are the focus of many online interactive resources and sites, it is worth considering the role of this strategy in today's contemporary mathematics classroom.

Van de Walle, Karp and Bay-Williams (2013, p. 71) distinguish between drill and practice, defining practice as 'different problem-based tasks or experiences, spread over numerous class periods, each addressing the same basic ideas', whereas drill refers to 'repetitive, non-problem-based exercises designed to improve skills or procedures already acquired'. While practice can provide students with increased opportunities to develop conceptual ideas and make connections, and develop alternative and flexible strategies, drill is more limited in that it focuses on singular methods and provides a rule-oriented view of what mathematics is about (Van de Walle, Karp and Bay-Williams 2013). While there is a place for drill in mathematics, it need not occur as frequently as it does in today's classrooms, and usually only when automaticity with the skill or strategy is the desired outcome (Van de Walle, Karp and Bay-Williams 2013).

The role of technology in drill and practice

Many websites promote drill and practice as being effective for improving students' learning and recall of facts. While some websites undoubtedly engage students and provide practice with the recall of facts, it is important to be discerning when selecting websites for your students to use. The websites in Activity 2.7 are just two examples of those that use drill and practice to 'teach' number facts. The first website is similar in layout to a page from a textbook, with the added advantage of the user being able to select an operation on which to focus, together with the upper and lower numbers. The user is given 60 seconds to complete each challenge. The second website is visually more appealing, as a picture is gradually revealed as the correct answers to simple addition questions are entered.

ACTIVITY 2.7

Visit the following websites and complete one or two challenges on each site:

- Math.com: Basic Math: <http://www.math.com/students/practice/arithmeticpractice.htm>
- APlusMath: Hidden Picture: <http://www.aplusmath.com/Games/Hidden Picture/HiddenPicture.php>

1 How engaging did you find the activities? Which website appealed to you more and why?
2 What happened on each site when you got an answer wrong?
3 Which activity capitalised more on the affordances that technology can offer?
4 Contrast both activities to the traditional page of 'sums' contained in many worksheets and textbooks. What alternatives are there for consolidating children's knowledge of number facts? How important do you think 'speed' is when completing such activities?

Alternatives to traditional drill and practice

Although currently catering mostly for the early primary years and beyond, HOTmaths contains a number of online activities that offer more than the traditional drill and practice routines. One example of an activity that links multiplication with addition and encourages skip counting is outlined in Activity 2.8.

ACTIVITY 2.8

The following activity, Balloon Bunches, is suitable for Year 3 students. You can access this widget by logging into HOTmaths and entering its name in the search field.

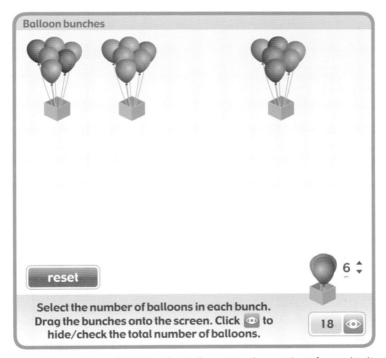

Figure 2.11 Screenshot of HOTmaths Balloon Bunches widget for multiplication

- Open Balloon Bunches.
- Show the total number of balloons. *Example:* Create three bunches of five balloons.
- Record as repeated addition or three groups of five, then introduce or review the multiplication symbol. *Example:* 3 × 5.

- Hide the total value and predict what will happen if another bunch of balloons is added.
- Write this as an expression using the multiplication symbol. *Example:* 4 × 5 = ?
- Discuss strategies to work out the total number of balloons. *Example:* Skip count, count balloons, add 5 to 3 × 5.
- Model a range of multiplication examples, and record them using multiplication number sentences and the multiplication symbol.

How does this activity contrast with the two examples you accessed in Activity 2.7? To what extent do you think it places multiplication into a context to which children might relate? How would you use it in the classroom?

Multiplication as arrays

Arrays are useful models for moving children from additive thinking to multiplicative thinking. An array is an arrangement of rows and columns, such as a rectangle of square blocks. The array can be turned around to demonstrate the commutative property of multiplication:

HOTmaths has an Arrays widget suitable for Year 3 students that can be used to visually demonstrate the concept of arrays and explicitly draw attention to the commutative aspect of multiplication (see Figure 2.12). Through

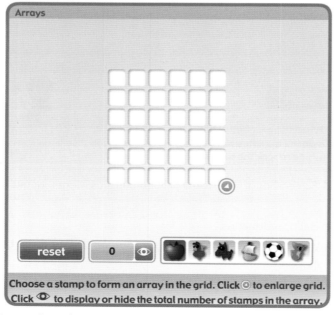

Figure 2.12 Screenshot of HOTmaths Arrays widget

placing objects in the grid squares, children can set up their own arrays and describe them in different ways (e.g. five rows of four can be changed to show four rows of five). Again, you can access this widget by logging into HOTmaths and entering its name in the search field.

Place value

Early number development activities such as those outlined in this chapter can provide the foundation for understanding the base-10 numbers, or that collections of 10 (hundreds, thousands and so on) can be counted. For example, a collection of 25 counters can be seen as 25 individual counters or as two groups of 10 counters, and five additional counters. According to Siemon (2007, p. 3), place value ideas should be introduced as follows:

- Introduce the 'new' unit – that is, 10 ones is one 10 via bundling and counting 10s.
- Introduce the names for multiples of 10 (language only, no symbols).
- Make, name and record numbers 20–99 using appropriate models (for example, make six tens and three ones, read and write 63, record using a place value chart).
- Make, name and record numbers 10–19, pointing out inconsistency in language (should be onety-seven, onety-eight, etc.).
- Consolidate place value knowledge.

Siemon (2007) also advocates that place value should be reinforced through regular activities:

- Make, name, read and record numbers using appropriate models.
- Compare two numbers. Which is bigger? Why? How do you know?
- Order and sequence numbers.
- Count forwards and backwards in place-value parts.
- Rename numbers in as many different ways as possible.

Using technology to support development of big ideas in place value

HOTmaths has a number of widgets that can be used to reinforce these ideas. For example, Jersey Numbers (see Figure 2.13) requires users to place the jerseys in correct ascending or descending order, thereby reinforcing the concept of order and sequencing numbers.

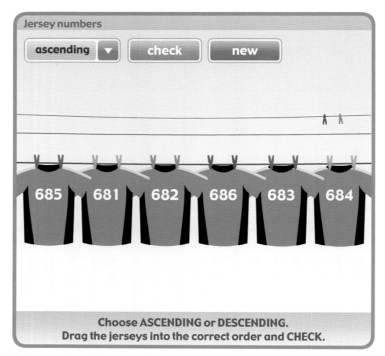

Figure 2.13 Screenshot of HOTmaths Jersey Numbers widget

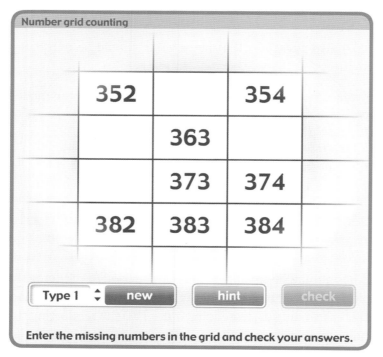

Figure 2.14 Screenshot of HOTmaths Number Grid Counting widget

The sequencing of numbers can also be reinforced through the Number Grid Counting widget, which requires the placement of numbers into empty cells to complete a grid (see Figure 2.14). You can access either of the widgets shown in Figures 2.13 and 2.14 by logging into HOTmaths and entering their respective names in the search field.

Global

Arguably one of the best known models for demonstrating the base-10 system is multi-arithmetic blocks (MAB). Blocks come in joined groups of thousands, hundreds, tens and ones, and can be used to represent numbers into the thousands. Many interactive sites have **virtual manipulative** versions of these blocks, which can be used to make and represent numbers and to operate with numbers.

ACTIVITY 2.9

HOTmaths has a Number Machine widget in which blocks can be clicked and dragged into the appropriate column to represent the numbers displayed on the screen. This is a suitable resource for Year 3 students. Again, you can access this widget by logging into HOTmaths and entering its name in the search field.

Figure 2.15 Screenshot of HOTmaths Number Machine widget

Use the widget to represent some different numbers using concepts of grouping.

- What did you notice happened to the representations of thousands as the numbers increased? How do you think this would affect children's understanding?
- Why do you think MAB materials are used so extensively to teach place value concepts? Can you see any limitations with these materials?

While technology should not drive the curriculum, it can influence what mathematics is taught. According to Goos, Stillman and Vale (2007), teachers' confidence about and use of technology can result in some content or skills being omitted, extended or altered in sequence. With reference to the teaching of early number, are there any aspects of number development that you consider would be more accessible to early learners as a result of technology?

PAUSE AND REFLECT

- According to Perry, Dockett and Harley (2012), the Australian Curriculum: Mathematics suggests that a child's mathematical life begins at school entry, and that what has occurred to the child beforehand is largely irrelevant. Do you agree or disagree with this statement?
- How important do you think it is to consider children's earlier mathematical experiences? How can early childhood educators capitalise on the mathematical knowledge that children bring to school?
- Think about your own early schooling and the approaches your teachers used to teach mathematical concepts. How much focus was placed on engaging the learner? What role do you see technology playing in engaging the learners of today?
- In this chapter, you visited a number of websites aimed at developing early mathematical concepts. Think about which website appealed to you most. What features made it appealing? Did it incorporate mathematics that was important to learn? Did it offer affordances beyond what could be achieved with traditional classroom materials?

Further reading/websites for exploration

Further reading

This chapter provided information about the different problem types associated with addition and subtraction. For further reading about these, and problem types for multiplication and division, see Carpenter et al. (1999).

During recent years, there has been very little research published in Australasian contexts that focuses on early childhood mathematics education and technologies (MacDonald et al. 2012). The following research articles may be of interest, as they document some innovative ways in which technology has been used to promote rich mathematical thinking and sustained engagement for young children:

Highfield, K 2010, 'Robotic toys as a catalyst for mathematical problem solving', *Australian Primary Mathematics Classroom*, vol. 15, no. 2, pp. 22–7.

Yelland, N & Kilderry, A 2010, 'Becoming numerate with information and communication technologies in the twenty-first century', *International Journal of Early Years Education*, vol. 18, no. 2, pp. 91–106.

Throughout 2012, the *Australian Primary Mathematics Classroom* (APMC) has included a regular feature, 'Teaching with Technology'. In vol. 17, no. 2 (2012), Catherine Attard talks about the use of robotics to teach mathematics.

Websites

- ABCya.com: <http://www.abcya.com/one_hundred_number_chart_game.htm>.
- Ambleside Primary: Interactive Counter Square: <http://www.amblesideprimary.com/ambleweb/mentalmaths/countersquare.html>.
- Ambleside Primary: Interactive Number Grids: <http://www.amblesideprimary.com/ambleweb/mentalmaths/interactivenumbergrids.html>.
- APlusMath: Hidden Picture: <http://www.aplusmath.com/Games/HiddenPicture/HiddenPicture.php>.
- Apples 4 the Teacher: Interactive 100 Chart: <http://www.apples4theteacher.com/math/games/100-number-chart-one.html#interactive100chart>.
- Australian Curriculum: Mathematics: <http://www.australiancurriculum.edu.au/Mathematics>
- Butterfly Ten Frame: <http://www.curriculumsupport.education.nsw.gov.au/countmein/teachers_teaching_ideas_-_butterfly_ten_frame.html>
- Count Me in Too: <http://www.curriculumsupport.education.nsw.gov.au/countmein>
- Count Me In: Hundred Chart: <http://www.curriculumsupport.education.nsw.gov.au/countmein/children_hundred_chart.html>
- Crickweb: <http://www.crickweb.co.uk>
- Illuminations: <http://illuminations.nctm.org/ActivityDetail.aspx?ID=73>
- Math.com: Basic Math: <http://www.math.com/students/practice/arithmeticpractice.htm>
- New Zealand Number Framework: <http://nzmaths.co.nz/new-zealand-number-framework>
- Nursery Rhymes 4 U: Counting: <http://www.nurseryrhymes4u.com/NURSERY_RHYMES/counting.html>

CHAPTER 3

Exploring measurement

LEARNING OUTCOMES

By the end of this chapter, you will:

- be familiar with the measurement sequence used to introduce and develop an understanding of each attribute
- have an understanding of the attributes of length, area, volume and capacity, angle, mass, time, temperature, and value and money
- understand the importance of developing estimation skills, spatial awareness and meaningful benchmarks
- understand the principles involved when measuring with units
- understand how technology can be integrated and utilised to support, develop and extend students' measurement experiences.

Measurement is an aspect of the mathematics curriculum that has wide usage in everyday life. A basic level of knowledge, skills and confidence in measurement is very much part of being 'numerate'. An analysis of the measuring process suggests that children learn to measure first by becoming aware of the physical **attributes** of objects and how they compare with other objects. **Estimation** is a significant aspect of measurement, and should be seen as an integral part of the measurement process. The ability to estimate is enhanced when students have strong **spatial awareness** and are able to visualise and represent measurement situations in their head. Students therefore need to be given plenty of opportunities to engage in measurement activities that focus on developing a sound understanding of the attribute being measured, along with the act of measuring.

KEY TERMS

- **Angle:** A measure of the amount of turn between two lines; a figure made of two rays with a common end-point
- **Area:** The amount of space contained within a closed shape
- **Attribute:** A property or characteristic of something
- **Capacity:** The measure of how much a three-dimensional object can hold
- **Estimation:** An approximation or judgement about the attribute(s) of something
- **Hefting:** Estimating the mass of objects by holding them in the hands
- **Length:** A measure of something from end to end
- **Mass:** The amount of matter in an object
- **Non-standard units/informal units:** Everyday materials or objects that can be used to measure various attributes – for example, hands, feet, straws, tiles, marbles
- **Spatial awareness:** The ability to mentally visualise objects and spatial relationships
- **Tangrams:** An ancient Chinese dissection puzzle comprising seven flat shapes
- **Temperature:** The measure of how hot or cold something is
- **Time:** The duration of an event from its beginning to its end
- **Value:** The measure of a cost placed on something
- **Volume:** The amount of space occupied by a three-dimensional object

Learning sequence for measurement

It is generally accepted that there is a learning sequence for measurement that can be used to develop effective measuring processes (e.g. Van de Walle, Karp & Bay-Williams 2013; Booker et al. 2010). It enables the building of understanding in all the measuring topics (**length**, **area**, **volume** and **capacity**, **angle**, **mass**, **time**, **temperature**, and money and **value**).

1 Identify the attribute

The first and most important stage in the sequence is for the attribute to be identified – that is, are we interested in finding out about the length of an object, the area of an object, how heavy something is or how much space it takes up?

Children need to be given lots of opportunities to explore what different attributes mean before being asked to measure in standard ways. For example, if asked to find the height of something, do children realise that this essentially means length? Confusion often occurs between the attributes of area, perimeter (length) and volume, so it is vital that a robust understanding of the attribute is developed before moving on in the sequence.

2 Comparing and ordering

While comparing and ordering will no doubt be part of children's experiences of identifying the attribute, the second stage in the sequence focuses on these aspects and will help to consolidate an understanding of the attribute. Comparing and ordering activities should focus on the attribute and initially involve direct comparison: Who is taller? Whose pencil is the longest? Which tile would cover the most area? Which ball is heavier? Indirect methods may be necessary for some attributes, such as comparing the volume of two boxes, and this can lead to the introduction of non-standard, or informal, units.

3 Non-standard units

Non-standard, or **informal, units** are essential for developing the concept of measuring and for children to see the need for standard units. These units should utilise everyday objects that are familiar to children, and should primarily be uniform in nature and resemble the standard unit. For example, straws, paper clips and pencils are all suitable materials for measuring the length of objects, as they can be laid end to end along the length of something. Playing cards, dominoes and tiles are useful for measuring area, as they can be fitted together to cover a surface without leaving gaps. Most attributes have suitable non-standard units that can be used, although non-standard units for time and temperature are not as readily interchangeable. When students measure with non-standard units, important aspects about measuring can be introduced and consolidated – for example, principles such as that the unit must not change, lining up the units correctly, and leaving no gaps need to be explicitly taught and practised. It is also helpful to measure the same objects with different non-standard units to develop the understanding that there is an inverse relationship between the size of the unit chosen and the number of units required to measure that object. Non-standard units are also useful in their own right – as adults, we often use them as benchmarks for estimating in everyday situations.

4 Standard units

After students have had considerable experience with using non-standard units and understand the principles behind measurement, then they can be introduced to standard units. Encourage students to discuss and think about the limitations of non-standard units and the usefulness of having standard ways by which to communicate measurements. For example, it might be interesting to see that it takes 100 pieces of A4 paper to cover the floor in a room, but this would not be helpful if we were ordering carpet to cover the floor. Similarly, discussion around the different sizes of students' feet can highlight the need for standard units if measuring and communicating the length of the tennis court. The use of measuring instruments needs to be taught explicitly, and their construction discussed. A useful task involves students constructing their own ruler (see Activity 3.1), whereby students are encouraged to make connections between the iteration marks (centimetres) and the unit, and view the ruler as a convenient tool that assists in the measuring process. Students also need to develop an understanding of the relationships between units such as millimetres, centimetres, metres and kilometres, and later to explore the relationships among different units such as millilitres, cubic centimetres and grams, and cubic metres and litres.

5 Application of formulae

When students become proficient at measuring correctly and with selecting and using appropriate standard units, they can be provided with opportunities to use measurement applications and formulae. Students should be encouraged to construct their own methods for calculating areas, perimeters and volumes before being introduced and/or instructed to use 'rules'. For example, covering the surface of a desk with playing cards can demonstrate the relationship between arrays and multiplication and the concept of multiplying the units along the length by the units placed along the width (see Figures 3.1 and 3.2).

Figure 3.1 Cards covering desk

Figure 3.2 Cards partially placed

ACTIVITY 3.1

- Decide on a convenient unit to use (e.g. pen top, pen, mobile phone).
- Cut out pieces of paper that are the length of your unit.
- Stick these along a strip of cardboard to make a ruler.
- How would you allocate numbers on your ruler?
- Use your ruler to measure:
  - the length, width and height of your table
  - the length of an A4 page
  - the length of someone's arm.
- Compare your answers with those of another group/friend. What do you notice?

Measurement topics

The measurement sequence can be used to develop an awareness and knowledge of the attribute being measured, and can be applied to the measurement of any attribute, including geometric measures of length, volume and angle, physical measures of mass, time and temperature, and money and value.

Measurement is included with Geometry as a content strand in both the Australian Curriculum: Mathematics and in the New Zealand Mathematics Curriculum. Throughout their schooling, students are expected to develop an increasingly sophisticated understanding of size and shape, including making meaningful measurements of quantities and choosing appropriate metric units of measurement. In the New Zealand Mathematics Curriculum, for example, students at Level 1 are expected to 'order and compare objects or events by length, area, volume and capacity, weight (mass), turn (angle), temperature, and time by direct comparison and/or counting whole numbers of units'. By Level 6, students are expected to measure at a level of precision appropriate to the task, apply relationships between units and calculate volumes of 3D shapes using formulae.

Beginning in the early primary years, and extending through to Year 10, the Measurement and Geometry strand in the Australian Curriculum: Mathematics (ACARA 2011) includes a sub-strand, 'Using Units of Measurement', which lists the following aspects as being indicative of the teaching emphases in the primary years:

- Use direct and indirect comparisons to decide which is longer, heavier or holds more, and explain reasoning in everyday language (ACMMG006).
- Connect days of the week to familiar events and actions (ACMMG008).
- Measure and compare the lengths and capacities of pairs of objects using uniform informal units (ACMMG019).
- Compare masses of objects using balance scales (ACMMG038).
- Tell time to the quarter-hour, using the language of 'past' and 'to' (ACMMG039).

Interestingly, while many descriptors are general in nature and particular attributes are not mentioned every year (e.g. only length and time are referred to in Year 1), time is mentioned frequently, and there is an emphasis on telling the time and using units of time. Metric units are first mentioned in Year 3, with calculation of perimeter and area (of rectangles) referred to in Year 5. Formulae are first mentioned in Year 7. Money (along with financial mathematics) is included under the Number and Algebra content strand from Year 1 through to Year 10.

Within the Australian Curriculum: Mathematics (ACARA 2011), the proficiency strands should also be used to engage students in understanding measurement concepts and solving problems involving measurement contexts. Many of these contexts require students to engage in spatial thinking, such as indirectly comparing the lengths or areas of different objects.

Each of the measurement attributes mentioned earlier and referred to in the Australian Curriculum: Mathematics will now be discussed.

Length

Length refers to the measurement of something from end to end, and is usually the first attribute introduced to children. Children should be given frequent opportunities to investigate length measures, initially with straight lines, then extension into curved distances and length around plane shapes. Because length is often the first attribute encountered by children, teachers need to capitalise on both developing the concept and building an understanding of what it means to measure and estimate. Activities that help children to develop an appreciation of comparing and ordering, and discussion of appropriate vocabulary, are particularly valuable in the early years of schooling. The following are examples of length activities that could be used with a younger class to develop the concept of length:

- In small groups, ask each child to make a play dough worm. Compare the length of each worm to determine who made the longest worm.
- Make class books based around the attribute of length (for example, the title of the book could be: 'I am tall enough to …'; or 'The gate is wide enough to …') (Department of Education and the Arts, Tasmania 1994).
- After reading *The Gingerbread Man*, provide students with gingerbread family members and ask them to place them in order from shortest to tallest (see the NZmaths website, <http://www.nzmaths.co.nz/resource/gingerbread-man?parent_node=>).

PAUSE AND REFLECT

A class of Prep children (ages 5 and 6) were asked to bring in their teddies from home, with the aim being to find out who had the 'biggest' teddy.

- What discussions might be stimulated by such an activity?
- Would this be an appropriate activity for developing an understanding of the attribute of length?

Area

Area is a two-dimensional concept that refers to the amount of space within a closed region. Experiences that encourage children to investigate covering surfaces should be provided early on and, where possible, direct comparisons should be made. Where such direct comparisons cannot be made, Van de Walle, Karp and Bay-Williams (2013) recommend activities in which one area is rearranged. Cutting a shape into two parts and reassembling it into a different shape can show that the 'before' and 'after' shapes have the same area, even though they are different shapes. This notion is often not readily apparent to younger children, and relates to the ability to conserve area. Students tend to confuse the concepts of area and perimeter. This may be attributable to the tendency to teach the two concepts in tandem – particularly in the older grades – and a heavy reliance on memorising formulae to calculate them both. The potential for confusion and further discussion of students' misconceptions about area is discussed later in this chapter. The following are examples of activities that could be undertaken to develop an understanding of the attribute of area.

Hand prints

Give each child a piece of A3 paper and ask them to estimate how many of their hand prints it would take to cover the paper. Have children make prints of their hand and cover the surface of the paper. Discuss how many it took, compare with other children's answers and discuss the notion of 'gaps' remaining between the hand prints.

Icing the cake

Give children three different-shaped 'cakes': rectangle, square and circle. Ask them to predict which cake would need the most icing to cover its top. Discuss ways in which the areas could be compared and/or the use of non-standard units.

Tangrams

Tangrams, or dissection puzzles, are a useful resource for investigating size and shape concepts. Provide students with the outline of several shapes made with tangram pieces and encourage them to use the pieces to decide which shapes have the same area.

ACTIVITY 3.2

Visit the tangram site at Math Playground, <http://www.mathplayground.com/tangrams.html>. Complete three or four of the activities. How could you use the experiences to develop students' conservation of area?

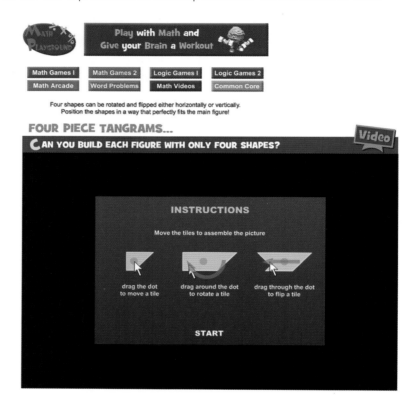

Figure 3.3 Tangram screenshot

Volume and capacity

Volume refers to the three-dimensional space that a figure or object occupies, while capacity is a measure of how much that three-dimensional object can hold. In primary school, most of the experiences children encounter are really about measuring capacity, and involve the comparison of different containers. As with area, some children will not have developed conservation of volume – believing, for example, that a taller container will always hold more than a shorter container, even if the amount of liquid in each is the same.

Water, sand, rice and beans are all appropriate non-standard units that can be used to compare and measure the capacity of various containers. Activity 3.3 is indicative of the types of experiences in which young children should be involved to investigate capacity.

ACTIVITY 3.3

Provide students with a series of five or six labelled containers of different sizes and shapes, with the task being to place them in order from least volume (or capacity) to most. This can be quite challenging. Children should be encouraged to work in groups to come up with a solution and to explain how they worked it out.

Figure 3.4 Measuring liquids

Angle

The attribute of angle can be described as 'the spread of the angle's rays', as angles are composed of two rays that are infinite in length with a common vertex. In practice, there is a tendency to leave the study of angles until later grades, where students' first encounter often involves the use of protractors, rather than activities that help them to identify the concept. Young children can be introduced to the attribute of angle and be involved in comparing and ordering activities that may then result in a more conceptual understanding of the attribute when exposed to more formal experiences further on in their schooling. Activity 3.4 is an example of an activity that would be appropriate to use with younger children to explore the concept of angles.

ACTIVITY 3.4

Angles in the environment

Take the class on an angle hunt. Discuss what an angle is and where we see them in the environment.

Angle turner

Using two different-coloured circles, make an angle turner Use the angle turner to represent different angles – for example, 'Show me what angle is made when it is three o'clock on an analogue clock,' or 'Show me what angle the door is open to.'

As with the other attributes, the use of standard units and formulae should occur later in the measurement sequence. In a similar way to using non-standard units to measure the length and area of different objects, students should also be given the opportunity to measure angles with non-standard units. Van de Walle, Karp and Bay-Williams (2013) suggest that a wedge can be made from a piece of cardboard and used to measure different angles by counting the number of times it will fit into a given angle. Students could then use their individual wedges to measure similar angles and discuss the reasons for different results that are related to the size of their unit.

PAUSE AND REFLECT

Many students have difficulties with measuring angles and reading a protractor. What do you think contributes to these difficulties? There are a number of websites that provide online protractors and require the use of protractors to measure angles. One example, at Teacher Led, <http://www.teacherled.com/resources/anglemeasure/anglemeasureload.html>, provides a protractor and activities suitable for use on the Interactive Whiteboard (IWB). Identify how you would model the use of the protractor through this site.

Chapter 3: Exploring measurement 57

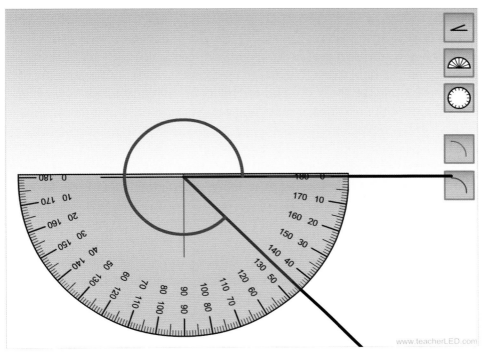

Figure 3.5 Interactive protractor from Teacher Led

ACTIVITY 3.5

Follow the directions below to make your own wax protractor. What would be the benefits of having students complete such an activity? Use your protractor to measure some different angles drawn on a piece of paper. What do you notice?

Sources:

- Schoolwires: <http://shusd.schoolwires.com/cms/lib6/CA01000647/Centricity/Domain/161/4thgd_unit_6.pdf>
- Everyday Math Online: <https://emccss.everydaymathonline.com/em-crosswalk/pdf/4/g4_tlg_lesson_6_6_mm_pages_190_and_507_509.pdf>

Mass

Although the terms 'mass' and 'weight' are often used interchangeably, it is important to remember that they are different. Mass is the amount of a substance, while weight is the pull of gravity on that substance. Having said this, on earth the measures of mass and weight are about the same, and measurement experiences in primary school generally involve measuring weight. One of the earliest understandings to develop with students is the notion that mass is not necessarily proportionate to volume, and that an object's mass cannot be calculated by looking at it. Early experiences should require children to compare the mass of two objects by holding one in each hand (**hefting**) in order to demonstrate what 'heavier' means. Balance scales can then be used to demonstrate how one side of the scale lowers when the object is heavier. Non-standard units such as marbles, cubes or washers can be used to 'calculate' the mass of different objects before introducing children to standard units such as grams and kilograms. As mass is a physical measure, children need opportunities to be physically involved in weighing different objects, and comparing and ordering the weights of these objects. The following activity is an example of a suitable experience to engage young children in focusing on the attribute of mass.

Play dough animals

In small groups, have each child construct a worm using play dough. Discuss whether or not the longest worm would also be the heaviest. Try different ways to determine this. Place the worms in order of lightest to heaviest.

Ask the children to now make their worm into a cat. Will their cat be the same weight? Why/why not? Now get them to make the cat into different-sized balls. Which ball is the heaviest? What would happen if all the balls were placed on the balance scale? Would they weigh the same as the cat? The worm? (see NZmaths, <http://www.nzmaths.co.nz/resource/seesaws?parent_node=>).

Time

Time can be thought of as the duration of an event from its beginning to its end, and it differs from other attributes in that it cannot be perceived through sensory experiences in the way that other measurement concepts can. Time is also subjective according to context – something can take a long time or a short time, depending upon the situation and people's perceptions; similarly, one hour can seem a long time when you are waiting for a bus or a short time if you are playing a game.

There is a tendency in classrooms to focus instruction on time around reading the time or a clock face. As the clock represents an instrument for telling time using the standard units of minutes and hours, focus on this aspect should occur later in the measurement sequence. We do not, for example, expect young children to measure with a ruler or scales, yet many children are expected to be able to tell the time from a very early age. As with the other attributes, children need to be given plenty of opportunities to develop an understanding of the attribute of time, as well as experiences in comparing and ordering time and measuring time with non-standard units, before being introduced to standard units. Experiences that require students to order the duration of events can be used to establish a feel for how long something takes, and activities such as placing events in their correct order can help develop the concept of time as a sequence. Non-standard units can involve the use of hand claps, pendulum swings and sand running through a bottle. Students often show confusion with reading analogue clocks, which can be related to the different actions and functions of the two hands. Van de Walle, Karp and Bay-Williams (2013) recommend the use of a one-handed clock, which allows a focus on approximate language (for example, it's about seven o'clock; it's a little past nine o'clock).

ACTIVITY 3.6

There are hundreds of apps and interactive websites that have a focus on clock reading. Visit the following sites:

- Bang on Time: <http://resources.oswego.org/games/BangOnTime/clockwordres.html>
- On Time!: <http://www.sheppardsoftware.com/mathgames/earlymath/on_time_game1.htm>.

The first site requires users to stop the clock when the required time is reached, while the second site requires users to move the hands on the clock to match given times. Explore both sites and take note of what happens when incorrect answers are given. Would you recommend the use of either site by students? What understandings would you expect students to develop after completing the activities? Why do you think so many websites focus on clock reading, rather than trying to develop the concept of time as an attribute?

ACTIVITY 3.7

Look at the screenshot below from the HOTmaths Analog Clock widget. What do you notice about the way the clock is numbered? What do the blue numbers show? How helpful do you think this would be in assisting students to tell the time? This widget can be found by logging into HOTmaths and entering its name in the search field.

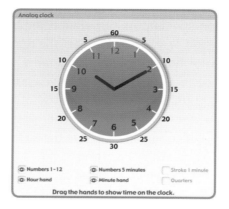

Figure 3.6 Screenshot of HOTmaths Analog Clock widget

Temperature

Temperature is the measure of how hot or cold things are, with the judgement of this being dependent upon people's perceptions or feelings. Children will typically use terms such as 'cold', 'warm' or 'hot' to describe the weather, food and drinks, and it is likely that the interpretation of these terms will vary among individuals. There are no non-standard units for temperature, and the use of standard units to describe temperature is complicated by the use of the different scales (Celsius or Fahrenheit) and the inclusion of negative numbers. Students' experiences with temperature should begin with a focus on the attribute and developing the understanding that describing temperature is based on sensory experiences – for example, a surface may feel hot because your hand is cold. Before introducing students formally to the thermometer, it is useful to have them design their own thermometers and include descriptions of their own developed scales to designate a range of temperatures.

Money and value

Money is the unit used to measure the value or cost of something. As with the other attributes, it is important to establish an understanding of what value means rather than beginning with a focus on operating with currency. In the primary grades, students typically are required to:

- recognise coins and notes
- know the value of coins and notes
- create equivalent coin collections
- make change
- round money amounts up or down.

In the early years, many worthwhile experiences and learning opportunities can be provided through the establishment of a class shop, which can be used for free and directed play. The relative value of items can also be discussed, and games such as 'The Price is Right' can be played to compare and order the cost of various items. Non-standard units can take the form of counters or stickers before formal units are introduced. In the Australian Curriculum: Mathematics, money and value are included under the Number and Algebra strand, in 'Money and Financial Mathematics', and are first mentioned at Year 1: Recognise, describe and order Australian coins according to their value (ACMNA017). Where possible, provide lots of opportunities for students to calculate and give change using real money before expecting them to carry out purely written computational tasks. Interesting investigations can be carried out with older students on currencies used in other countries and exchange rates. A currency converter can be found at XE Currency Converter, <http://www.xe.com/ucc>.

Estimation

Estimation is an integral part of everyday life, as many measurement situations do not require exact measurements (Hodgson et al. 2003). The significance of estimation as an everyday and natural aspect of measurement needs to be conveyed to students, yet many students tend to view it as a difficult technique where success is dependent upon how close their estimate is to that of the teacher (Muir 2005). Muir outlines eight principles that can be used in classrooms to enhance the value of measurement estimation experiences for students:

- Estimation is useful.
- Estimation should be related to real life.

- Estimation saves time.
- Estimation experiences should be purposeful and relevant.
- The ability to estimate is enhanced through practice and over time.
- Personal benchmarks are useful referents for estimation.
- Sometimes it is better to over- or under-estimate, depending on the context.
- Estimation is used to validate measuring tools and methods.

CLASSROOM SNAPSHOT 3.1

The following is adapted from an article (Muir 2012) based on a lesson conducted with a Year 5/6 class.

The lesson began with the whole class sitting in front of the whiteboard. It was introduced by posing three questions, one at a time, to the class and soliciting three different responses to each question. The questions were:

- How far is it from Launceston to Hobart?
- How far is it from the earth to the sun?
- How old is your teacher?

Three responses for each question were recorded on the board and students were then asked to vote for the most reasonable answer. Discussion then occurred on the reasons for the vote, what referents were used to make estimates, which questions were 'easier' to estimate and why. The activity was particularly useful for gauging which referents were used by students. For example, students indicated that they found it easier to estimate the teacher's age because they could gauge whether or not she was older than their parents, whereas distances posed more of a problem.

How big, how tall, how many?

The next activity required the students to provide (and receive) answers to a number of questions that did not have an obvious answer and involved larger numbers (e.g. How high is Mount Everest? What is the population of China? How tall is the world's tallest man? How long would it take you to count to a million?). *The Guinness Book of World Records* (Glenday 2007), Google and the Australian Bureau of Statistics website, <http://www.abs.gov.au>, are all useful resources for questions. Cards containing the questions were stuck on the students' backs without the question being revealed to its wearer (see Figure 3.7). Each student then asked three people to provide an answer to the question, and in turn provided answers to others' questions. The responses were recorded on a pro forma. After everyone had recorded three responses, the class regrouped and answers were discussed. Before the students were allowed to look at their questions, they first had to decide on an appropriate answer, based on the responses received, and second identify an appropriate question that would 'fit' the answer. The students were very curious about what their questions were, and also about the 'correct' answers.

One student, Susan, received three responses to her question, 'How high is Mount Everest?' Is it 1000 m, 2000 m or 500 000 m?' She chose '2000 m' as a reasonable answer (the correct answer is 8848 m) and predicted that her question was 'How far is it from somewhere to somewhere?'

Figure 3.7 Reading the question

PAUSE AND REFLECT

When the activities above were conducted with the class, the teacher was surprised that many students did not demonstrate a 'sense of the relative and absolute magnitude of numbers', and did not utilise a 'system of benchmarks' with which they could operate to make reasonable estimates (Muir 2012). How would you capitalise on the experiences from this lesson to develop these aspects? How could ICT be utilised to demonstrate the magnitude of numbers?

ACTIVITY 3.8

Classroom snapshot 3.1 referred to sources such as *The Guinness Book of World Records* and the ABS website as useful resources for generating questions and investigations. Visit *The Guinness Book of Records* website, <http://www.guinnessworldrecords.com>, and then use either the 'Explore Records' or 'Search Records' tab to locate records related to size. Search for the world's tallest man and view the linked YouTube clip. Either during or after watching the clip, try to answer the following questions:

- How tall do you think Sultan Kösen is?
- How tall do you think the waiter is? The driver of the car?
- What would be the height of an average doorway?
- How close would Sultan's head be to the ceiling in your house?

Now pause and reflect on how you came up with your answers. What benchmarks did you use to make your estimates?

How did you use your ability to visualise situations to help you make reasonable estimates? How do you think not having this ability would affect students' capacity to make reasonable estimates?

Further explore the site to identify three examples of pictures or reports that could be used as source questions for a lesson similar to that outlined in Classroom snapshot 3.1.

Making estimation meaningful

As mentioned earlier, many students tend to view estimation as a 'guess', and fail to see it as a useful or meaningful experience. One way to make estimation more meaningful is to allow students to revise their estimates. Try the following:

- Choose a distance to estimate – for example, the length of your classroom.
- Choose an informal unit to measure with – for example, your foot.
- Estimate how many of your feet (end to end) it would take to measure the classroom from one end to the other and write it down.
- Next, take five steps (placing feet touching), then pause. Look at your estimate and adjust if necessary.
- Take five more steps in the same way. Pause and adjust your estimate if necessary.
- Keep going in this way – you can revise your estimate as many times as you like.

Through engaging students in similar experiences, you can maintain their interest throughout the measuring process and facilitate the understanding that estimation is useful, purposeful and has real-life applications.

ACTIVITY 3.9

Go to the *The Guinness Book of World Records* website, <http://www.guinnessworldrecords.com>, and look up the world's longest snake. Think about how you could use the picture of the snake to develop the first three stages in the sequence of measurement: the attribute of length; comparing and ordering length; and measuring with informal units.

Measuring with non-standard units

It is worth spending time on this aspect of the sequence whenever a new attribute is being taught or formally introduced. Students of all ages should be given opportunities to engage in measuring experiences with informal units prior to being introduced to standard units. These experiences should be aimed at developing measurement 'principles', and can help students see how and why standard units are necessary. The following understandings should be developed with students as they participate in measurement activities using informal units:

- The unit must not change – for example, we should select one type of informal unit, such as straws, to measure the length of the table, rather than a straw, a pencil and a rubber.
- The units must be placed end to end (when measuring length), with no gaps or overlapping units.
- The units need to be used in a uniform manner – that is, if dominoes are being used to find the area of the top of a desk, then each domino needs to be placed in the same orientation in order to accurately represent the standard unit.
- There is a direct relationship between the size of the unit and the number required – that is, the smaller the unit, the bigger the number, and vice versa.

While students should be involved in physically selecting units and measuring with them, there are a number of interactive sites and applications that can be used to supplement and consolidate classroom experiences, and to focus on developing the understandings associated with measuring with informal units.

Although mostly focused on developing children's understanding of number, the Count Me in Too website at <http://www.curriculumsupport.education.nsw.gov.au/countmein/index.htm> does contain some measurement activities that are appropriate for younger children. Plasticine Snakes (see Figure 3.8), for example, requires children to:

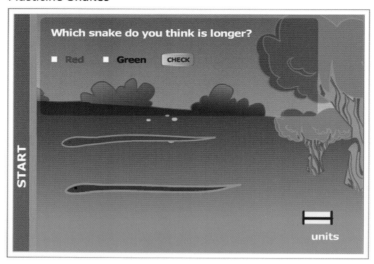

Figure 3.8 Screenshot from the Count Me in Too website showing Plasticine Snakes

Estimate and select the snake that you think is the longest, either the red or green. Drag the top snake to the starting line. Click and drag a measuring unit below the snake. Continue to click and drag measuring units to measure the length of the snake. Type the number of units used in the white box. Drag the other snake to the starting line and use the units to measure the length of the snake. Write the number of units used in the white box and press 'check'.

Although this is an independent activity, the teacher still plays a vital role in encouraging discussion around the principles associated with using informal units efficiently, as previously highlighted. Interesting conversations could occur around the reasons for incorrect answers, such as leaving gaps between the units.

HOTmaths offers widgets that include opportunities to measure length, mass and capacity using informal units. Figure 3.9 shows a screenshot from an activity that requires users to order the objects from lightest to heaviest, using a pan balance and marbles as informal units. While not advocating this as a substitute for gaining a sense of the mass of an object through **hefting**, the widget does provide a visual representation of how the scales are balanced when the mass of both sides is the same. This widget can be accessed by logging into HOTmaths and entering its name in the search field.

Global

Figure 3.9 Screenshot of HOTmaths Lightest to Heaviest widget

Measurement misconceptions and difficulties

Conservation

Many of the common misconceptions that students display in relation to measurement involve the concept of conservation. Young children are said to conserve the attribute of length if they recognise that the length of a given piece of string does not change according to whether or not it is curled up or laid straight. Similarly, the conservation of capacity can be demonstrated by recognising that four cups of water is still the same amount, regardless of the height and width of a container. Many middle primary and older students do not readily understand, for example, that rearranging areas into different shapes does not affect the amount of area. Cutting a shape into two parts and then reassembling it into a different shape can show that the before and after shapes have the same area, even though they are different shapes. The tangram activity in Activity 3.2 could also be used to help develop students' awareness of this. NZmaths also has an activity where students can develop spatial awareness skills through making pentominoes from 5 centimetre cubes (see 'Outlining Area', <http://nzmaths.co.nz/resource/outlining-area>).

ACTIVITY 3.10

Global

HOTmaths has a widget called Measuring Capacity (Figure 3.10), where users are required to fill three different-sized containers with cups of liquid. This widget can be found by logging into HOTmaths and entering its name in the search field. Do the activity three times. What do you notice about the number of cups taken to fill the containers? In the screenshot, the tallest container actually holds the least amount of liquid. How could you capitalise on this teaching opportunity with children?

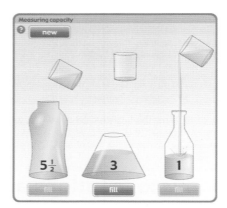

Figure 3.10 Screenshot of HOTmaths Measuring Capacity widget

Reading scales

Ryan and Williams (2007) found that confusion about the 'ticks' and 'intervals' on measuring devices was a common error made by students across all ages. Other common errors involved always interpreting the end number on the ruler as being the 'answer', regardless of whether or not the ruler was lined up with 0 to begin with, and misinterpreting the scale markings, believing that each interval or marking was one complete unit.

There are many interactive sites that provide opportunities for children to focus on reading instruments and scales. The ICT Games website, <http://www.ictgames.com/weight.html>, for example, includes a Scale Reader activity in which children can read and record the weight indicated by the scale. It is quite challenging in that children need to work out what scale is being used and then accurately read and record the measurement (which is sometimes between intervals) (see Figure 3.11).

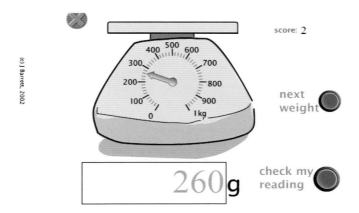

Figure 3.11 Scale Reader screenshot
Source: Printed with permission from <http://www.ictgames.com/weight.html>.

ACTIVITY 3.11

As a follow-up, students could access the HOTmaths Mystery Mass widget (Figure 3.12), which gives practice with scale reading. It is supported by the Estimating and Measuring Mass HOTsheet (Figure 3.13). You can access both of these resources by logging into HOTmaths, entering their respective names in the search field and selecting the appropriate results tab.

Figure 3.12 Screenshot of HOTmaths Mystery Mass widget

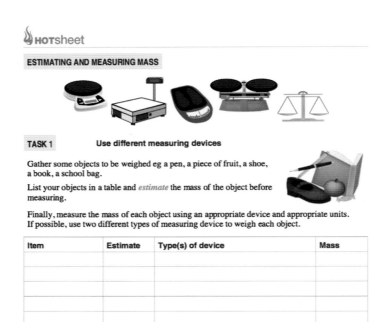

Figure 3.13 Task 1 of the Estimating and Measuring Mass HOTsheet

Broken ruler

Figure 3.14 shows a broken ruler and a segment to be measured. This is a task that is commonly used on test items, and it requires students to use the ruler to measure the length of the segment. It is a useful assessment item in that students who do not have a solid understanding of what the intervals on a ruler represent are likely to indicate that the segment is a bit more than 7 cm long. It would be useful to project a 'broken ruler' on the IWB and discuss how it could be used to measure different objects. It would also be valuable to provide students with a number of incorrect responses to the question and to have a class discussion on possible reasons for the incorrect responses.

There are a number of web-based resources that feature broken rulers, and these vary in terms of providing useful learning opportunities. The Length Strength: Centimeters website, <http://www.hbschool.com/activity/length_strength1_centi>, provides an activity that uses a broken ruler to measure a

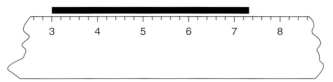

Figure 3.14 Broken ruler

number of different objects. Unfortunately, the ruler is only 'broken' at the end and does not require students to count intervals from different starting points. A better example is provided at the Broken Rulers: Measure Me website, <http://maine.edc.org/file.php/1/tools/BrokenRulers_MeasureMe.html>, but is limited in that the ruler uses inches and the measurements are of pictures of pets, and hence provide unrealistic results.

Area and perimeter confusion

In the Mathematics Assessment for Learning and Teaching (MALT) project, researchers found that 32 per cent of 13-year-old students calculated the perimeter of a shape, rather than the area, to find a missing dimension (Ryan & Williams 2007). They also found that while 60 per cent of 14-year-olds could calculate the 'distance a referee ran around a rugby pitch 90 m long and 60 m wide', 14 per cent of them calculated the area and another 12 per cent simply added 90 and 60. In her seminal study, Ma (1999) found that many practising teachers believed there was a constant relationship between the area and perimeter of a rectangle, and that whenever the perimeter of a rectangle increases, the area also increases. Livy, Muir and Maher (2012) also found that this was a misconception held by many pre-service teachers.

ACTIVITY 3.12

Mary says that whenever you increase the perimeter of a rectangle, the area also increases. John says this is not true. Who is correct – Mary or John?

Visit the National Library of Virtual Manipulatives website at <http://nlvm.usu.edu/en/nav/vlibrary.html> and access the virtual geoboard (see Figure 3.15).

Investigate making a number of different rectangles that have a perimeter of 16 centimetres.

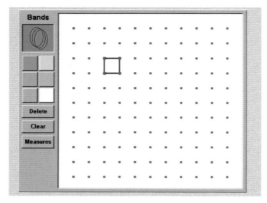

Figure 3.15 Virtual geoboard

- Which one has the largest area?
- What do you notice about the shape of the rectangles (remember that a square is a rectangle)?
- What conclusions can you draw about the relationship between the area and perimeter of a rectangle?

A focus on activities that involve covering a shape with tiles and counting the tiles within it can help to reinforce and consolidate students' understanding of the attribute of area.

CLASSROOM SNAPSHOT 3.2

After reading *Zack's Alligator* (Mozelle 1989), the teacher showed her Year 3/4 class a 'growing creature' that she had purchased from a novelty shop. On the packet it claimed that the starfish would increase 600 per cent when placed in water for 72 hours. This provided an ideal opportunity to investigate area as an attribute, and to introduce students to square centimetres as the standard unit for measuring area. Each day, one of the students placed the starfish on the grid paper and traced around it. If the claim was correct, after three days the starfish should have reached its maximum size. The class decided to keep measuring for five days to ensure maximum growth had been achieved. The tracings were compared informally and then students counted the squares to more accurately record the results. The tracings were projected on an IWB, which clearly showed the grid squares and raised the issue of how to count partial squares (see Figure 3.16). Different colours were then used to match up partial squares with the counting and recording of squares modelled with the whole class. The class found that the starfish grew in area from approximately 44 cm^2 to 204 cm^2 – an increase of about 400 per cent rather than 600 per cent.

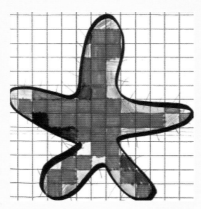
Figure 3.16 Colouring partial squares to determine the area

Classroom snapshot 3.2 demonstrates the practical nature of measurement and how technology can be incorporated to enhance measurement experiences. In addition to investigating the change in the area of the starfish, the class was also immersed in other experiences that focused on the attribute of area. As the result of a number of practical activities and experiences, students were able to make connections between the multiplicative nature of arrays to discover the formula for area for themselves. Through covering objects with tiles, for example, students were able to determine that a quick and efficient way of counting all the tiles was to count the tiles across the top and bottom and multiply them together (see Figure 3.17).

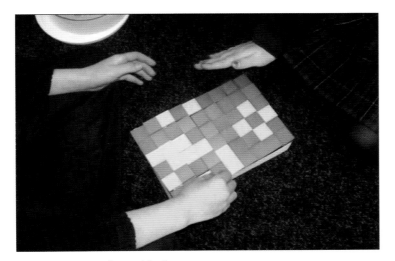
Figure 3.17 Covering a surface with tiles

Extending measurement concepts into older years

CLASSROOM SNAPSHOT 3.3

Having taken digital photos of 'Mr Splash' (Figures 3.18 and 3.19), Mrs Jones, a Year 7 teacher, projected the images on the IWB and introduced the following problem to the students:

> 'I have a photograph of Mr Splash. I wonder if we can work out how tall he might be?'

Figure 3.18 Mr Splash
Source: Photos courtesy of Sharyn Livy.

Figure 3.19 Measuring Mr Splash

Students were asked to volunteer their estimates of the height of Mr Splash, and encouraged to explain how they decided upon their predictions. They were then asked to compare their predictions with their own heights and those of other people in the class. In order to more accurately determine class heights and measurements, they then worked in small groups to measure heights and some body dimensions. The data were entered into TinkerPlots (Konold & Miller 2005), with data recorded for every student (Figure 3.20).

Each group was then allocated one relationship to investigate and subsequently to produce a graph – for example, height compared with head length, arm span or leg length (this can be done by hand or using technology). One group produced the TinkerPlot graph in Figure 3.21, showing the relationship between height and foot length.

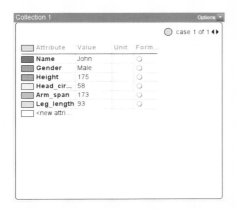
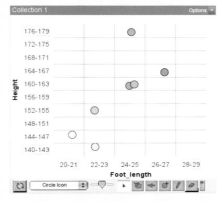

Figure 3.20 TinkerPlots – entering data   **Figure 3.21** TinkerPlots – interpreting data

PAUSE AND REFLECT

Using the data collected in Classroom snapshot 3.3, what other relationships could be investigated? In what ways do you think the use of TinkerPlots contributed to students' ability to complete the task? How do you think students could have undertaken the investigations without the use of technology? Would it have been as effective?

TinkerPlots is an example of a program that can be incorporated into older years to enhance and extend measurement concepts. It is further referred to in Chapter 8, and would be a useful resource for the upper primary and early secondary years. Data can be sourced from students' own investigations (for example, 'People say you are as tall as your arm span – investigate') or from the Australian Census at School site at <http://www.abs.gov.au/censusatschool>.

Establishing formulae for areas and volumes

According to the Australian Curriculum: Mathematics (ACARA 2011), students in Year 7 are expected to use units of measurement to:

- establish the formulae for areas of rectangles, triangles and parallelograms, and use these in problem-solving (ACMMG159)
- calculate volumes of rectangular prisms (ACMMG160).

In Year 8, these expectations are expanded to include:

- choosing appropriate units of measurement for area and volume and converting from one unit to another (ACMMG195)
- finding perimeters and areas of parallelograms, rhombuses and kites (ACMMG196)
- investigating the relationship between features of circles such as circumference, area, radius and diameter, and using formulae to solve problems involving circumference and area (ACMMG197)
- developing the formulae for volumes of rectangular and triangular prisms and prisms in general, and using formulae to solve problems involving volume (ACMMG198)
- solving problems involving duration, including using 12- and 24-hour time within a single time zone (ACMMG199).

These expectations are consistent with the New Zealand Level 6 expectations.

There are a number of interactive sites and resources that can be used to develop an understanding of the above concepts. The Illuminations website, <http://illuminations.nctm.org>, contains area, volume and capacity tools that can be used to investigate, for example, how the base and height of a figure can be used to determine its area. Figure 3.22 shows a screenshot of an interactive figure that can be manipulated to determine the area.

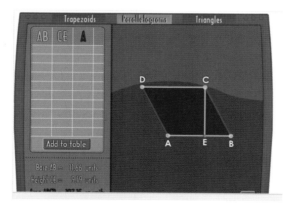

Figure 3.22 Screenshot showing area tool from Illuminations website

HOTmaths has a number of useful widgets that provide interactive opportunities to reconstruct figures much more easily than can be done by hand. Figure 3.23 shows a screenshot of a widget that can be manipulated to demonstrate how a section of a parallelogram can be sliced off and repositioned to resemble a rectangle and the steps involved for calculating the area. This widget can be accessed by logging into HOTmaths and entering its name in the search field.

Chapter 3: Exploring measurement

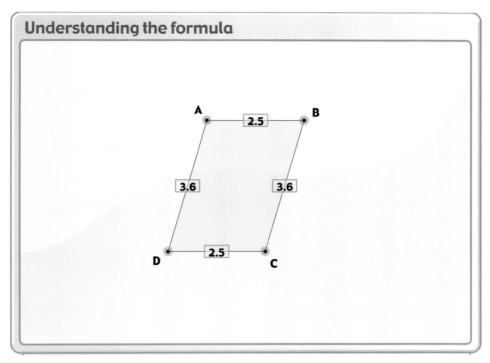

Figure 3.23 Screenshot of HOTmaths Understanding the Formula widget

ACTIVITY 3.13

Drill and practice widgets are also available to provide practice with calculating volume. One example is a game called Meltoids (see Figure 3.24). It is important to note that such activities should only be accessed once students have a solid understanding of the relevant attribute, have been immersed in the appropriate measurement sequence and have had the opportunity to derive and investigate formulae for themselves. As with all interactive online activities, learning will be maximised through input and guidance from the teacher.

You can access this resource by logging into HOTmaths and clicking on the Games icon on the Dashboard.

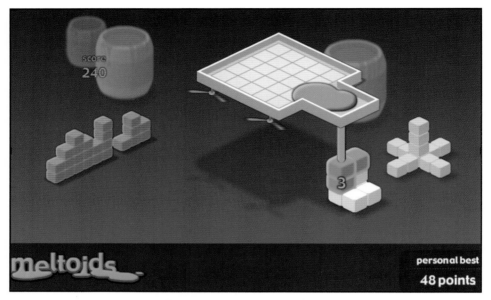

Figure 3.24 Screenshot of HOTmaths Meltoids game

Angles and triangles

Although generally accepted as a topic within measurement, in the Australian Curriculum: Mathematics, **angle** is included under 'Geometric Reasoning'. As discussed earlier in this chapter, angle can be considered an attribute about which understandings can be developed through applying the sequence to teach measurement. In relation to angle, the final stage in the sequence, application of formulae, can be demonstrated through investigating the sum of the angles in a triangle.

ACTIVITY 3.14

Students are often told that the sum of all the angles in a triangle totals 180 degrees, but how many of them have developed an understanding of why this is the case? HOTmaths has a useful widget that demonstrates clearly why this occurs, without focusing on manipulating the numbers (see Figure 3.25). This widget can be found by logging into the HOTmaths site at <http://www.hotmaths.com.au> and typing its name into the search field. The activity can also be undertaken in class, using paper triangles and tearing off the corners.

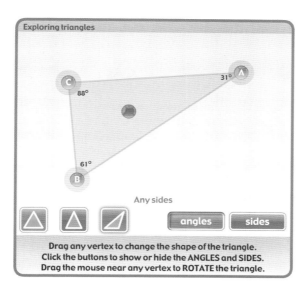

Figure 3.25 Screenshot of HOTmaths Exploring Triangles widget

Area and circumference of a circle

The relationship between the circumference of a circle (distance around the outside, or perimeter) and the length of the diameter can be investigated at a number of levels. The circumference and diameter of circular items such as lids, cans and jars can be measured and entered into a table. Plots can be made of the data using TinkerPlots or graphic calculators. The plot should show that most ratios would be approximately 3:1 or 3:2, with a straight line showing through the origin. The exact ratio (pi) is an irrational number that is about 3.14159. There are a number of online resources that demonstrate the origin of pi and how it was derived. It is useful to share some of these descriptions with students, particularly if they demonstrate the rearrangement of a circle's segments into eight or more parts.

Khan Academy is a website that is popular with students in the older years. It contains a number of video presentations in the form of tutorials demonstrating how to carry out various operations and procedures. The following link is to a tutorial based on finding the radius, diameter and circumference of a circle: <http://www.khanacademy.org/math/geometry/circles/v/circles--radius--diameter-and-circumference>. Again, it needs to be emphasised that the application of formulae occurs late in the measurement sequence, and students need to be provided with lots of opportunities to discover formulae for themselves.

Investigating times and time zones

World time zones can be accessed through The World Clock website, <http://www.timeanddate.com/worldclock>. This site can be used to determine the respective times of cities throughout the world. Cross-curricular links could be encouraged through mapping the zones on a world map and using some of the information to predict which cities would be 'ahead' or 'behind' Australia in time. Further investigations could involve the use of travel websites to plan trips, and to calculate distances and travel rates. Other activities could involve research into other aspects of time, such as why there are 60 minutes in an hour, why October is the tenth month instead of the eighth, and how other cultures measure their years.

ACTIVITY 3.15

Global

HOTmaths has some excellent time zone widgets and associated activities. You can access these widgets by logging into HOTmaths and entering their respective names in the search field (see Figure 3.26).

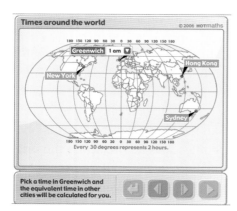

Figure 3.26a Screenshot of HOTmaths Times Around the World widget

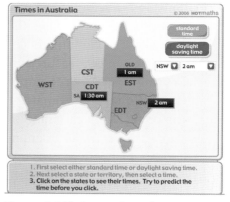

Figure 3.26b Screenshot of HOTmaths Time in Australia widget

Making and using a scale

Bathroom and kitchen scales are compact devices that produce a number to indicate the weight of an object placed on them. Simple spring scales can be made using rubber bands, paper cups and weights to calibrate the scale. Full instructions are provided at the Krieger Science website, <http://kriegerscience.wordpress.com/2011/09/29/how-to-make-a-spring-scale>. Learning opportunities here should focus on choosing appropriate informal units and constructing a scale that accurately distinguishes between the mass of different objects.

Reflection

This chapter has provided an overview of the main factors that need to be considered when teaching measurement concepts. Examples of how ICT can be used to introduce, develop and extend students' understanding of the different attributes have been incorporated throughout. As measurement is such a practical activity, think carefully about how ICT can be used to enhance students' spatial thinking skills and learning experiences, and how it can complement what can be done physically. As with the other learning areas, there are endless resources that are accessible online, and as teachers we need to be aware of the affordances and constraints of these resources and select from them accordingly.

Websites for exploration

Bang on Time: <http://resources.oswego.org/games/BangOnTime/clockwordres.html>
Census at School: <http://www.abs.gov.au/censusatschool>
HOTmaths: <www.hotmaths.com.au>
Khan Academy: Basic Geometry: <http://www.khanacademy.org/math/geometry/circles/v/circles--radius--diameter-and-circumference>
Length, Strength: Centimeters: <http://www.hbschool.com/activity/length_strength1_centi>
Math Playground: <http://www.mathplayground.com/tangrams.html>
NZmaths: <http://www.nzmaths.co.nz/resource/gingerbread-man?parent_node=>
Sheppard Software: Time: <http://sheppardsoftware.com/mathgames/menus/time.htm>
TeacherLed.com: <http://www.teacherled.com/resources/anglemeasure/anglemeasureload.html>
Time and Date: <http://www.timeanddate.com/worldclock>
XE Currency Converter: <http://www.xe.com/ucc>

CHAPTER 4

Exploring geometry

LEARNING OUTCOMES

By the end of this chapter, you will:

- understand the breadth of concepts included in the geometry section of curriculum documents generally, particularly the Australian Curriculum: Mathematics and the New Zealand Mathematics Curriculum
- be familiar with a theoretical framework used as a lens through which to view students' geometrical thinking, known as the van Hiele theory
- have a pedagogical framework that is useful for designing sequential student tasks to assist students to grow in their understandings of geometrical concepts
- understand the important role of language and maintaining 'student ownership' of the geometrical ideas
- be familiar with the use of technological tools to enhance our teaching of geometrical concepts for the e-generation.

KEY TERMS

- **Bisect:** Divide into two equal parts
- **Congruent figures:** Figures that are exactly the same size and shape
- **Dissect:** Divide into two parts
- **Edge:** The interval where two faces of a solid meet
- **Face:** A flat surface of a polyhedron
- **Isometric projection:** A corner view of an object

- **Spatial awareness**: the ability to be aware of oneself in space. It involves the organised knowledge of objects in relation to oneself in that given space and an understanding of the relationship of these objects when there is a change of position.
- **Three-dimensional (3D) objects:** Having three dimensions, requiring three coordinates to specify a point
- **Transformation:** Shifting or modifying a shape, including reflecting, enlarging, translating and rotating
- **Two-dimensional (2D) figures:** Having two dimensions, a flat surface with no depth, requiring two coordinates to specify a point
- **Vertices:** Plural of vertex; a point on a 3D shape where three or more straight edges meet to make a corner
- **Visualisation:** A mental image that is similar to a visual perception

Educationally, we are in an exciting time in terms of geometrical investigations in the classroom. While the manipulation of concrete materials to enable student construction of 2D figures and 3D objects has been readily available for many years, there are a growing number of mathematics classrooms that have access to dynamic geometry software and interactive sites that enable real-time creation and exploration of geometric figures and their properties. There is sometimes confusion when comparing curriculum documents in relation to using the words 'space' or 'geometry' or 'shape'. Geometry is often described as the exploration of space. This involves an investigation of shape, size and place. This chapter explores the development of geometrical concepts and the manner in which we can facilitate exploratory experiences to assist students in their development.

Geometrical concepts

While the Geometry and Measurement content areas are combined in the Australian Curriculum: Mathematics (ACARA 2012) and in the New Zealand Mathematics Curriculum (New Zealand Ministry of Education 2010), a complete chapter is devoted to the principles and practicalities of teaching geometry in the primary years (approximately 5–12 years of age) and moving into lower secondary (approximately 12–13 years of age). While the issues are dealt with separately, it is not possible to ignore the specific links that geometry has with other content areas, in particular measurement (as it is clearly linked within the Australian and New Zealand Mathematics Curricula) and number and algebra. The following activities are not age-specific, and should be used in the classroom according to students' levels of understanding.

The geometrical content in primary curriculum documents tends to include an investigation of relationships among and within:

- **two-dimensional figures** and their properties
- **three-dimensional objects** and their properties
- relevant positions, movements and **transformation** of shapes
- representations and interpretations of locations.

Each of these concept groups is explored through recognising, describing, comparing, constructing, classifying and developing geometrical arguments.

Visualisation is a key component of this strand, as the visual perception of shapes and their properties changes as the learner's conceptual focus develops. The relationship-building among these geometrical concepts builds to a focus on the interrelationships among the properties and their figures through student-centred tasks. To get to this point, learners have a long journey to take with specific hurdles along the way. The journey begins with 'sort, describe, and name familiar two dimensional shapes and three dimensional objects in the environment' (ACMMG009), similar to GM1-2 achievement objectives in NZmaths, collectively described as 'sorts objects by their appearance'.

Figure 4.1 Geometry and everyday objects

The shapes to be sorted and described are squares, circles, triangles, rectangles, spheres and cubes; however, this is not an exhaustive list, and should include any shapes of interest.

Taking geometry outside the classroom

There are many geometry activities that are suited to the outdoors. If we are going to build on children's everyday experiences, a big part of this will be in the context of outdoor activities. They are as simple as:

- going for a shape walk around the school and asking students to record the shapes they see and where they found them
- making a circle outside using a string and chalk
- designing and constructing a mini-basketball court
- designing and drawing a map for an outside obstacle course
- exploring the common shapes used in bridge and building designs (this leads nicely into an indoor activity where the students work in pairs using spaghetti or drinking straws and Blu-Tack to make a bridge that could hold a certain object between two tables)
- making wet tennis ball angles (see Figure 4.2).

To assist in the exploration of geometrical concepts, a developmental model will mould the discussion.

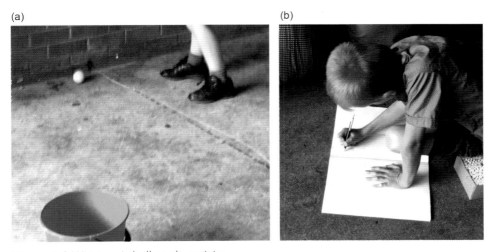

Figure 4.2 Wet tennis ball angle activity

Theoretical framework

A range of theoretical frameworks assist in interpreting students' understanding of mathematical concepts. Teachers have at their disposal a developmental theory particular to geometry that has been researched (Serow 2007a) since the 1950s, known as the van Hiele theory (van Hiele 1986). As in most areas of mathematics education, there is considerable debate concerning the validity and nature of the developmental models, such as the van Hiele theory, and associated features. Putting this debate aside for now, this model provides a practical lens that enables teachers to view their students' conceptual development and the targeted design of their teaching and learning tasks.

The van Hiele theory characterises five hierarchical levels and two transitional levels of geometric thought that provide a window for viewing students' growth and understanding in geometry concepts. The chosen descriptions of three of the five van Hiele levels of thinking (evident in primary and secondary settings), alongside more recently identified transitional levels of thinking (Pegg & Baker 1999; Serow 2007b), are explained within the context of students' growing understanding of the relationships among figures and their properties. As the context of the curriculum begins with describing and sorting, and spirals through any mathematics curriculum, sorting, classifying and describing properties and figures constitute the type of student task referred to.

While NZmaths also uses the van Hiele theory to inform the content and pedagogy of the Geometry strand, Geometry Information (NZmaths 2010) has included a Level 0, known as the Pre-recognition Stage, where the students are characterised as only being able to distinguish between some of the familiar shapes, such as, 'they may be able to distinguish between a square and a circle but not between a square and a triangle' (NZmaths 2010). The Level 0 described by NZmaths is an earlier development within Level 1 of the van Hiele levels of thinking described in Table 4.1.

Table 4.1 Van Hiele levels of thinking in the context of grouping and describing figures

Level 1	Definition
	Figures are judged by their appearance. A figure is recognised by its form or shape. The properties of a figure play no explicit role in the identification of a figure.
	Sample student responses at this level 'The pyramid is a pointy shape.' 'A square is like a box.' 'When a square is rotated with the diagonals vertically and horizontally positioned, it will be called a diamond.' 'A rhombus is a pushed over square.'

Chapter 4: Exploring geometry   87

	Teaching focus to promote development Any activity that involves constructing, building, sorting, drawing and discussing 2D and 3D figures will assist students in moving to Level 2.
Transitional level 1/2	**Definition** Properties are perceived as 'features'. When grouping shapes, students complete this spontaneously upon the identification of similar features. For example, students at this level will be aware of the number of sides and angles of a shape. The students may look at a shape and tell you which sides are equal, etc., but it is merely a feature of that specific shape and not something that is recognised as a signifier of a class of shapes. This is what distinguishes a 'feature' from a 'property'.
	Sample student responses at this level 'A square has four sides and four corners.' When prompted to discuss the sides the student may add, 'All the sides are the same.' A student will be happy to group pyramids and triangular-based prisms together because 'they both have triangles on them'.
	Teaching focus to promote development Students require a selection of hands-on activities that explore each of the properties of figures. Based upon the properties investigated, at this level the students require classification activities where the focus is on the selection of properties which are characteristic of each group, and the identification of group (class) names. Class refers to the group of shapes according to their properties.
Level 2A	**Definition** Figures are classified on the basis of one property. This property is treated as a unique signifier of the shape and is used to describe groupings.
	Sample student responses at this level 'The equilateral triangle has three sides equal.' 'I have put all the pyramids together because they all have triangles going up to the point.'
	Teaching focus to promote development At this stage, the focus is on the collective class of figures, as opposed to specific examples. This begins with a complete isolation of classes, and gradually shifts to recognition of similar properties across classes that are supported by visual cues. To enable this shift, students need to be encouraged to communicate recognised property differences that prevent relationships, and to be directed to find similarities across classes of figures.

(Continued)

Table 4.1 (cont.)

Level		
Level 2B	**Definition**	Figures are classified on the basis of more than one property. Students will make links among different groups of shapes when the links are supported by visual cues. Students will use more than one property as unique signifiers when attempting to provide minimum descriptions.
	Sample student response at this level	'I think the square and the rectangle are linked to one another because they both have four right angles and opposite sides are parallel. They can't go in the same group because the rectangle has to have two different sides to each other.'
	Teaching focus to promote development	It is essential to focus on the domains of student language use at this level. Students would benefit from the introduction of inclusive property descriptions. An example of an exclusive property description is 'two sides are equal and the other two sides are equal, but the two pairs are different from each other' (when describing the properties of a rectangle). An inclusive property description for the same property would be 'the rectangle has opposite sides equal'. Exclusive descriptions hinder relationships among classes of figures, whereas inclusive descriptions enable relationships among classes of figures. Activities that involve materials with moveable parts, and student-created figures using dynamic geometry software showing different figures that may belong in the same class of shapes, are suitable here. The classroom environment should be conducive to focusing upon relationships that exist between classes that are not supported by visual cues.
Level 3	**Definition**	Thinking at this level is characterised by a focus on the relationships among the properties and figures. Links are made between classes of figures that are not supported by visual cues. At this level, the links are often described in a manner that is 'one way'. In other words, there is an ordering between two properties.
	Sample student response at this level	'Yes, the rectangle and rhombus are linked to each other because they both have opposite sides equal and opposite sides parallel.'
	Teaching focus to promote development	Students would benefit from tasks requiring them to formulate descriptions/definitions of figures incorporating sub-sets. This may include tasks requiring the development and discussion of geometrical concept maps and flowcharts.

Transitional level 3/4	**Definition** Students focus on a single relationship that is two directional. Students will begin at this level by making tentative statements with regard to class inclusion. As students focus on multiple property relationships, they will accept notions of class inclusion; however, their justifications may not be consistent. Levels 3/4 and 4 are included here, although they will rarely be targeted at the primary level, to flag the next level of geometrical understanding.
Level 4	**Definition** Further conditions are placed upon class inclusion and general overviews of different relating concepts are formed. Students are able to succinctly and spontaneously use interrelationships among properties and figures to solve deductive problems. This level is evident in the secondary setting, and student responses involve deductive reasoning. Students are able to find their own way through a deductive proof when thinking at this level. An example of Level 4 thinking could involve applying congruency theorems and relationships among quadrilateral properties to solve a problem.

CLASSROOM SNAPSHOT 4.1

A casual teacher, Mr Walker, was allocated a Year 4 class for two teaching days. The usual teacher of the class left the following lesson notes for the mathematics lesson. 'We are finishing a unit on comparing and describing shapes. Year 4 had finished quadrilaterals. In the mathematics lesson time, please ask the students to identify a square, parallelogram, rhombus, kite, trapezium and rectangle by holding up the shape cards on my desk. The students can then draw each shape and write a description of each one.' Mr Walker found the cards, sat the students on the floor at the front of the room to have a class discussion and began what he thought would be a quick recap of quadrilateral identification and some descriptions using features and properties. As the cards were held up, students volunteered the following responses:

MING-LE: That is a rectangle.
MR WALKER: What can you tell me about it?
MING-LE: It has two sides the same and the other two sides the same but different to the other two.
MR WALKER: Can anyone tell me anything else about the rectangle?
ABDUL: It has four right angles like the square does.
[When the rhombus card was held up, all the children in the class put their hands up as high as they could.]

Mr Walker: Wow, we are all keen to answer this one! What is it?

Jessica: It is a rhombus because it looks like a square that has been 'rhommed' by a bus.

Mr Walker: Can you tell me anything else about this shape?

Jessica: If you straightened it up it would be a square again.

The teacher turned the card to a different orientation.

Mr Walker: What can you tell me now?

Leon: When you have it that way it is a diamond.

PAUSE AND REFLECT

The classroom discussion concerning the quadrilaterals provided a window through which to view students' conceptual understanding.

- What do you know about Jessica and Leon's understanding?
- What activity would you choose to target the properties of the rhombus?
- What technology might you use to assist you, and how could you use it?

There are seven key features that directly impact on the pedagogical practices we use when targeting geometrical concepts in the classroom. Teaching points that stem from the van Hiele theory are discussed below.

Hierarchical nature

The levels of thinking are hierarchical in nature – that is, a student cannot proceed to a particular level of thinking without understanding the previous level. For the teacher, this means it is not possible for a student to skip a level. Progression to the next level depends more on learning experiences than it does on biological maturation. These experiences require investigation, exploration and discussion.

Different level, different language

Each of the levels of thinking has its own language. While the words we choose to use may be common at different levels, the meaning attached to the words will be different. Once we have achieved a particular level of thinking, it is not possible to return to that level. This highlights a barrier that can exist in the classroom – not only between students at different levels, but also between students and the teacher. One positive lies in the fact that students' conceptual understandings can be determined through the uniqueness of the language used at each level. The students' classroom talk is a window through which to view their conceptual development.

ACTIVITY 4.1

Consider the following four student descriptions of a pair of parallel lines. How would these align with each student's level of thinking in geometry?

- *Student 1:* Parallel lines are like railway tracks.
- *Student 2:* Parallel lines are lines that will never meet if they keep going.
- *Student 3:* Parallel lines are always the same distance apart.
- *Student 4:* The perpendicular distance between a pair of parallel lines is the same at any point.

Crisis of thinking

The movement from one level to the next is not a simple process. When students make the transition in one area of geometry, they need to pass through what is called a crisis of thinking (van Hiele 1986). This requires a mental reorganisation. For example, a student moving from Level 2 to Level 3 needs to move from viewing the properties of shapes as isolated bits of information to focusing on the relationships among the properties and their figures. The human race has a tendency to avoid a crisis at all costs. To assist students in surmounting this crisis of thinking, as opposed to avoiding it, teachers can implement a pedagogical framework to assist in lesson and unit design. This is known as the van Hiele teaching phases, and will be explored further in this chapter. A sample teaching sequence incorporating technology will also be provided.

Level reduction

Level reduction involves introducing knowledge via a procedure or trick that allows the learner to solve a task without reaching the required level of thinking for the task. When geometrical concepts are taught in this fashion, ownership of the mathematical idea is not maintained by the students, and they are not equipped to complete unfamiliar questions at a similar level. Introduction of level-reduction techniques at an inappropriate time in the classroom will usually set students up for failure every time they come across problems relating to the particular concept that require variations of the process to be applied. Level reduction takes only one meaning when initiated by the student.

Progression requires instruction, exploration and reflection

To move students from one level to the next requires the teacher to construct an environment in which the students move to the next level and leave behind the structure of the previous level – something about which they have come to feel comfortable and secure. It is necessary for students to make more and more links to the next level, which means providing opportunities that place the students in a situation where it is necessary to make the change. This requires an exploration of the new structure and a general exposure to the language necessary to communicate effectively within the structure. Students need time to reflect upon the generalisations formed.

Implicit and explicit understanding

Growth through the levels requires learning experiences that facilitate the analysis of elements of the lower levels. Van Hiele (1986, p. 6) describes the attainment of the higher level as evident when 'the rules governing the lower structure have been made explicit and studied, thereby themselves becoming a new structure'.

Discontinuity

This feature has always been controversial, particularly when students have been identified as transitional, between the original five discrete levels of understanding. It is evident that each new level requires a new language, which implies a sudden leap rather than a gentle progression.

Geometry in the primary classroom

Three-dimensional (3D) objects, also known as solid shapes, require the same amount of attention as two-dimensional figures. There are many opportunities for extended explorations from the early years of primary to upper primary. According to many curriculum documents, including the Australian Curriculum: Mathematics, students should cover aspects of the following:

- Sort, describe and name familiar two-dimensional shapes and three-dimensional objects in the environment (ACMMG009).
- Recognise and classify familiar two-dimensional shapes and three-dimensional objects using obvious features (ACMMG022).
- Describe the features of three-dimensional objects (ACMMG043).
- Make models of three-dimensional objects and describe key features (ACMMG063).
- Connect three-dimensional objects with their nets and other two-dimensional representations (ACMMG111).
- Construct simple prisms and pyramids (ACMMG140).
- Draw different views of prisms and solids formed from different combinations of prisms (ACMMG161).

The NZmaths outcomes cover the same content and are summarised as:

- Identify and describe the plane shapes found in objects (GM2-4).
- Classify plane shapes and prisms by their spatial features (GM3-3).
- Represent objects with drawings and models (GM3-4).

The spiralling nature of this process, or revisiting the same geometrical concepts over a period of time, is lost if the outcomes are interpreted literally and in an unrelated fashion. For example, it would not be appropriate to leave the drawing of 3D objects until the upper primary years. Instead, the outcomes need to be viewed holistically. It makes better sense to ask students in the first year of school to build interesting towers with a range of concrete materials and then ask them to draw what they have built. It is also a lost opportunity if the students are not asked to name and describe the shapes that they have used to build their tower. This activity could also link to a measurement activity, where the students compare the heights of the different towers and how they are represented in their drawings.

Students may begin describing the features of 3D objects as those that are pointy, those that roll, those that slide, those that have curved **edges**. They will begin to identify the 2D shapes on the **faces** of 3D objects. As with our 2D discussions, it is essential to use the correct terminology for 3D features when referring to them. In relation to 3D objects, you may hear students describe the cube as having six sides. After affirming the student in identifying this, the discussion could move to the fact that they are called faces on a 3D object. In fact, the cube has six *faces*, eight **vertices** and 12 *edges*. These concepts are continually revisited in their explorations, even in the upper primary years, where students may complete a task where they tabulate the number of faces, edges and vertices on various 3D objects and search for any patterns and relationships they can find. It is a pleasant surprise when students identify Euler's theorem in various forms. A nice connection to patterns and algebra is forged in this type of activity. HOTmaths has a useful widget called Describing 3D Objects (Figure 4.3) that supports this kind of exploration. This widget can be accessed by logging into HOTmaths and entering its name in the search field.

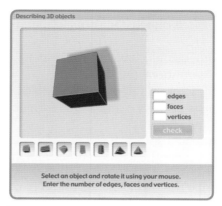

Figure 4.3 Screenshot of HOTmaths Describing 3D Objects widget

ACTIVITY 4.2

Tabulate the number of faces, edges and vertices on 10 known 3D objects (e.g. cube, rectangular prism, triangular-based pyramid). Can you find a rule that connects the number of faces, edges and vertices on any 3D object? In how many different ways can you write the relationship?

As in all areas of mathematics, it is important to keep the explorations as open as possible. For example, instead of asking students to match the cube with its corresponding net from a selection of nets, ask them to come up with as many different nets of the cube as they can on 1 cm² grid paper. There are several HOTmaths widgets and HOTsheets that could be used following this exploration to consolidate the concept of the net as a 2D representation of a 3D object. Such a widget is illustrated in Figure 4.4. This Looking at Nets widget, along with alternative resources, can be accessed by logging into HOTmaths and selecting 'Australian Curriculum' as the Course list and then 'AC Year 5' as the Course. Then choose '3D objects' as your Topic, and finally 'Nets & models' as your Lesson.

AC

(a)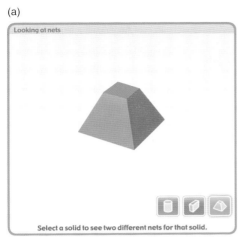

(b)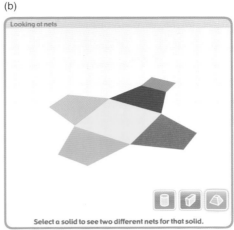

Figure 4.4 Screenshot of HOTmaths Looking at Nets widget

ACTIVITY 4.3

Find as many different nets for the cube as you can. How do you know when you have found them all? What do you need to be careful about?

Try this same activity with other 3D shapes. How can you use the IWB to assist you with a class discussion of the nets found by the students?

CLASSROOM SNAPSHOT 4.2

Mrs Baker was looking for an interesting way to introduce her students to **isometric projections**. She searched high and wide for a contextual teaching strategy. She came across an activity on the Maths300 website, <http://www.maths300.com>, which is a mathematics teaching resource where you can find complete interactive lessons for early primary to upper secondary mathematics classes.

The activity was called 4-Cube Houses, and involved the students taking on the role of an architect to design as many house designs as possible using four cubes with various defining instructions. An important component of the lesson involved the students drawing their designs on isometric paper. At the end of the first lesson, the students were thoroughly engaged in finding all possible designs. As they were packing up the items on their desks ready for recess, one student was heard telling another, 'I thought we were doing maths before morning tea today.'

In the upper primary years, as students are constructing prisms and pyramids using various commercial and everyday materials, they will benefit from tasks that draw their attention to the cross-sections of these objects. Figure 4.5 shows an example of a HOTmaths widget targeting this concept. Again, this widget can be accessed by logging into HOTmaths and entering its name in the search field.

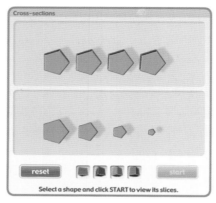

Figure 4.5 Screenshot of HOTmaths Cross-sections widget for exploring cross-sections of prisms and pyramids

This is the key to higher-order classifications that begin to move away from a focus on the shape of the faces only. When objects are considered in conjunction with their cross-sections, students will begin to differentiate between the classes of prisms, pyramids, cones, cylinders and spheres.

ACTIVITY 4.4

Go to the Maths300 website at <http://www.maths300.com> and take the sample tour. This site has an annual subscription to enable access to the complete lessons and associated downloadable software.

PAUSE AND REFLECT

How would you best describe the characteristics of groups of figures named prisms, pyramids, cones, cylinders and spheres? How would you describe the cross-sections of each of these groups of shapes?

IWB technology can be found in many primary classrooms across the world. The levels of IWB use in the mathematics classroom will be addressed in Chapter 11; however, the affordances of IWB use in developing geometrical understandings need to be flagged now. Many diagrams can be used as stimuli for student discussion and activity. Figure 4.6 provides examples of diagrams that could be used from a HOTmaths lesson. You can access these diagrams by logging into HOTmaths and selecting 'Cambridge Primary Maths' as the Course list and then 'Aus Curric 5' as the Course. Then choose '3D objects' as your Topic, and finally 'Different views' as your Lesson. This can be used as an interesting introductory activity to a lesson by using the curtain facility on IWB software to either cover up the structure or cover up the views, and asking the students to create with the structure or the views. The HOTmaths Drawing Views of Solids widget that accompanies this lesson may also be helpful, although it is important that students experience the different views using concrete materials.

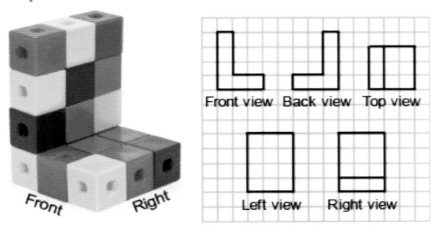

Figure 4.6 Extract from HOTmaths Different Views lesson notes

Figures 4.7–4.9 tell the story of an experienced teacher integrating IWB technology into the mathematics classroom for the first time. It is evident that the teacher integrated IWB technology into the teaching/learning sequence within myriad practical and discussion-promoting tasks. This teacher acknowledged the gradual introduction and development of mathematical language and followed the van Hiele teaching phases as a guide to do this. The students did not perceive the IWB as the teacher's tool, and instead viewed it as a class tool that was shared freely.

Figure 4.7 Using concrete materials to construct pyramids

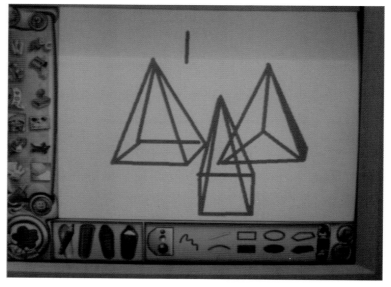

Figure 4.8 Representing pyramids using IWB technology

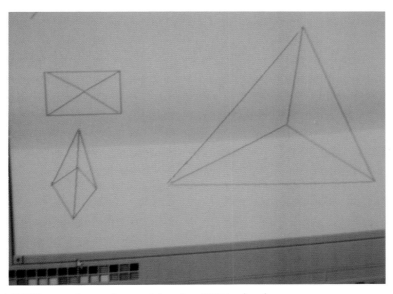

Figure 4.9 Different views of pyramids using IWB technology

In upper primary, some children will begin describing the similarities and differences of cross-sections of groups of 3D figures. Once these are observed and explored by the students, it is reasonable to begin using definitions such as:

- Prisms and cylinders have congruent cross-sections when cut on a plane parallel to the base.
- Pyramids and cones have similar cross-sections when cut on a plane parallel to the base.

While technological tools provide interesting avenues to explore geometrical concepts, it is important to utilise the affordances of everyday materials and contexts that do not involve ICT tools. The structure of the New Zealand Mathematics Curriculum provides elaborations to the unit level, and many of these use everyday materials throughout the lesson sequences. Within the shape units of work for GM2-4, titled Foil Fun (NZmaths 2010), which targets 'explore and describe faces, edges, and corners of 2D and 3D objects' and 'make, name and describe polygons and other plane shapes', the activities require the students to manipulate packages, various solid shapes found in everyday life, nets, foil covers, and 3D representations from magazines.

Two-dimensional shapes

The van Hiele framework clearly articulates a developmental pathway that shapes our teaching of geometry. The level structure, which moves from identifying figures based on overall appearance to focusing on the properties as unrelated elements to focusing on the relationships that exist among the properties, is evident in many national mathematics curriculum documents. The Australian Curriculum: Mathematics is no exception and, as already discussed, centres on the following outcomes in the primary context:

- Sort, describe and name familiar two-dimensional shapes and three-dimensional objects in the environment (ACMMG009).
- Recognise and classify familiar two-dimensional shapes and three-dimensional objects using obvious features (ACMMG022).
- Describe and draw two-dimensional shapes, with and without digital technologies (ACMMG042).
- Compare and describe two-dimensional shapes that result from combining and splitting common shapes, with and without the use of digital technologies (ACMMG088).
- Connect three-dimensional objects with their nets and other two-dimensional representations (ACMMG111).
- Classify triangles according to their side and angle properties, and describe quadrilaterals (ACMMG165).
- Demonstrate that the angle sum of a triangle is 180 degrees and use this to find the angle sum of a quadrilateral (ACMMG166).

It is not until the early secondary years that students are expected to explore concepts concerning **congruent figures** that involve transformations, conditions for congruency of triangles and solving problems using congruency as a tool.

As discussed earlier, while recognition and naming of 2D shapes in the environment is fine in the early years, 2D shape explorations gradually build a focus on the relationships among the properties and figures. Table 4.2 illustrates the properties of the parallelogram that students need to begin to know before the relationships become the target. The findings are described below using inclusive property descriptions.

When exploring diagonals, you will notice that students often confuse the words '**bisect**' and '**dissect**'. These need to be made explicit. As discussed earlier, students will progress from exclusive to inclusive property descriptions. Students who are operating at Level 2, where the properties are known but exist in isolation, will often include no axes of symmetry for the parallelogram, hence not allowing the sub-sets of square, rectangle and rhombus into

Table 4.2 Quadrilateral property descriptions

Shape	What can you tell me about the ...	Your findings!
Parallelogram	Equality of sides	Opposite sides are equal.
	Equality of angles	Opposite angles are equal.
	Parallelism	Opposite sides are parallel.
	Diagonals	Diagonals bisect each other.

the class of parallelograms. This is a difficult hurdle to overcome, and doing so is a target outcome in the secondary setting. It is important for primary teachers to be aware of the importance of the early property explorations; otherwise concepts such as symmetry and diagonal properties could possibly be overlooked.

ACTIVITY 4.5

Design a similar template for exploration of all the triangle and quadrilateral figures. What would be the differences between students' descriptions of their findings at Level 1, 2 and 3 thinking?

It is common for some students at Level 1 and Level 2 to describe the square as if it is the 'king of the quads'. At Level 1, students may describe all other quadrilaterals as a morphed version of the square. This is evident in Classroom snapshot 4.3.

CLASSROOM SNAPSHOT 4.3

Peter's teacher showed him all the quadrilaterals, one after another, on flashcards. He very quickly told the teacher the name of all the quadrilaterals. When she asked Peter to describe the shapes he replied:

> The rectangle is a stretched out square, the parallelogram is a pushed over rectangle, the rhombus is a pushed over square, the kite is a stretched diamond and the trapezium is a rectangle with bits chopped off.

The use of the word 'diamond' is an example of the use of inappropriate language that hinders students' progression in identifying the properties of figures. Teachers should take every opportunity to explore the same figures in different orientations to reinforce the notion that a change in position does not alter the properties of a figure. It is not appropriate to use the term 'diamond' for a square in a specific orientation. In fact, there is no need to use the term 'diamond' for the square or the rhombus. After exploration of the properties of 2D figures, it is useful to ask students to complete a table where they choose all the properties that belong to each class of figures. An example of this can be found in one of the HOTmaths HOTsheets, titled Summarising Quadrilateral Properties (Figure 4.10). You can access this resource by logging into HOTmaths, entering its name in the search field and then selecting the HOTsheets tab.

AC

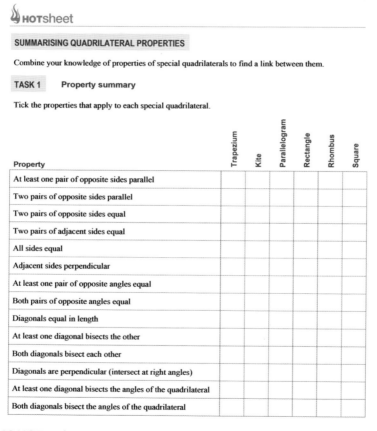

Figure 4.10 HOTmaths Summarising Quadrilateral Properties HOTsheet

This HOTsheet also includes an activity in which students complete a quadrilateral family tree (see Figure 4.11). Another online resource that is similar to the HOTmaths activity is Property Chart from nrich, <http://nrich.maths.org/2927>, where students work in pairs to draw quadrilaterals that meet the designated properties.

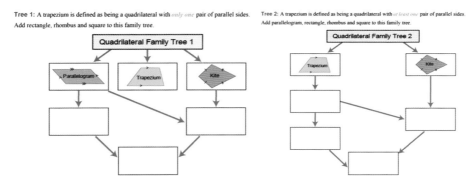

Figure 4.11 Summarising quadrilateral properties

Teachers in the primary years may have the rewarding experience of teaching students who begin describing classes of quadrilateral figures with sub-sets. Figure 4.12 shows an actual higher-order student response to Activity 4.6. Note the inconsistent response where the student states that the 'rhombus is a special square'.

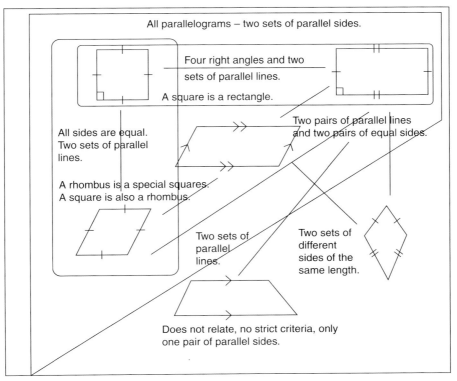

Figure 4.12 Student work sample of quadrilateral relationships and justifications

ACTIVITY 4.6

Draw a tree diagram that shows how all the quadrilaterals are related to each other. You must justify the groups and links that you make.

There are many interactive online games that enable manipulation of figures in an engaging environment. HOTmaths has a growing number of interactive games that can act as catalysts for further classroom discussion of geometric ideas. The first of these, named Peanut Bridge, requires the students to rotate and alter the sizes of shapes to complete the bridge so that a trio of elephants can cross it safely (Figure 4.13). This is an excellent example of an ICT tool that explores aspects of shape orientation in a context that is of interest to young learners. This game is suitable for individual, small-group or whole-class situations. The platforms of accessibility enable it to be explored on desktops, laptops, tablets or IWB technology. You can access this resource by logging into HOTmaths and clicking on the Games icon on the Dashboard.

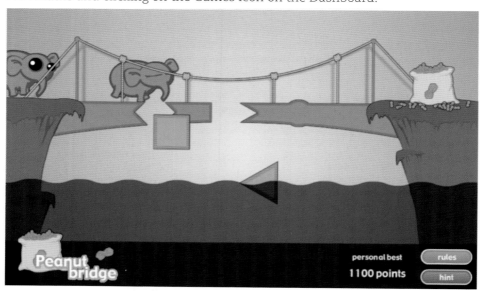

Figure 4.13 Screenshot of HOTmaths Peanut Bridge game

Another interesting game included in the HOTmaths material is called Shape Sketch (Figure 4.14). This game is suitable for upper primary students, and is one to keep revisiting as it targets the language associated with triangle and quadrilateral figures and properties. You can also access this resource by logging into HOTmaths and clicking on the Games icon on the Dashboard.

Figure 4.14 Screenshot of HOTmaths Shape Sketch game

In response to the recognised difficulties caused by the cognitive hurdles encountered by many students through the characterisation of the development of relationships among figures and those among properties, we need to consider suitable teaching strategies to assist students to meet and rise above the hurdles.

One readily available tool is known as dynamic geometry software (DGS). This began with Cabri, <http://www.cabri.com>, and Geometer's Sketchpad, <http://www.keycurriculum.com/products/sketchpad>, and has moved into freeware with the production of GeoGebra, <http://www.geogebra.org>. DGS tools provide teachers with the opportunity to explore the relationships among figures and properties, both intuitively and inductively. Primary-age students have found DGS to be accessible and engaging as the multiple (and formerly tedious) drawings required to identify relationships among figures and properties are carried out with just the drag of the cursor.

GeoGebra is described as a form of dynamic mathematics software (DMS) that is open-source. GeoGebra combines DGS tools and Computer Algebra Systems (CAS) that together have the potential to target the relationships among the concepts of geometry and algebra. Originally, the tools were designed for the middle-school and secondary years; however, further developments have resulted in a primary version, known as GeoGebraPrim, that is more accessible to primary-aged students. Many DMS programs, including GeoGebra, can be used in a variety of ways in the classroom. For example, the student activity can solely involve student constructions and explorations, or the teacher can initiate exploration through design of templates before the lesson.

When introducing any form of dynamic geometry software into the classroom, it is usual to start with the segment tool. This becomes a lesson on its own, as the students, through their need to choose the appropriate tool, work through the terms 'line', 'ray' and 'segment'.

ACTIVITY 4.7

Go to the GeoGebra website at <http://www.geogebra.org> and download the dynamic geometry software. You will notice that there is a primary version. You may also use any form of DGS to complete this activity, such as Geometer's Sketchpad, and even the iPad version of Sketchpad. To familiarise yourself with DGS, complete the tasks outlined below in separate windows or on separate pages. In this case, we learn by getting our hands dirty.

- Write your name using the segment tool.
- Draw a house using all the quadrilaterals you know.
- Create a picture that uses the reflection tool. You might like to start with a person performing a certain action.
- Construct a robust square (one that is constructed using the properties of the square and that, when dragged, will remain a square). This task can be repeated with any quadrilateral, triangle or regular polygon.
- Construct an irregular quadrilateral, then mark the mid-points on each side of the quadrilateral and join the mid-points to form another quadrilateral. What shape have you made? Drag your shape to see whether it changes. How can you prove what you have made?
- Use the circle tool to construct an equilateral triangle.

Figure 4.15 shows a student work sample of a rectangle constructed using GeoGebra software.

The student work sample in Figure 4.16 is part of a larger sample created in response to designing a quadrilateral starter game using Geometer's Sketchpad. The students were asked to design the diagonal structure of each of the quadrilaterals they know for younger children to guess the figure and put in the required sides. It proved to be a wonderful way to engage students in diagonal explorations. As with many constructions, it is important to ask students to use the textbox facility to consolidate their ideas and gradually formalise the language used.

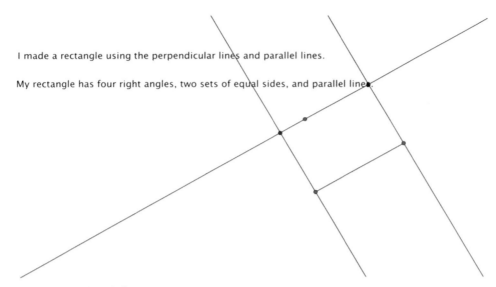

Figure 4.15 GeoGebra construction

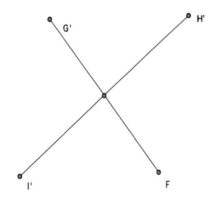

Figure 4.16 Diagonal starter construction
Source: Serow (2007a, p. 396).

Angles

In any exploration of their world, students are faced daily with angles around them. Angles are used to explain location, directions, slopes or gradients – even movement in dance. Keeping this in mind, it is essential that we include elements of the environment as stimuli for our investigation of angles in many different contexts. This is where the challenge lies when teaching about angles. The concepts explored in the middle primary years have been described by the Department of Education and Training (2004, p. 8) in several main areas:

- identifying two-line angles
- identifying the arms and vertex of one-line angles
- comparing angles using informal means such as angle testers
- describing angles using everyday language and the mathematical terms 'right', 'acute' and 'obtuse'.

It is not until the upper primary years that we begin to do the following:

- Estimate, measure and compare angles using degrees. Construct angles using a protractor (ACMMG112).
- Investigate, with or without digital technologies, angles on a straight line, angles at a point and vertically opposite angles. Use results to find unknown angles (ACMMG141).

As with mental computation requirements in the Number and Algebra strand, it is essential that students are provided with the opportunity to develop 'angle sense'. This is where they become automatic in their ability to visualise, estimate and compare angles, and eventually relate angle properties to other properties of 2D and 3D shapes. Time needs to be devoted to identifying angles in everyday life, such as on furniture, body parts, scissors, clock hands, body turns and opening doors, and in classroom resources such as pattern blocks. Students may construct angles at a point using straws, an activity that relates nicely to the development of fraction concepts.

The New Zealand curriculum structure also spirals the conceptual development of angles using a pedagogical framework that is closely linked to the measurement framework discussed in Chapter 3. The framework is:

1. Identify the attribute of an angle.
2. Compare and order angles.
3. Non-standard units (informal units).
4. Standard units (formal units).
5. Applying and interpreting (strong connection to the measurement strand).

Location and transformations

The exploration of location concepts begins well before young children start compulsory education. This strand of the curriculum targets **spatial awareness**, a term used to describe children's understanding of their location and the location of objects in relation to their bodies. This awareness begins to develop as soon as children are able to explore the world around them. It can be observed from a very young age, such as when a baby begins to reach for objects and becomes mobile. As an adult, we rely on spatial awareness on a daily basis, whether it be navigating, finding your car in a large car park, or maximising space when packing, to name a few examples. In the primary classroom, it is essential for teachers to seize opportunities to discuss locations of objects/people/places, to use comparative terms such as 'closer to' or 'further than', to talk about relationship amongst items such as 'under' or 'beside', to measure distances (which is different from measuring the length of an object), and to practise giving directions. The development of spatial structuring links to the Measurement, Number and Algebra strands through the structuring of rectangular figures, which also links to arrays. Like mathematics curriculum documents elsewhere, the Australian Curriculum: Mathematics begins this area of geometry with the following concepts:

- Describe position and movement (ACMMG010).
- Give and follow directions to familiar locations (ACMMG023).
- Interpret simple maps of familiar locations and identify the relative positions of key features (ACMMG044).
- Investigate the effect of one-step slides and flips with and without digital technologies (ACMMG045).
- Create and interpret simple grid maps to show position and pathways. Identify symmetry in the environment (ACMMG065).
- Use simple scales, legends and directions to interpret information contained in basic maps (ACMMG090).
- Create symmetrical patterns, pictures and shapes with and without digital technologies (ACMMG091).
- Use a grid reference to describe locations. Describe routes using landmarks and directional language (ACMMG113).
- Describe translations, reflections and rotations of two-dimensional shapes. Identify line and rotational symmetries (ACMMG114).
- Apply the enlargement transformation to familiar two-dimensional shapes and explore the properties of the resulting image compared with the original (ACMMG115).
- Investigate combinations of translations, reflections and rotations, with and without the use of digital technologies (ACMMG142).
- Introduce the Cartesian coordinate system using all four quadrants (ACMMG143).

NZmaths (2010) includes similar outcomes; however, the activity and unit example exemplify contextualisation and student-centred activities. The initial outcome related to the sub-strand of location is 'give and follow instructions for movement that involve distances, directions, and half or quarter turns' (GM1-3). A sequence of activities is provided that takes the students on a 'follow me' journey using the children's picture book *A Lion in the Night* by Pamela Allen as a stimulus to explore early location concepts. The unit also makes use of digital cameras, with the young students taking on the role of 'director' as they photograph various positions described in the story. Visual representations of vocabulary are available for the teacher, which illustrate 'across', 'into', 'over', 'under', 'through', 'out', 'past' and 'around'. This is a good example of literacy activities across the mathematics curriculum involving children's literature.

It is important to note the inter-connectivity in the mathematics outcomes of any syllabus. The notion of half- and quarter-turns relates closely to this topic, as well as to angles and fractions. All the *transformations* (translations, reflections and rotations) are also part of the exploration of 2D and 3D shapes. The interconnectivity is evident in the three Year 1 children's floor plans of their homes in Figures 4.17–4.19. Each of the diagrams provides information about where the students are situated on their developmental journey in geometry.

ACTIVITY 4.8

- What are the characteristics of each child's floor plan in Figures 4.17–4.19?
- What do the diagrams indicate to you about the developmental pathway leading to the ability to draw accurate floor plans of a well-known place? (See Callingham 2008 for further information.)

The teaching of geometry provides some wonderful opportunities to integrate with visual arts in the mathematics classroom. One example of this is the exploration of tessellations. While it is not explicitly identified as an outcome of the Australian Curriculum: Mathematics, the study of tessellations does draw together concepts such as properties of 2D shapes, including interior angles, angles at a point and transformations of 2D shapes, thus providing an opportunity to further explore a group of upper primary geometry concepts. It is not unusual to walk into primary classrooms around the world and be entertained by beautifully coloured tessellating designs made with pattern blocks and other templates. It would be a real shame to spend so much time exploring shapes that create regular and semi-regular tessellations and to never take the time to explore why they do so. For some students, it is a light-bulb moment to realise the connection between the interior angles of the tessellating shapes and the factors of 360 degrees. In addition to the beautiful display, one needs to walk into the classroom and see the students' responses to finishing the statements:

- These patterns use one shape to tessellate because …
- These patterns use two shapes to tessellate because …

One of the most significant roles of the mathematics teacher is their ability to make the mathematical ideas developed by the children explicit.

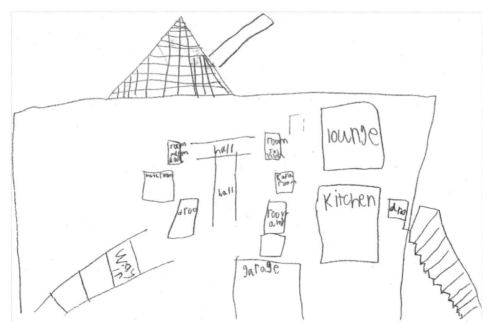

Figure 4.17 Sample floor plan 1

Figure 4.18 Sample floor plan 2

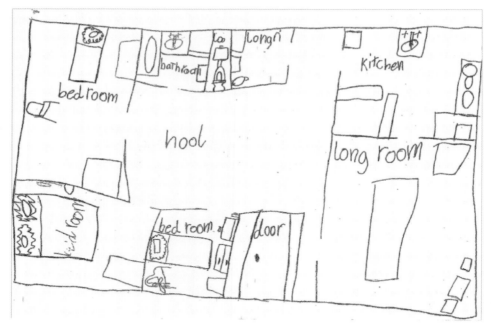

Figure 4.19 Sample floor plan 3

The van Hiele teaching phases

While many experienced teachers confidently use technology as display tools in the mathematics classroom, it is not uncommon for teachers to struggle with finding ways to effectively use ICT as a teaching/learning tool. One teaching framework, based on the work of Dina van Hiele-Geldof, supports teachers in using ICT to promote conceptual development through a scaffold of five teaching phases. The phases aim to assist in moving students from one level of understanding to the next due to the premise that 'help from other people is necessary for so many learning processes' (van Hiele 1986, p. 181).

The structure of the van Hiele teaching phases facilitates students' opportunities to display insight, described as essential in van Hiele's geometrical thinking framework. These opportunities to display insight enable students to maintain ownership of their mathematical ideas. Just as the phase framework supports students in transitioning from one level to the next, it also provides a structure to address the concern that 'teachers often feel reluctant or uncomfortable because their pedagogical knowledge perhaps does not include a framework for conducting technology-based activities in their lessons' (Chua & Wu 2005, p. 387).

The phase approach promotes 'maths talk' in the classroom, and students are encouraged to seek clarification from each other and their teacher as they gradually move from teacher-directed tasks to student-directed tasks. Interestingly, despite the initial phases being teacher-directed, they remain student-centred throughout the five phases. A key component is the development of language, which gradually becomes more technical and formal.

A description of the phases is provided below with an emphasis on the changing role of language as the student progresses through them. The phase descriptions in Table 4.3 are taken from Serow (2007a, p. 384).

Table 4.3 Descriptions of the van Hiele teaching phases

Phase	Description of AIM of phase
1 Information	For students to become familiar with the working domain through discussion and exploration. Discussions take place between teacher and students that stress the content to be used.
2 Directed orientation	For students to identify the focus of the topic through a series of teacher-guided tasks. At this stage, students are given the opportunity to exchange views. Through this discussion, there is a gradual and implicit introduction of more formal language.
3 Explicitation	For students to become conscious of the new ideas and express these in accepted mathematical language. The concepts now need to be made explicit using accepted language. Care is taken to develop the technical language with understanding through the exchange of ideas.
4 Free orientation	For students to complete activities in which they are required to find their own way in the network of relations. The students are now familiar with the domain and are ready to explore it. Through their problem-solving, the students' language develops further as they begin to identify cues to assist them.
5 Integration	For students to build an overview of the material investigated. Summaries concern the new understandings of the concepts involved and incorporate language of the new level. While the purpose of the instruction is now clear to the students, it is still necessary for the teacher to assist during this phase.

The phases provide a means for defining and aiding progression from one level of understanding to the next. This does not mean that each time a student passes through the five-phase process within concept development they have reached the next level. The phases do, however, provide students with the opportunity to come closer to moving to the next level. It is interesting to note that this teaching process is not centred upon one specific form of instruction. The five-phase process lends itself to many teaching styles, and each phase has a specific and important purpose.

The five-phase teaching approach provides a structure on which to base a teaching/learning sequence using a variety of available classroom tools. It promotes the integration of technological tools while still enabling the use of other hands-on materials. As can be seen, the phase approach begins with clear teacher direction involving exploration through simple tasks, and moves to activities that require student initiative in the form of problem-solving.

Sample teaching sequence

The following teaching sequence was designed with two main elements in mind: the developmental framework and the embedding of technology. This sequence focuses on exploring triangle properties. Please note that the teaching sequence is in order of delivery, with students moving both forwards and backwards within van Hiele phases during the various activities presented for the learning and teaching of this geometric concept. The teaching phases are spiralling in nature. While it is not appropriate to skip levels, it is appropriate to go back to previous phases and continue from each phase before finishing at Phase 5. This may occur more than once.

You can access any of the HOTmaths resources referred to in the activities in this teaching sequence by logging into HOTmaths and entering the respective resource's name in the search field.

Phase 1: Information

- *Activity:* Play a game of celebrity heads with three triangle cards.
- *Activity:* Class IWB activity where the students need to move shapes into two groups: triangles and not-triangles.

Phase 2: Directed orientation

- *Activity:* Students make 12 different triangles on geoboards (pinboards) using rubber bands, and draw them on dot paper. Students may use 1 cm grid and isometric dot paper if needed.

Phase 3: Explicitation

- *Activity:* Students share examples of the triangles they have made using the electronic geoboard (HOTmaths Pinboard widget) (see Figure 4.20).

Chapter 4: Exploring geometry

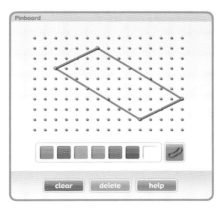

Figure 4.20 Screenshot of HOTmaths Pinboard widget

Phase 2: Directed orientation

- *Activity:* Finding triangles in tile patterns (HOTmaths Triangles in Shapes HOTsheet) (see Figure 4.21).
- *Activity:* Folding squares and rectangles to make triangles (HOTmaths Triangles in Shapes HOTsheet) (see Figure 4.22).

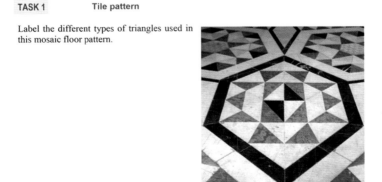

Figure 4.21 Finding triangles in tile patterns from HOTmaths Triangles in Shapes HOTsheet

Phase 3: Explicitation

- *Activity:* Students cut out the 12 triangles they have made. Using a protractor and a ruler, they measure the sides and angles on each triangle. Students are encouraged to use paper folding to find any axes of symmetry and to mark them on the appropriate triangles.
- *Activity:* Students complete a spreadsheet with four columns headed 'triangle', 'sides', 'angles' and 'symmetry'.

> **TASK 2**   **Folding a rectangle and a square**
>
> Cut out this square and rectangle separately. Make folds along each diagonal then open out the shapes so they are flat.
>
> Describe the number and types of triangle you can now see in each shape.
>
> **Square:** _____
>
> _____
>
> **Rectangle:** _____
>
> _____

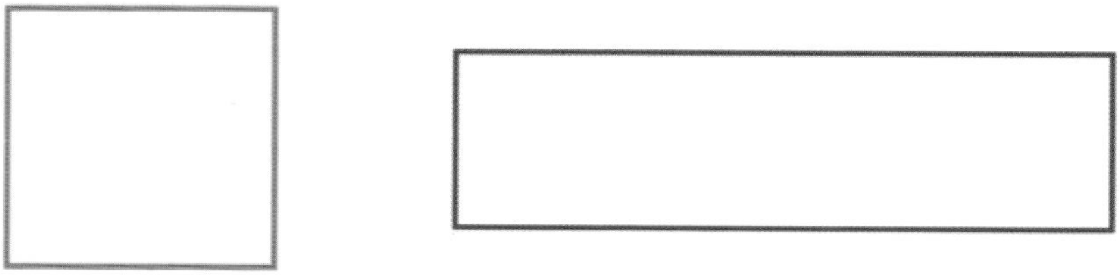

Figure 4.22 Folding a rectangle and a square from HOTmaths Triangles in Shapes HOTsheet

Phase 4: Free orientation

- *Activity:* In pairs, with 24 triangles for each pair, students are asked to sort their shapes into appropriate groups and record their groups. They need to provide a detailed description of each group and a description of any relationships across groups.
- *Activity:* Draw a detailed concept map indicating how the triangle groups are related to one another.
- *Activity:* Students use the circle tool in dynamic geometry software (DGS) to create an equilateral triangle.
- *Activity:* Use DGS to create an example of each of the triangle types possible. Students could come up with: equilateral, right-angled isosceles, acute-angled isosceles, obtuse-angled isosceles, right-angled scalene, acute-angled scalene and obtuse-angled scalene. The HOTmaths Sorting Triangles HOTsheet has an emphasis on students describing different triangles and can provide useful consolidation (see Figure 4.23). Some students may provide a higher-order response and begin noting that the equilateral triangle is a sub-set of the isosceles class of triangles with additional properties. If this happens, it is the right time to ask the child why, and to congratulate them on their mathematical thinking.

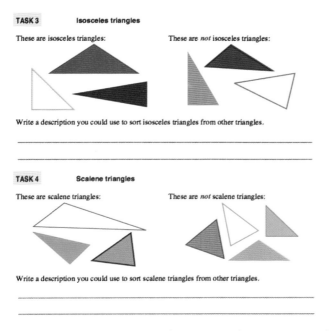

Figure 4.23 Sample describing triangles tasks from HOTmaths Sorting Triangles HOTsheet

Phase 5: Integration

- *Activity*: Students present their DGS diagrams and description of each triangle type and create an information pamphlet. Students add their own summary page at the end of the pamphlet explaining what they know about triangles.

It is important to note that, throughout this teaching sequence, assessment is embedded seamlessly into teaching/learning activities. There is ample opportunity to observe and gather information concerning students' individual understanding of geometrical concepts. The notion of assessment for learning will be addressed in detail in Chapter 10.

Suggested geometry tablet applications to explore

There is an expanding wealth of mathematics applications available on mobile devices. Geometry is one content area that is particularly open to manipulation of objects. To make the most of the application as an opportunity for conceptual growth in terms of geometrical understanding, it is essential for it to be placed thoughtfully within the teaching and learning sequence. Examples of applications include:

- Geometry Pad by Bytes Arithmetic LLC (available via searching on iTunes, <https://itunes.apple.com>)
- Sketchpad Explorer by KCP Technologies, http://www.dynamicgeometry.com>
- Tangram XL Free by NG (available via searching on iTunes, <https://itunes.apple.com>).

In addition to free and purchasable applications on mobile devices, many of the available tools provide media for investigation and catalysts for discussion. One example of this is the digital camera on many devices – even the most basic mobile phone. Classroom snapshot 4.4 describes an interesting teaching moment in a Year 3 classroom where a pre-service teacher was beginning a lesson targeting investigating symmetry in the environment.

CLASSROOM SNAPSHOT 4.4

It was Week 2 of a three-week practicum, and Miss Hope was about to teach a mathematics lesson targeting symmetry. She decided to begin the lesson with an introductory activity that involved taking a photo of one student on a mobile device, displaying it via the IWB and using the curtain tool to cover up half the face with a vertical line down the centre of the nose. She requested a couple of students to draw in the outline of the missing half of the face and any of the main features of the face. Miss Hope then asked a student to reveal the whole face and see how well it matched up. The teacher went on to ask the students, 'What did you notice?'

A very rich discussion followed concerning symmetry, axes of symmetry, whether or not a human face has symmetry, where we find symmetry in our environment and what makes a shape symmetrical. The entire discussion flowed from the students' observations during the introductory activity.

Miss Hope then proceeded to state that today they were going to explore items in their environment to determine whether they were symmetrical. As a whole class, the students did one example of a flower together.

The students were provided with six laminated diagrams, including tile patterns. The students were asked to record any symmetry they could find in any way, to say how they knew this, and to illustrate in a diagram the symmetry that was observed. The following day, the students constructed their own nature-inspired design using dynamic geometry software (DGS) (see Figure 4.24).

Chapter 4: Exploring geometry

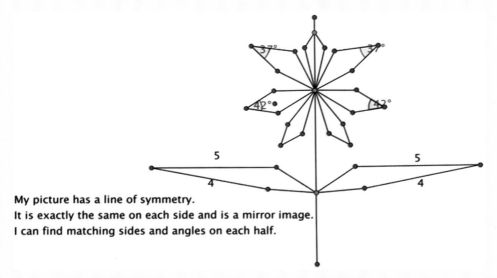

My picture has a line of symmetry.
It is exactly the same on each side and is a mirror image.
I can find matching sides and angles on each half.

Figure 4.24 Sample of reflection using dynamic geometry software

Teachers will find that the geometry tasks in the test generator have a range of applications in the classroom. While they may be used to generate a sample of questions to identify where the students are at and store this information, the questions can be used to facilitate class discussion on the IWB, in an introduction to a lesson or even altered and extended to create more open-ended tasks. Often just adding the question 'Why?' makes all the difference to a geometry task. These ideas will be addressed in Chapter 10.

Reflection

In reading this chapter and completing the activity tasks, you will have explored samples of the breadth of content in the geometry strand. This can be placed broadly in overlapping 2D, 3D, angles, location and transformations categories. The van Hiele theory has been described as a means for viewing students' level of thinking in geometry, and providing a window to more accurately target the next student task in an attempt to meet individual student needs. ICT tools have been discussed; however, this is an ever-expanding field that needs to be considered in a flexible manner to enable new and emerging technologies to be placed in our teacher tool kit. To assist planning in geometry, the van Hiele teaching phases have been presented as a suitable pedagogical framework that is open to a wide range of teaching strategies and available resources.

Websites for exploration

Cabri: <http://www.cabri.com>

GeoGebra: <http://www.geogebra.org>

Geometer's Sketchpad: <http://www.keycurriculum.com/products/sketchpad>

Geometry Pad by Bytes Arithmetic LLC (available via searching iTunes: <https://itunes.apple.com>)

Google Sketchup: <http://www.sketchup.com>

HOTmaths: <http://www.hotmaths.com.au>

National Council of Teachers of Mathematics, Illuminations: Resources for Teaching Maths, Isometric Drawing Tool (available via searching at <http://illuminations.nctm.org>)

Maths300: <http://www.maths300.com>

NRICH Maths: <http://nrich.maths.org/2927>

Scootle: <http://www.scootle.edu.au>

Sketchpad Explorer by KCP Technologies: <http://www.dynamicgeometry.com/General_Resources/Sketchpad_Explorer_for_iPad.html>

Tangram XL Free by NG (available by searching iTunes: <https://itunes.apple.com>)

CHAPTER 5

Exploring whole number computation

LEARNING OUTCOMES

By the end of this chapter, you will:

- recognise the difference between additive and multiplicative thinking
- be able to plan for appropriate use of drill and practice activities
- choose appropriate representations to illustrate different ways of thinking about multiplication and division
- use technology effectively to develop understanding of whole number computation.

KEY TERMS

- **Algorithm:** A step-by-step procedure for undertaking a computation
- **Computation:** All operations on numbers that are used to calculate a result or answer
- **Distributive property**: When multiplying one number by another number, the result is the same as multiplying its addends and summing the products. It can be expressed as $a(b + c) = ab + ac$ (e.g. $5 \times 17 = 5 \times 10 + 5 \times 7 = 50 + 35 = 85$).
- **Factor**: A whole number that divides exactly into another number or a whole number that multiplies with another whole number to make a third number (e.g. 3 and 4 are both factors of 12)
- **Multiplicative thinking:** Thinking and reasoning about more than one quantity or value at once (e.g. doubling the side length of a square means that the area quadruples)

- **Place value**: The value of a digit depends on its place in the number. For example, in the number 361, the 3 has the value of 300 or three hundreds, 6 has a value of 60, or six tens, and there is one unit.
- **Whole number:** A number that has no fractional parts; an integer (e.g. 71 is a whole number but 71.5 is not)

Developing computational skills and the concepts that underpin proportional reasoning is a large component of the primary mathematics curriculum. Moving children's thinking towards proportional reasoning will be covered in more detail in Chapter 6. There is a very large research base about the development of these aspects of number and algebra. In this chapter, the focus is on the effective use of technology to enhance learning: developing computational skills, the relationships between different operations and moving from additive to **multiplicative thinking**.

Across the primary years, the mathematics curriculum places considerable emphasis on **computation** and the development of more complex conceptual understandings of the ways in which numbers work. Computation includes all those actions we use to add, subtract, multiply and divide numbers of many different types. We compute mentally, by using technology such as calculators or computers, and by using a variety of pen-and-paper methods. These activities are part of the domain of mathematics we call arithmetic. With the development of sophisticated technology tools, it is more important than ever to ensure that children receive a good grounding in the operations, and can compute effectively and efficiently, choosing methods and tools appropriate to their situation. Children will need a well-developed sense of **place value** and a conceptual understanding of parts of a whole to deal with the ideas in the curriculum.

Technology provides a range of useful tools to help develop students' understanding. These tools should be used in conjunction with concrete materials and traditional pen-and-paper approaches to provide a balanced program that addresses both procedural and conceptual understanding. Students throughout the primary years need opportunities to 'play', or work informally, with numbers of all types: large and small whole numbers, fractions, decimals and percentages, expressed in a wide variety of ways and presented in diverse contexts. The more opportunities students have to become familiar with numbers, the more fluent they will become when working with them, and this in turn will help develop understanding, improve their mathematical reasoning and support their problem-solving. McIntosh (2004) stresses the importance of children developing their own informal understanding, which can be built on to develop the foundational ideas that are needed to progress mathematically.

There are many sources of information about ways to develop students' understanding of the four operations of addition, subtraction, multiplication and division. Initially, students learn to use these with whole numbers; later, the operations are applied to part-whole numbers such as fractions and decimals; and, in the early part of high school, negative numbers are introduced. In this chapter, the focus is less on different approaches to computation; instead, we consider how technology of various kinds can support and aid students' understanding.

This chapter addresses **whole number** operations, building on the early number work from Chapter 2. Key ideas are developed about appropriate pedagogy for working with primary-age children using technology effectively to enhance understanding.

Operations with whole numbers

In the middle primary years, there is a focus on developing children's computational skills, beginning with addition and subtraction and moving to multiplication and division. The New Zealand mathematics curriculum states that number and algebra should constitute about 80 per cent of the curriculum in Years 3 and 4, and this also constitutes an important part of the Australian Curriculum: Mathematics in these years. Children need to experience a variety of approaches to operations, and technology can be a very useful addition to a teacher's tool kit. Today, there is less emphasis on developing standard procedures, or algorithms, but children need a well-developed sense of number to understand the different processes, including the commonly encountered algorithms – step-by-step mathematical processes to arrive at a solution – for written computation.

Addition and subtraction

Addition involves joining two or more numbers or quantities to get one number (called the sum or total) (<http://www.amathsdictionaryforkids.com>). Addition problems are of two types. In *merger* problems, two groups are combined to create a new group. An example of a merger problem is 'I have 12 green marbles and I am given 7 red marbles. How many marbles do I have now?' In *parts-of-a-whole* problems, the questions are about parts of a group. A typical problem of this type is 'I have a bag of marbles. Twelve of these are green and seven are red. How many marbles do I have altogether?' Mathematically, the solution approach is the same: 12 + 7 = 19. From a representation point of view, however, these two problems look different. The merger type requires two separate sets of objects to be combined, such as in the HOTmaths FUNdamentals Add Fish widget (Figure 5.1). You can access this resource by logging into HOTmaths and clicking on the FUNdamentals icon on the Dashboard. Next, click on 'Addition and subtraction', then 'Addition skills', and finally 'Add fish'.

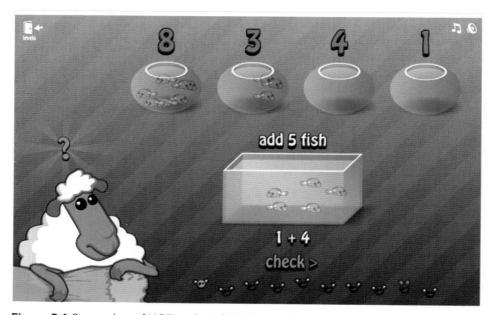

Figure 5.1 Screenshot of HOTmaths Add Fish FUNdamentals

The parts-of-a-whole problems require the whole to be identified, and can be represented by numbers bars or Cuisenaire Rods, such as those in the HOTmaths Number Bars widget (see Figure 5.2). This widget can be accessed by logging into HOTmaths and entering its name in the search field. A set of interactive Cuisenaire Rods is available from the nrich website at <http://nrich

Figure 5.2 HOTmaths Number Bars widget

.maths.org/4339>. Both of these resources can be used very effectively with an IWB, with children asked to drag the bars to complete the appropriate problem.

Addition is commutative – that is, it may be done in any order. For example, the mathematical sentences could be written as 4 + 3 = 7 or 3 + 4 = 7; and 4 + 6 = 10 or 6 + 4 = 10. This important idea can be developed using tools such as the Number Bars widget. The commutative law is an important principle for children to recognise because it can make addition more efficient by starting with the largest or most convenient number. An example of this convenience is when summing data, as shown in Classroom snapshot 5.1.

CLASSROOM SNAPSHOT 5.1

A Year 5 class collects some data about the amount of litter in the schoolyard. Students record the numbers in a table on the IWB. They add the numbers of different pieces of litter together to find the total number of pieces of litter.

Kind of litter	Number of pieces
Icy pole sticks	11
Chocolate wrappers	12
Milk cartons	20
Chip packets	18
Hamburger wrappers	29
Fruit box straws	14

Figure 5.3 Table from IWB

Several strategies could be used to sum together the data shown in Figure 5.3 to obtain a count of all the pieces collected. One is to start with the largest number (29) and find compatible numbers to build to groups of 10. In this instance, 29 + 11 makes 40, which is a 'friendly' number on which to build because it is a multiple of 10. Adding on 20 (another 'friendly' number) to the 40 makes 60 pieces. It is worth noticing that 18 + 12 also makes a 'friendly' number: 18 + 12 = 30. Add this on to 60 to give 90. Now the only number left to add is 14. So 90 + 14 = 104. In total, 104 pieces of litter were collected.

This problem could have started with noticing that 18 + 12 = 30, and building from there in a similar fashion. It could also have been completed using a standard **algorithm**, adding all the units, regrouping into tens and then adding the tens. Knowing that addition can be done in any order provides a starting point for developing a variety of ways of solving a problem.

Subtraction is the opposite of addition. This means that for every addition fact there is a related subtraction fact. Children need to experience these related number facts and to be reminded about them. The number bars can be very useful ways to represent the related facts. In Figure 5.2, it is clear that if 6 is removed from 10, then 4 will be left, and similarly if 4 is removed from 10, then 6 will be left. This can be powerfully and effectively demonstrated using an IWB. Consider, for example, a lesson in which a teacher leads the students through a simple problem using the HOTmaths Number Bars widget. Only the teacher's voice is presented in Classroom snapshot 5.2, so you can focus on the teaching approach and the teacher's language.

CLASSROOM SNAPSHOT 5.2

Michael, please will you come and make a number bar with the value of 10? You can choose any colour.

Sally, choose another colour and make a number bar underneath to the 10 bar. You can make it any value.

Now we have a blue 10 bar and a pink 5 bar. What is the difference between 10 and 5? How can we find out?

Let's hide the values – Lisa, tap on the value sign.

Who thinks they know the answer?

James, you think that the difference between 10 and 5 is 5. Come and make a bar that would show whether or not you are correct.

Wait a minute, James, before you drag the bar out. Who thinks that James made a good choice by choosing the same colour as Lisa?

Ben, you agree with James and think the answer should be 5 so the bars should be the same colour. Okay James, let's check it.

From this starting point, the teacher could go on to discuss 'doubles', different meanings of subtraction, strategies used by the children to arrive at their answers or other ways of recording the problem. The choice of what the teacher does next will depend on the class, their prior learning and what the objective of the lesson is.

Subtraction can also have several meanings. The common 'take away' includes problems of the type 'I had 16 marbles but I lost 5. How many marbles are left?' Other meanings are 'difference between', such as 'I had 16 marbles and now I have 11. How many marbles have I lost?' Finally, there is the 'how many more' meaning, such as 'I have 11 marbles. How many more will I need to have 16 marbles?' This last meaning initially looks like an addition problem, but is actually solved by undertaking a subtraction. It is a missing addend problem: 11 + 5 = 16, and makes use of the interrelationship between addition and subtraction. Examples of widgets that reinforce these different meanings can be found in the FUNdamentals section of HOTmaths, which can be accessed through the Dashboard (for example, Balloon Pop and Complete the Sentence).

Children may also 'over-generalise' and apply the commutativity principle to subtraction. This misunderstanding leads to the common error of subtracting the smaller number from the larger one in every situation. Teachers may inadvertently add to the difficulty by making remarks like, 'You can't take a larger number from a smaller one' when children are setting up their own subtraction problems. One way to begin to address this thinking is to set up all of the related number facts using the HOTmaths Expression Calculator widget, which can be accessed by logging into HOTmaths and entering its name in the search field. For example:

$$6 + 7 = 13 \qquad 7 + 6 = 13$$

Related subtraction facts are:

$$13 - 6 = 7 \qquad 13 - 7 = 6$$

But

$$7 - 13 = -6 \text{ and } 6 - 13 = -7$$

Even though children are not expected to understand negative numbers until they are older, children in the primary years can understand that the answer to the reversed subtraction problem is different. Complex explanations are not necessary to illustrate this point. For instance, by using the examples of temperature, or going below the sea, children can understand that commutativity principles cannot be applied to subtraction. This activity also lends itself to the use of an IWB, with children involved in dragging the digits to create the equation.

It is vital to stress the place value aspects of the numbers involved in addition and subtraction. Learning to decompose numbers into component parts is important. Empty number lines are particularly useful (Figure 5.4). They provide strong visual imagery for children. For example:

$$453 + 167$$

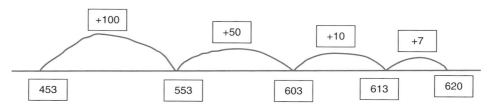

Figure 5.4 Using a number line

Global

Again, such an activity can be used to develop discussion and class involvement, and this may be enhanced by using an IWB. The interactive empty number line widget (Number Line Creator) in HOTmaths is a useful tool, and several number line activities are also included in FUNdamentals.

ACTIVITY 5.1

Log into HOTmaths, enter 'Number Sentence Builder' in the search field and click on this title in the results. Build number sentences that show all of the different meanings of addition and subtraction. Turn your number sentences into 'word problems'. What words did you use that meant addition or subtraction? Make sentences of the type 23 = 10 + 13 as well as the more common 10 + 13 = 23. Why is it important to show different ways of making number sentences?

Figure 5.5 Screenshot of HOTmaths Number Sentence Builder widget

Curriculum aims for computation are to develop confident, flexible thinkers who can take a variety of approaches to completing any computation.

Underpinning computational strategies must be a sound understanding of place value. Being able to use place value partitioning of numbers is a powerful strategy. The unit of work called 'Addition and Subtraction Pick n Mix' available from NZmaths at <http://www.nzmaths.co.nz> provides a framework for developing place value understanding and efficient computation using a variety of approaches. Sharing children's ideas is very important and use of a document camera with an IWB provides an efficient and effective way of sharing children's work. Another useful open online resource is the Australian Association of Mathematics Teachers Top Drawer, <http://topdrawer.aamt.edu.au>, particularly the mental computation drawer. Underpinning the mathematics curriculum are the proficiencies of understanding, fluency, problem-solving and reasoning. These are explicitly named in the Australian Curriculum: Mathematics, and are implicit in the approaches taken to learning mathematics in the New Zealand curriculum. It is important when teaching mathematics to keep these in mind, and also to consider ways in which they can be developed further.

ACTIVITY 5.2

AC

Log into HOTmaths and select 'Australian Curriculum' from the Course list. Choose one activity from an Addition and Subtraction topic for each year level from Year 3 through to Year 6. Choose different types of activities for each year level, or explore the same activity at different year levels.

- Which activities support the fluency proficiency? Do any activities help to develop reasoning, understanding or problem-solving?
- How would you use any of these activities in your classroom to develop the proficiencies?

Multiplication and division

Young children often have an intuitive sense of multiplication and division through grouping and sharing activities, although these activities may not necessarily lead to useful mathematical ideas without some intervention. A good teacher will identify and build on children's intuitive understanding.

Multiplication is the process by which one number is scaled up (or down) by another number – that is, it is a stretching (or contracting) process.

ACTIVITY 5.3

Use a word processing or drawing program. Go to 'Insert shapes' and choose a square. Insert the square into your document. Copy the square exactly (click on the square, then copy and paste). You have *added* one square. You can do this as many times as you wish in order to create, ultimately, an infinite number of squares. Now select one of your squares. Grab one corner of the square and drag it to make the square bigger. You have now *multiplied* the square size by increasing the length of both sides. The square has been scaled up.

Additive thinking is characterised by a series of repeated operations in which the same element is added on each time. Multiplicative thinking involves stretching (or contracting) the element – the relationship between the quantities is the crucial determining characteristic. When whole numbers are involved (as in tables facts), repeated addition – adding on the same number each time – will lead to the same answer as 'stretching' the number that is, 5 + 5 + 5 + 5 gives the same result as 4 x 5.

Multiplication is often associated only with 'times tables', but these are simply a summary of a specific set of number patterns that are useful to aid computation. Learning these facts is important, however, and the Australian Curriculum: Mathematics indicates that by the end of Year 4 children should be able to 'recall multiplication facts to 10 × 10 and related division facts' (ACMNA075). The New Zealand Maths site (<nzmaths.co.nz>) describes a game for times tables practice, similar to Rock, Paper Scissors, that could be used by parents, teacher aides or a more knowledgeable peer. There are several ways in which multiplication can be approached in the classroom, and technology provides a variety of resources that can help.

One useful representation of multiplication is found in arrays. An array provides a picture of the ways in which a number can be shown: objects are arranged in rows and columns, such that every row is the same length and every column is the same height. The number 12, for example, could be represented as shown below. It is important that children have opportunities to see and experience all of these arrangements, and to realise that they represent the same number in different ways. These kinds of opportunities also emphasise the commutativity of multiplication. Commutativity is often referred to as 'spin-arounds' when helping children to learn 'tables' facts.

```
        12 × 1                              3 × 4                or 1 × 12
  ☺ ☺ ☺ ☺ ☺ ☺ ☺ ☺ ☺ ☺ ☺ ☺              ☺ ☺ ☺                    ☺
         6 × 2                             ☺ ☺ ☺                    ☺
    ☺ ☺ ☺ ☺ ☺ ☺                           ☺ ☺ ☺                    ☺
    ☺ ☺ ☺ ☺ ☺ ☺                           ☺ ☺ ☺                    ☺
         4 × 3                             2 × 6                    ☺
      ☺ ☺ ☺ ☺                              ☺ ☺                     ☺
      ☺ ☺ ☺ ☺                              ☺ ☺                     ☺
      ☺ ☺ ☺ ☺                              ☺ ☺                     ☺
                                           ☺ ☺                     ☺
                                           ☺ ☺                     ☺
                                           ☺ ☺                     ☺
                                                                   ☺
```

Ask children to make arrays of numbers chosen by them using a variety of concrete objects such as bricks, pebbles, leaves, shells or counters, then ask them to record their arrays on squared paper, together with the related multiplication fact. Activities such as this will also lead to discussion about the kinds of numbers that can be represented as multiple arrays in contrast to prime numbers, for which there is only the horizontal or vertical array – for example, 1 × 7 and 7 × 1. HOTmaths has an Arrays widget and a Using Arrays HOTsheet for practising the use of arrays. These can be accessed by logging into HOTmaths and entering their respective names into the search field and then selecting the appropriate results tab. As children move through school, they should be encouraged to think about arrays not as a collection of single objects organised in a particular way, but rather as a rectangle, and using squared paper representations can help with this development.

A different approach to thinking about multiplication is to use repeated equivalent groups. Multiplication is represented as a repetitive addition of the same group. This representation can foster skip counting, which can be a useful strategy for small numbers (up to 5) of groups. For example, skip counting by 5 to 5 × 5 = 25 is a sensible strategy, but attempting to skip count with larger numbers – for example, 8 × 7 – can be problematic. Often children using this strategy lose count of the number of skips and are out by a **factor** – for example, answering 49 or 64 to 8 × 7. The Balloon Bunches widget in HOTmaths (see Figure 5.6) provides an opportunity to develop ideas about equivalent groups and skip counting. There are also resources in FUNdamentals that could be used to reinforce the different representations of multiplication (arrays or equivalent groups and skip counting).

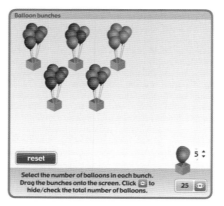

Figure 5.6 Screenshot of HOTmaths Balloon Bunches widget

Another online resource is Pobble Arrays (L2058) from the e-content available from Scootle, <http://www.scootle.edu.au>, or the New Zealand digistore at <http://digistore.tki.org.nz>. This resource could be used to develop the notion of factors. Older primary-age children should become familiar with using the term 'factor', and be able to explain what it means. The idea that 24 has factors of 1, 2, 3, 4, 6, 8, 12 and 24 provides powerful support for dealing with a range of computations, and can be developed from array representations. This understanding is important for later algebra activities. Two helpful interactive web-based activities are Factorize on the Illuminations website of the National Council of Teachers of Mathematics, <http://illuminations.nctm.org> and Factor Tree from softschools.com, <http://www.softschools.com>.

Division has the same relationship with multiplication as subtraction has with addition. It is the inverse operation, and 'undoes' multiplication operations. This understanding is important for children, and when learning tables facts they should also experience all the related division facts. For example, 7 × 6 = 42 should also be written as 6 × 7 = 42, 42 ÷ 7 = 6 and 42 ÷ 6 = 7 to emphasise the interrelated nature of the operations.

Division has two meanings. First, there is a *sharing* (or partition division) meaning, in which a given number is partitioned into equal groups. Second, there is a *grouping* meaning, in which a given number is divided into a given number or quota. This grouping meaning is sometimes called quotition division. Although the mathematical meaning and operation are the same, for children the different ways in which a problem is worded can make it difficult to recognise the operation needed as division. As an example, think about the subtle differences in these two problems.

There are 54 biscuits packed into 6 packets. How many biscuits in each packet?

This problem requires 54 to be portioned or shared among six equal groups. The answer is the number of items per group.

There are 54 biscuits which have to be divided into equal packets of 6 biscuits. How many packets would be needed?

This problem requires 54 to be divided into equal groups (packets), each of six biscuits. The answer is the number of groups. The mathematical solution in both instances is 54 ÷ 6 = 9.

ACTIVITY 5.4

AC

Log into HOTmaths and access the Teams widget by entering its name in the search field. Try the activity several times. Look particularly at the wording of each problem, whether you are asked to divide the players into a given number of teams or to make teams of a given number. Decide whether the problem requires *sharing* or *grouping*. Make up some problems of each type for yourself. Explain in your own words the difference between sharing and group conceptions of division. Scootle and the Digistore also have a number of activities, fact sheets and information about multiplication and division. Try, for example, Divide it Up: Grouping tool (L2810), Divide it Up: Sharing tool (L2809) or some of the arrays learning objects.

A word about calculators

Few topics have engendered as much discussion and dissent as the use of calculators in the classroom. This technology is widely used in everyday life – think of cash registers or petrol pumps, for example. In the classroom, however, there is a concern that using a calculator will reduce arithmetical skills. Research into calculator use has repeatedly emphasised the potential benefits of using calculators, although much of that research was completed in the 1980s and 1990s (e.g. Ruthven 1998). In the Australian and New Zealand curriculum documents, there is an expectation that technology – including calculators – will be available in schools. As with all tools, outcomes depend on what kind of use is made of calculators.

Many people claim that using a calculator, or other forms of technology, makes children lazy and that it will cause them to avoid learning basic number facts. These views are most common among parents and systems rather than schools and teachers (Banks 2011). Using a calculator does not replace the need for children to develop fluency and instant recall of basic number facts. Rather, calculator use emphasises the need to have mental tools for simple, straightforward computations, and high-level estimation skills in order to decide whether the answer arrived at is reasonable. Students should be able to make informed choices about whether an answer is best obtained using a mental, written or technology-based approach (Higgins 1990; Sowder 1990).

PAUSE AND REFLECT

A Year 5 teacher has given a worksheet to her class so they can practise number operations. Alex has provided this answer to the problem: 13 × 6.

13 × 6 = 52   I doubled it three times.

The teacher looks at the answer and says, 'You've got that wrong. Use a calculator to check.'

What has Alex done wrong? What message is Alex getting when he is told to use a calculator? How could the teacher have built on Alex's incorrect answer to develop the understanding needed? Discuss these questions with another person, and write down what you would do to help Alex.

Many commercially produced worksheets instruct students to 'Use a calculator to check your answer'. On the face of it, this appears to be sound advice, but when teaching school children, it is easy for them to avoid the thinking processes that are so important.

If children know that they can 'check' their answers, they may simply fill in the worksheet with little thought, knowing that they can write in the correct answers after they have 'checked'. In addition, the implicit message, reinforced by the teacher's 'You've got that wrong' comment, is that the calculator is always correct and that computation is not a personal skill that matters.

Calculators must be used appropriately. They are a very useful form of technology to develop understanding of patterns and skip counting, and they can be used to remove the computational load when the lesson focus is on some other concept, such as interpreting data or solving a complex problem involving measurement.

Drill and practice

Along with calculators, there is considerable discussion about the role of practice when learning new skills. Many people (e.g. Allen 2011) lament the passing of 'chanting' tables and think that today's children are not expected to memorise number facts. The research of Cowan (2011) clearly indicates that rapid recall of number facts aids mathematics proficiency, but that conceptual understanding of the ways in which numbers work is critical for continued success. The Australian Curriculum: Mathematics expects children to 'recall number facts' in both Year 3 and Year 4, and in New Zealand children should be able to apply addition, subtraction and simple multiplication facts to a variety of mathematical situations. Recall of number facts has to go beyond rote recitation to a deeper understanding of the four operations with whole numbers to develop the proficiencies of fluency, reasoning, understanding and problem-solving. It is not that drill and practice are not needed; rather, the focus should be on the nature of that practice, which should develop deep, connected knowledge of the number system.

Practice and developing efficient computational strategies should not always consist of drill. Consider the ways in which Veena in Classroom snapshot 5.3 is using technology to work with her class to develop number sense, based on multiplication.

CLASSROOM SNAPSHOT 5.3

Veena is a teacher with a Year 5 class that is developing strategies for multiplication with larger numbers. The class works in small groups selected by Veena so that there are different abilities in every group. She reminds the children that they may not tell anyone how to do a problem but should help each other by asking questions. Each group is given a set of three word problems involving three-digit by two-digit multiplication. The questions are graded in difficulty and each group can choose the question that they think will challenge them best. The questions are set in the context of a computer game where for every treasure collected there is a reward. The numbers were deliberately chosen to provide messy problems: 127 × 15, 254 × 36, 598 × 87. The aim is to develop further understanding of the use of the **distributive property** that all the children know about and can use confidently with smaller numbers. The children work on their problems for about 15 minutes while Veena walks around and asks questions of the children, such as, 'What is the meaning of the 1 in 127?" and, 'What are the different ways that you could split up 254?' She calls the class together and asks each group to come to the front with its work. The work is placed under a document camera and projected onto the IWB. Several different solutions are shown, including a standard written algorithm and different ways of using the distributive law. Each solution is explained by one member of the group. Veena takes a screenshot and saves each solution. When the range of solutions has been shown, Veena brings them all onto the screen and there is a class discussion of the ways in which the approaches work, and how they rely on the distributive property. Each child is then challenged to try the problem again in their maths book using a different method from the one that their group put up. Veena collects the books at the end of the lesson for later review and ends the lesson with a quick round-up, during which the children say what they have learned during the lesson.

There are other ways of using technology effectively to provide practice, develop understanding or consolidate learning. It is not the technology itself that enhances learning, however, but the nature and quality of the interactions between the children themselves, and the teacher and children, as exemplified in Classroom snapshot 5.3. Problems need to be posed carefully and with thought to provide different levels of challenge. It is important to have discussion both during the learning activity and particularly at the end of the lesson. Asking the children to solve the same problem in a different way ensures that the students engage with new ideas. The brief review at the end helps both the children and Veena reflect on the lesson, and sets up a platform for the subsequent mathematics lesson.

PAUSE AND REFLECT

There are sound reasons, backed by research, for students needing to be able to recall number facts automatically (e.g. Pegg, Graham & Bellert 2005). What is the best way to achieve this?

Some common approaches include:

- 10 quick facts at the start of a lesson
- timed tests of basic facts
- competitive games such as Sherrif (Swan & Marshall 2009)
- worksheets with many repetitions of a previously taught procedure
- conceptual games such as Numero™ (Asplin, Frid & Sparrow 2006)
- the use of programs such as *Math Blaster* (<http://www.mathblaster.com>)
- Scorcher activities in HOTmaths.

Discuss which of these approaches you have seen in classrooms. What approaches do you think are helpful to students? What is the role of technology in providing practice opportunities? What resources will you use when you are teaching?

Reflection

Operating with whole numbers is a key aspect of the primary curriculum. In parallel with the development of these skills, it is important to ensure that students build foundations for proportional reasoning through developing multiplicative thinking and understanding number relationships, including those dealing with parts of a whole, such as fractions, decimals and percentages. Children need time to develop understanding and consolidate new concepts, and teaching should focus on both conceptual understanding and procedural fluency. Providing a range of problem-solving practice activities and opportunities for discussion supported by good use of technology will help to support children's mathematical development.

Further reading/websites for exploration

Further reading

The readings suggested here are all available electronically, and provide some additional background.

Allen, E 2011, 'Scandal of the primary pupils who can get full marks in maths without even knowing their times tables', *Mail Online*, 7 September 2011, <http://www.dailymail.co.uk/news/article-2034442/Pupils-passing-maths-exams-good-marks-dont-know-times-tables.html>.

Asplin, P, Frid, S & Sparrow, L 2006, 'Game playing to develop mental computation: A case study', in P Grootenboer, R Zevenbergen & M Chinnappan (eds), *Identities, cultures, and learning spaces* (Proceedings of the 29th annual conference of the Mathematics Education Research Group of Australasia, Canberra), Adelaide: MERGA, <http://www.merga.net.au>.

Banks, S 2011, 'A historical analysis of attitudes toward the use of calculators in junior high and high school math classrooms in the United

States since 1975', MA thesis, University of Cedarville, Cedarville, OH, <http://digitalcommons.cedarville.edu/cgi/viewcontent.cgi?article=1030&context=education_theses>.

Cowan, R 2011, *The development and importance of proficiency in basic calculation*, London: Institute of Education, <http://www.ioe.ac.uk/Study_Departments/PHD_dev_basic_calculation.pdf>.

Higgins, J 1990, 'Calculators and common sense', *The Arithmetic Teacher*, vol. 37, no. 7, pp. 4–5.

Jacob, L & Willis, S 2001, 'Recognising the difference between additive and multiplicative thinking in young children', in J Bobis, B Perry & M Mitchelmore (eds), *Numeracy and beyond* (Proceedings of the 24th Annual Conference of the Mathematics Education Research Group of Australasia), Sydney: MERGA, <http://www.merga.net.au>.

McIntosh, A 2004, 'Developing computation', *Australian Primary Mathematics Classroom*, vol. 9, no. 4, pp. 47–9.

Pegg, J, Graham, L & Bellert, A 2005, 'The effect of improved automaticity of basic number skills on persistently low-achieving pupils', in HL Chick & JL Vincent (eds), *Proceedings of the 29th Conference of the International Group for the Psychology of Mathematics Education, Vol. 4*, edited by HL Chick & JL Vincent, Melbourne: PME, pp. 49–56, <http://www.emis.de/proceedings/PME29/PME29RRPapers/PME29Vol4PeggEtAl.pdf>.

Ruthven, K 1998, 'The use of mental, written and calculator strategies of numerical computation by upper primary pupils within a "calculator-aware" number curriculum', *British Educational Research Journal*, vol. 24, no. 1, pp. 21–42.

Siemon, D, Bleckly, J & Neal, D 2012, 'Working with the big ideas in number and the Australian Curriculum: Mathematics', in B Atweh, M Goos, R Jorgensen & D Siemon (eds), *Engaging the Australian Curriculum: Mathematics – perspectives from the field* [e-book], Mathematics Education Research Group of Australasia, pp. 19–45, <http://www.merga.net.au/node/223>.

Sowder, J 1990, 'Mental computation and number sense', *The Arithmetic Teacher*, vol. 37, no. 7, pp. 18–20.

Swan, P & Marshall, L 2009, 'Mathematics games as a pedagogical tool', in *CoSMEd 2009 3rd International Conference on Science and Mathematics Education Proceedings*, Penang, pp. 402–6, <http://www.recsam.edu.my/cosmed/cosmed09/AbstractsFullPapers2009/Abstract/Mathematics%20Parallel%20PDF/Full%20Paper/M26.pdf>.

Websites

Digistore: <http://digistore.tki.org.nz>
Illuminations: <http://illuminations.nctm.org>
Math Blaster: <http://www.mathblaster.com>
A Maths Dictionary for Kids: <http://www.amathsdictionaryforkids.com>
Nrich: <http://nrich.maths.org/4339>
Scootle: <http://www.scootle.edu.au>
Top Drawer: <http://topdrawer.aamt.edu.au>

CHAPTER 6

Part-whole numbers and proportional reasoning

LEARNING OUTCOMES

By the end of this chapter you will:

- understand the importance of part-whole numbers such as fractions, decimals and percentages
- choose appropriate representations to illustrate different ways of thinking about fractions, decimals and percentages
- understand the use of proportional reasoning in daily life
- use technology effectively to develop understanding of part-whole numbers.

KEY TERMS

- **Multiplicative thinking:** Thinking and reasoning about more than one quantity or value at once (e.g. doubling the side length of a square means that the area quadruples)
- **Part-whole numbers:** Those representations of number that relate a part to the whole, such as fractions (e.g. $\frac{3}{5}$ is three parts taken from five parts of a whole), decimals (e.g. 0.6 is six-tenths of a whole) and percentages (e.g. 35% is 35-hundredths of a whole)
- **Proportional reasoning:** Reasoning about relationships (e.g. 6 being two threes or three twos rather than one more than 5) or comparing quantities or values (e.g. 1 km is 1000 m).

We use **proportional reasoning** every day, often without being aware that we are reasoning in terms of two quantities that vary in relation to each other – that is, as one quantity increases or decreases, so does the other. I may decide to buy two tins of tomatoes. The price of each tin is the same, so if I purchase double the number of tins, the amount I pay also doubles. Despite using this thinking informally quite regularly, it is surprising how many people have trouble with this concept. Doubling or trebling a quantity is one thing, but what about wanting one-and-a-half times, or only needing one-fifth of something? These calculations can become very tricky. Often we make some kind of estimate, and either end up with too little or too much of something.

Proportional reasoning is used widely to solve a range of everyday problems from 'best buys' to understanding data presented in tables. It underpins scaling problems such as scale drawings of house plans and currency conversions, and appears in many other situations.

ACTIVITY 6.1

Think about what you do each day. Identify times when you may have needed to use proportional reasoning. Examples might be in cooking, making something, house decorating, shopping, going on a journey … there are many possibilities. Make a list and identify as many different situations as you can. For example, if you have to increase a recipe serving four people to feed 10 people, each of the ingredients will need to be increased $2\frac{1}{2}$ times because $10 = 2\frac{1}{2} \times 4$. Compare your list of everyday activities using proportional reasoning with that of another person, and comment on the similarities and differences.

Developing **multiplicative thinking** is fundamental to proportional reasoning, and the shift from additive to multiplicative thinking is a focus for the middle years of schooling. In this chapter, the underpinning conceptual understanding of numbers such as fractions, decimals and percentages is explored as an important precursor to reasoning proportionally. Key ideas are developed about the appropriate pedagogy for working with primary-age children and using technology effectively to enhance understanding.

Some background

Moving from additive to multiplicative thinking is one of the great challenges in the primary years of schooling. There is considerable research on this topic, and being able to recognise key behaviours and capitalise on 'teachable moments' is an important skill of all good primary teachers. Siemon, Bleckly and Neal (2012, p. 25) suggest that by Year 4 students should begin to demonstrate multiplicative thinking, which is characterised by:

> Capacity to work flexibly with both the number in each group and the number of groups (e.g. can view six eights as five eights and one more eight). Recognises and works with multiple representations of multiplication and division (e.g. arrays, regions and 'times as many' or 'for each' idea).

By Year 6, students should show:

> Ability to partition quantities and representations equally using multiplicative reasoning (e.g. a fifth is smaller than a quarter, estimate 1 fifth on this basis then halve and halve remaining part again to represent fifths), recognise that partitioning distributes over previous acts of partitioning and that numbers can be divided to create new numbers.

Such thinking lays the foundation for the development of proportional reasoning during the secondary years. The importance of fractions, decimals and percentages is endorsed by Australian and New Zealand mathematics curriculum documents. Both countries describe an almost identical progression of fraction understanding from understanding the nature of one-half as two equal parts in the early years of the primary school to recognising equivalence, moving between representations of fractions, decimals and percentages, and carrying out straightforward operations by the upper primary years. *Teaching Fractions, Decimals, and Percentages* is a useful downloadable resource available to all teachers from the NZmaths site at <http://www.nzmaths.co.nz>.

PAUSE AND REFLECT

What do you remember about learning fractions, decimals and percentages at school? Jot down all the words that come to mind. What helped you to comprehend these numbers? What was difficult to understand? From your reflection, think about how you can translate your experience into positive learning events for the children you teach.

Figure 6.1 Fraction representations

Parts and wholes

The fraction one-half ($\frac{1}{2}$), and its decimal (0.5) and percentage (50%) equivalents, are widely experienced by children, and most children develop intuitive understandings of half. Many children, however, don't have a lot of experience of other fractions outside the classroom, so care should be taken to ensure that teaching materials use fractions, decimals and percentages that go beyond halves and quarters.

Fractions are notoriously difficult for children to understand, and technology can play a part in helping to support the development of fraction understanding. In the early years of primary school, children need to understand that a fraction can represent part of a whole and part of a group, and can be a number in its own right.

For children in the early years, a key idea is that the parts of the fraction must be equal in size, whether represented as a whole or a group. Activities such as folding paper into halves, and half again, and sharing objects such as blocks into two, and then four, equal groups, together with use of the language of fractions, will help young children develop understanding. Note the links to division – both sharing and grouping. The HOTmaths Farm Fractions widget is a support for developing the idea of fractions of a group, and would lend itself to a lesson using an IWB.

ACTIVITY 6.2

Log into HOTmaths and access the Farm Fractions widget by entering its name in the search field. Play with the activity to become familiar with it.

Global

What happens to the number of sheep as you change the fraction required? What happens to the fraction as you change the number of sheep? This widget represents a fraction as a part of a group. Think about the strengths of this representation.

Now find the Exploring Equivalent Fractions widget. Play with the widget to explore what it can do. This widget represents a fraction as an area model – that is, as part of a two-dimensional figure.

Now go to the Counting with Fractions widget. As before, become familiar with what the widget does. This widget represents a fraction as a number on the number line. How are these three representations similar and different? Represent $\frac{3}{5}$ in all three representations. What do you need to know to show the same fraction in different ways? Why should students recognise and use all of these representations?

Interactive Maths at Wikispaces.com, <http://www.interactivemaths.wikispaces.com>, also has a number of interactive resources that could be used either to demonstrate ideas about fractions or as an activity for children to undertake as part of a lesson based on a round-robin format. Look at Pizza Fractions or Cross the River.

If you use interactive resources when you are teaching, it is essential that these are combined with whole-class and individual or small-group discussion. Developing the language is crucial. Any online or ICT-based resource must be integrated into a coherent lesson sequence. Once children are familiar with what is expected, small groups could play the games or do the activities at the IWB. This would allow you to be talking to other groups of children but also keeping an eye on what the ICT group is doing.

The difficulty that children have with representing fractions other than half was described by Gould, Outhred and Mitchelmore (2006), who found that children in Year 7 still used naïve representations of one-third and one-sixth, most likely based on seeing a fraction as two whole numbers unrelated to each other. This finding is a reminder that diagrams or drawings that adults take for granted may be interpreted quite differently by children. This is important to bear in mind when using technological resources.

Different meanings of fractions

There are different ways of thinking about fractional numbers. As children progress through school, they need to experience both part-whole and part-part representations. For example, in Figure 6.2, the red section can be thought of as one part out of four, a part-whole notion because the whole is the large rectangle that has been divided into four parts. Usually, this is written in traditional fraction notation as $\frac{1}{4}$ It could also, however, be described as one red part to three white parts, a part-part idea because it describes how the different parts of the large rectangle are related to each other. This idea can be written as a ratio such as 1:3.

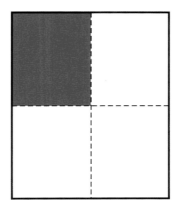

Figure 6.2 1:3 ratio

The same idea can be applied to fractions as represented by groups or sets (Figure 6.3). In the first array shown here, the box contains $\frac{1}{4}$ of the group of ☺. The second array shows the ratio of ♦ to ☺. For every one ♦ there are three ☺, and this can be written as 1:3.

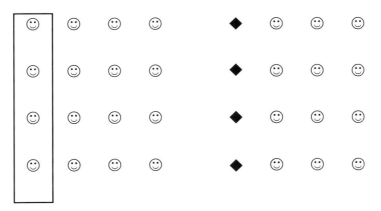

Figure 6.3 1:3 ratio sets array

Representations of fractions are the focus of the downloadable resource *Fractions, Pikelets and Lamingtons* (Department of Education and Training, NSW 2003). The difference between 'continuous' models, such as shapes and paper strips, and 'discrete' models, such as collections of objects, is important to understand because these may be useful in different situations.

ACTIVITY 6.3

Fractions can also be thought of as division. Consider, for example, dividing three pizzas among five people. How much pizza would each person receive? Solve that problem in any way you like. You can draw or use concrete materials if you want to. Now ask several other people of different ages and experience to solve the same problem. What different solutions did you see?

Each person gets $\frac{3}{5}$ of a pizza. Children will solve these kinds of problems in a variety of ways, and may represent their answer as a fraction. Often they will start by dividing the three pizzas in half and allocating half a pizza to each person. Then they have to deal with the leftover half of a pizza. Some children will divide this half into five parts and allocate each person one part. Often they cannot resolve this into a formal fraction solution. Other children will divide each pizza into five parts and share these parts among the five people so that each person gets three parts out of five from a pizza. These informal solutions provide important experiences for children and, more significantly, give teachers considerable insight into their thinking and understanding. There is an interactive game called Kids and Cookies available at <http://www.teacherlink.org/KidsAndCookies> that would be a useful follow-up after a class had tried the problem.

A key idea from the middle primary years onwards is that of equivalence. If this is well developed and understood, many of the difficulties with fraction operations that arise later can be avoided. Equivalent fractions are those that are the same size but have different denominators. The idea that $\frac{1}{3} = \frac{2}{6} = \frac{3}{9} = \frac{2}{71} = \frac{100}{300}$ and so on is powerful. Any fraction can be written in different ways. Children should be encouraged to explore equivalence and develop their own 'rule' to make equivalent fractions from any starting point.

One useful way of starting to develop ideas about equivalence is with a fraction wall. There are many different versions of the fraction wall available, but they are also easy to make using a table in Microsoft Word. Create a table with a single column and as many rows as you want. Then use the Split Cells feature under Table Tools – Layout to split each row into as many cells as you wish. Make sure that initially you have fraction families: halves, quarters, eighths, sixteenths and thirds, sixths, ninths and twelfths. Ask children to find and colour in all the different ways to make $\frac{1}{2}$ and $\frac{1}{3}$ and so on. Initially use only unit fractions – that is, those fractions with 1 as the numerator. Then extend into $\frac{2}{3}$, $\frac{3}{4}$ and more complex fractions. An interactive fraction wall that is available from <http://www.tes.co.uk> could be used with an IWB with small groups or a whole class. The AAMT Top Drawer website has a Fraction Wall game in the fractions drawer that could be used with the interactive wall as well as with concrete materials. This drawer also has a lot of useful information in it.

Older children can also be challenged to think about the relationships among different fractions, and how these change if a different whole is specified. In the fraction wall shown in Figure 6.4, if A = 1, then B = $\frac{1}{2}$, C = $\frac{1}{3}$ and so on. But if B is specified as the whole (that is, B = 1) then A = 2, D = $\frac{1}{2}$ and F = $\frac{1}{3}$. Naming C and E when B is the whole can become a challenge, and discussions with children about how they could find out how big C is if B = 1 can lead naturally to discussions about common denominators, making use of the equivalence between two lots of F and C, and three lots of F and B. The HOTmaths Comparing Fractions widget is also a useful tool to support the development of equivalence understanding. You can access this widget by logging into HOTmaths and entering its name in the search field.

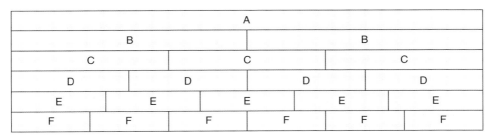

Figure 6.4 Fraction wall

Development of fraction understanding takes place over time, and through many different experiences. It is important that children have concrete experiences using activities such as paper folding and cutting. They should also use a variety of shapes – circles, rectangles, squares, hexagons – and paper streamers. By cutting and folding, matching shapes and placing one piece over another, children develop a feel for equivalence that can be transferred to a virtual environment.

If children understand equivalence, operating with fractions becomes much more intuitive, and children are less likely to try to rely on half-remembered rules. Adding $\frac{2}{3}$ and $\frac{1}{2}$, for example, can be visualised as a fraction wall. If A = 1, then $\frac{2}{3}$ can be seen as four lots of F or $\frac{4}{6}$, which has to be added to three lots of F or $\frac{3}{6}$, making $\frac{7}{6}$. It is easy then to see that this is one lot of A and one lot of F or $1\frac{1}{6}$. Not only does the addition become more intuitive, but understanding of mixed fractions can also be developed.

Fractions can also be thought of as operators. When used in this way, fractions can expand or contract a number, and are forms of mathematical functions closely associated with patterns and algebra. For example, $\frac{2}{3}$ of 15 could be seen as 'input 15, multiply that by 2 and divide the answer by 3 to get 10'. Making connections with early algebra concepts creates links and provides further rationale for learning about fractions in a metric world. The fraction $\frac{2}{3}$ is, in this instance, shrinking 15 down to 10.

This section has focused more on developing conceptual understanding of fractions than on using fractions to calculate new quantities. Developing a sound understanding provides a platform for future development.

Decimals

Decimals are particular fractions expressed in a different way. In some places they are termed decimal fractions, which emphasises the close relationship. Decimals can be written as fractions with denominators that are powers of 10 (10, 100, 1000, etc.).

A key understanding about decimals concerns the extension of place value understanding to parts of a number less than 1. You can find quite a good explanation of this for your own understanding at the Khan Academy website, <http://www.khanacademy.org/math/arithmetic/decimals/v/decimal-place-value>. To develop this understanding in children, however, requires more than an explanation – however good this may be. Children need to experience a range of ways of representing decimals.

Using a strip of paper divided into 10 equal parts, students can explore tenths and represent these as both fractions and decimals. This is a precursor to using squared paper where 10 by 10 squares represent one whole. Each small square is then $\frac{1}{100}$, or 0.01. Children need these concrete experiences that can be modelled using interactive representations such as the HOTmaths Representing Decimals widget. You can access this widget by logging into HOTmaths and entering its name in the search field.

Decimal number expanders (Figure 6.5) are another resource. There is an example of a decimal number expander at the Teaching and Learning about Decimals page of the Melbourne Graduate School of Education website, <https://extranet.education.unimelb.edu.au>, and this site also has templates for printing. Representations of this kind help children to visualise decimal numbers and begin to understand the relationships among the decimal parts of a number.

Multibase arithmetic blocks are also used to represent decimals. These should be used with some care. If children are used to seeing the small cube as a unit, they may have difficulty seeing the flat or large block as 1. There are various interactive representations available to show this approach.

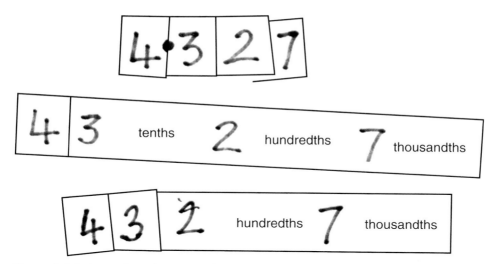

Figure 6.5 A decimal number expander showing different ways of presenting the same number

A better approach is to use Linear Arithmetic Blocks. This concrete model has a long rod as the unit, which can be covered by 10 pieces of poly pipe, or 100 smaller pieces of poly pipe, or 1000 washers (Figure 6.6). Commercial versions are available, but it is relatively easy to make your own. This model appears to be less confusing for children, and if it is made fairly large, it involves children in physically manipulating the equipment.

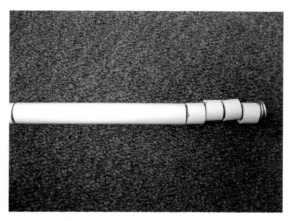

Figure 6.6 Linear arithmetic blocks

As with fractions, decimals can be represented on the number line. Many children do not believe that there are numbers between the counting numbers represented on a number line. They need practice and experience to understand that, for example, 2.7 lies between 2 and 3. As they move through school, children begin to experience decimal numbers that are more fine grained, to the 100th and 1000th place. Many children develop the idea that the longer the decimal, the larger it is. For these children, 0.678 is larger than 0.75. This notion needs to be addressed through place value activities and a range of representations.

Building a conceptual understanding of decimals is important before beginning to operate with decimals. The earliest operation is usually multiplication by powers of 10.

CLASSROOM SNAPSHOT 6.1

Kim wants to teach her Year 5 class about multiplying a decimal number by 10 with an emphasis on the place value aspects. She hands out cards with digits on them to some children, and one child receives a card with a large black dot – the decimal point. The children make the number 4.356. The child with the decimal point is asked to sit down and hold her card up high. The other children remain standing. When another child waves the magic ×10 wand, all the digits move one place to the left and the decimal point remains sitting. Students read the new number, 43.56, and discuss what they saw. They repeat this activity with different numbers. Then they return to their desks and make number slides (see Figure 6.7) to create different multiplications by 10 and then 100. The children write their own rules for multiplying decimals by powers of 10.

- Why did Kim first model the activity using the children?
- Why is it important for children to realise that the numbers move – not the decimal point?
- Where is this representation seen in daily life?

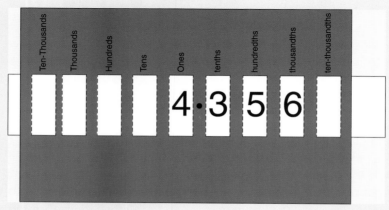

Figure 6.7 A number slide

Calculators are a useful technology for developing ideas about decimals. They could be used to reinforce Kim's lesson, for example. They are particularly valuable for developing ideas about the expansion and contraction of numbers when they are operated on by a decimal. Many children believe that multiplication always makes a number bigger. Challenging this notion by setting up situations where this is not true can be a powerful learning experience.

ACTIVITY 6.4

Use a calculator to complete these computations:

$$25 \times 3 \quad 25 \times 0.3 \quad 25 \times 0.03$$

Predict your answer to the problem 25 × 0.003. Were you correct?

Describe in your own words what is happening. Predict what you think would happen if you divided a number by a decimal. Set up some calculations of your own to test your conjecture.

Money and decimals

Money is often suggested as an introduction to understanding decimals. There is evidence, however, that money is of limited use when introducing decimals (see the Teaching and Learning about Decimals site at <http://extranet.edfac.unimelb.edu.au/DSME/decimals>).

Children see money as separate from the decimal numbers with which they work in classrooms. Money is concrete, whereas decimal numbers are abstract in nature. Today, money in Australia is rounded to the nearest 5 cents, and in New Zealand to 10 cents. Using money as a model for decimals may convince children that decimals can only be written to the nearest 0.05 or 0.10 or two decimal places (Steinle & Stacey 2001). Problems of the type 'I purchased a toy for $4.37 and some lollies for $1.24. How much money did I spend?' are unrealistic and do not provide context-based practice in adding decimals. There are many resources available that use such approaches, but they simply reinforce children's ideas that school mathematics is not related to their everyday lives. Money can be useful, but it should not be the only model used to introduce decimals.

Percentages

Percentages are another form of fractional number. Per cent means 'out of one hundred', and these numbers are probably more common in daily life than fractions. A percentage means that the number or quantity has been divided into 100 parts. In the Australian Curriculum: Mathematics, percentages are first found in Year 6, where they are associated with decimals and fractions. It is a logical and relatively straightforward step to recognise that percentages are decimal numbers and fractions of 100.

Children should begin to develop an intuitive sense of the size of a percentage, using 50% as a benchmark initially, and then 25% and 75% as halfway points between 0 and 50%, and 50% and 100%. As with fractions, it is essential to identify the whole. Describing something as 30% makes no sense unless the whole is defined. The Estimating Percentages widget in HOTmaths provides an opportunity to begin to develop a sense of the size of a percentage. You can access this widget by logging into HOTmaths and entering its name in the search field.

Recognising that a fraction may be interchangeable with a decimal or a percentage is useful because it makes computation easier later on. Percentage calculations are very common, and developing quick and easy equivalents in fractions and decimals is potentially powerful. In the primary years, the focus is on the common percentages and equivalents, such as 10%, 25%, 50% and multiples of these, such as 20% and 75%. Emphasise the interconnectedness: if children can recognise 50%, they can halve this to get 25%, in the same way that they deal with halves and quarters. People who have a well-developed sense of these relationships can move effortlessly between the different representations, choosing the one that makes most sense to them or that is most useful for a given purpose. Children should be encouraged to use mental strategies for common percentages. For example, finding 50% of a quantity is a halving activity, which most children deal with reasonably easily. In the primary years, developing these intuitive understandings is more important than learning algorithms.

There are many games and activities that can be used to develop a facility with using percentages, decimals and fractions interchangeably. These include Bingo-type games, matching games and puzzles. The Boxladders game in HOTmaths, accessed via the Games icon on the Dashboard, is another good example. These games provide practice and motivation for children. The HOTmaths Ordering Types of Fractions widget is a useful tool to develop a sense of the size of different representations.

The Illuminations site from the National Council of Teachers of Mathematics also has some interactive activities that could be used to support developing understanding. For example, try Fractions Models at <http://illuminations.nctm.org/ActivityDetail.aspx?ID=11>, which allows a variety of representations and shows the fraction, decimal and percentage equivalents.

The National Library of Virtual Manipulatives, <http://nlvm.usu.edu>, has a useful interactive tool that computes a part, a whole or a percentage, given any two of these quantities. It also provides some useful suggestions for using a tool of this kind. Use of a manipulative like this allows children to calculate quantities that might otherwise be too complex. From a teaching perspective, though, it is important to be purposeful about the use of such tools. Use them to help children to identify the relationships between the whole, the part and the percentage, and to relate these to real situations. Ask them to explain these relationships to develop reasoning and understanding.

Fractions, decimals and percentages are important numbers with which children need to come to grips. The underlying thinking is multiplicative, and lays the foundation for proportional reasoning that is a critical component of the curriculum in the middle years. These numbers are used frequently in other areas of mathematics, and also in other subjects where their mathematical complexity may be less well understood.

Proportional reasoning

Proportional reasoning can be understood in different ways, and is difficult to capture in a short definition. The key to thinking proportionally is to identify the relationships between numbers and quantities. For example, a piece of paper is three pegs or five USB drives long. Another piece of paper is two pegs long. How many USB drives long is the piece of paper? Try solving this problem before reading further.

A problem such as this one can be solved in a variety of ways. You could find out how many USB drives make one peg and then double the answer to find the length of the smaller piece of paper in USB drive units. You might say the new piece of paper is $\frac{2}{3}$ of the old one so the number of USB drives long is $\frac{2}{3}$ of 5. Children should be encouraged to find answers to problems of this type using whatever resources they have. With concrete materials, even children as young as Year 4 level can attempt problems like this – providing an opportunity to develop their problem-solving proficiency.

Proportional reasoning has strong links with algebra, and early activities working with patterns are important in building understanding.

CLASSROOM SNAPSHOT 6.2

Children in a Year 6 class were working on a pre-algebra task. They were using square tiles to model the number of people who could be seated at a series of square tables, as the number of tables was increased. They worked systematically through a series of questions that asked about the number of people that could be seated at two tables, four tables, five tables and ten tables. Then they were asked a challenge question: How many people could be seated at 99 tables?

The number of tables was too large to model with tiles, so the children had to find alternative approaches:

- Sam drew the picture shown in Figure 6.8 and counted the number of people.
- Min said, 'It's a pattern going up by two each time. I can skip count by two, 99 times, starting from 4 because that's the beginning number.'
- Jan said, 'It's easy. You just double the number of tables and add one for each end.'

All three children got the correct answer (200 people).

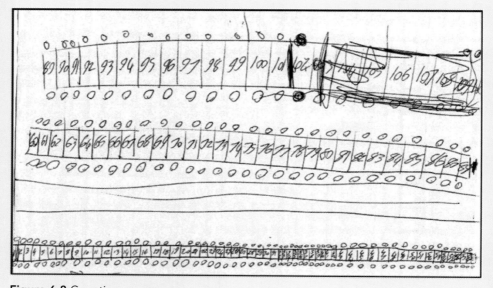

Figure 6.8 Counting on process

The three children in Classroom snapshot 6.2 demonstrate diverse ways of calculating their answer.

Sam is using counting strategies and needs to recognise that there are more efficient approaches. To provide suitable teaching for Sam, you may need to go back to early number work and provide age-appropriate resources to develop more effective strategies such as counting by fives or tens. The Leaping Frog widget in HOTmaths is one practice activity that Sam might find useful. You can access this widget by logging into HOTmaths and entering its name in the search field.

Min has recognised a pattern and has used that to help her skip count. However, skip counting is not efficient when large numbers are involved, and it is easy to lose count of the number of skips. She needs to begin to use patterns more effectively, recognising the underlying relationship. Her thinking is still essentially additive – she adds on two each time to get to the answer – but it is more sophisticated than Sam's approach. Min needs to begin to recognise the multiplicative relationship involved so that she can arrive at the answer more efficiently. She would benefit from activities such as making patterns and recording the growing patterns as a table (see the HOTmaths Toothpick Patterns HOTsheet for Year 6).

Jan is using multiplicative thinking. He has recognised a relationship between the ways in which the tables are arranged and the number of people who can be seated, and is using multiplication to arrive at the answer efficiently. He needs now to recognise and write rules for a variety of number pairs. Activities such as HOTmaths Determine the Rule widget or the Interactivate Function Machine, <http://www.shodor.org/interactivate/activities/FunctionMachine> would be helpful.

Multiplicative thinking is an underpinning concept for algebra, and activities such as writing rules for number pairs provide a useful base upon which to build. (See Chapter 2 for early number ideas that provide a foundation.)

The use of technology described in this chapter so far has had a focus on the mathematics. There are also useful resources that can be used to manage the lesson. In Classroom snapshot 6.3, the classroom teacher, Tina, is using a free online educational resource called Kahoot!, <https://getkahoot.com>, to allow the children to share their answers and then participate in a short quiz as a round-up to the lesson.

CLASSROOM SNAPSHOT 6.3

Tina is a beginning teacher, teaching a Year 5 class. She is one of a group of teachers in her school who are working together to explore and improve the ways in which they use technology in the classroom.

Before the lesson began, Tina asked the students to work in pairs. Each pair collected a laptop or a tablet from the class set. She asked them to log into Kahootit, <https://getkahoot.com>, and then to set the devices down.

The focus of the class was developing early proportional reasoning. Tina posed a starting problem on the IWB:

> In a Year 1 class, Ruby makes a line of bricks that is 7 bricks long. Walt wants to make one twice as long. How many bricks will he need? What if he made it three times as long? Or 10 times as long?

The children worked in pairs to solve the problem. Using an approach with which the children were familiar, they used a screen-sharing facility to put their solutions anonymously onto the whiteboard from their tablets or computers. Tina saw that most children were able to solve the problem, so she led a class discussion in which she encouraged children to share their approaches to solving it. Eventually the class came to the realisation that they were using the seven-times table. Tina posed a few problems based on other numbers such as 3, 5 and 8 to consolidate this idea, recording the children's answers manually on the IWB.

She then asked the children to return to the Kahootit page and gave them the necessary log-in PIN to access a quiz that she had prepared earlier. The quiz used multiple-choice questions and displayed the answer and the number of people who got the answer correct after 30 seconds' thinking time. The children were excited and engaged, talking about the problems and keeping track of what they had got right.

At the end of the lesson, the children returned their devices to the storage facility and the technology monitors checked that all the tablets and computers were correctly docked so that they could recharge.

Several discussion points arise from Tina's use of technology. All the sharing is anonymous, so children who are less confident can participate knowing that if they get an answer incorrect other children in the class don't know. Tina, however, knows her class well and by watching the children's reactions when the solutions are discussed, or their responses to the quiz answers, is able to identify those children who will need some additional help. The quiz is motivating but not competitive, because individuals are not identified. The children are working in pairs and their goal is to get the answer, not to beat the rest of the class. Tina can save the children's answers and the summary from the quiz for later thought and planning. There are good routines in place for using the technology. The class understands how to access the devices, how to log on to the sites requested and how to input their answers. Establishing good routines is an essential classroom-management technique when technology is involved, to avoid unnecessary delays or frustrations from laptops or tablets not being charged, and to minimise damage.

PAUSE AND REFLECT

In this chapter, several ways have been suggested in which technology can help you teach one of the more challenging domains of the primary mathematics curriculum. Look back through the chapter and identify as many different uses of technology as you can. What similar uses of technology have you seen or used yourself? For some uses of technology, you will need to develop new skills. Why should you invest your time and energy into this learning? What aspects of the TPACK framework do you need to develop to be an effective user of technology in the mathematics classroom?

Reflection

Fractions, decimals and percentages will continue to challenge teachers and children in the primary years. However, developing a sound understanding of the characteristics of these numbers is critical for mathematical development. You should have noticed that in this chapter the emphasis has been on concepts rather than procedures. In the primary years, operating on **part-whole numbers** should arise naturally from the discussions about the kinds of numbers that these are, rather than focus on computation and algorithms. Given a sound platform of conceptual understanding, the later complexities of proportional reasoning will be minimised and children will develop confidence in their mathematical capabilities.

Further reading

The readings suggested here are all available electronically, and provide some additional background.

Department of Education and Training, NSW 2003, *Fractions, pikelets and lamingtons*. Sydney: Department of Education and Training Professional Support and Curriculum Directorate, <http://www.schools.nsw.edu.au>.

Gould, P, Outhred, L, & Mitchelmore, M 2006, 'One-third is three-quarters of one-half', in P Grootenboer, R Zevenbergen, & M Chinnappan (eds), *Identities, cultures, and learning spaces* (Proceedings of the 29th Annual Conference of the Mathematics Education Research Group of Australasia, Canberra, Adelaide: MERGA, pp. 262–70, <http://www.merga.net.au>.

Jacob, L & Willis, S 2001, 'Recognising the difference between additive and multiplicative thinking in young children', in J Bobis, B Perry & M Mitchelmore (eds), *Numeracy and beyond* (Proceedings of the 24th Annual Conference of the Mathematics Education Research Group of Australasia), Sydney: MERGA, <http://www.merga.net.au>.

Ministry of Education 2008, *Teaching fractions, decimals, and percentages*, Wellington, NZ: Ministry of Education, <http://www.nzmaths.co.nz>.

Siemon, D, Bleckly, J & Neal, D 2012, 'Working with the big ideas in number and the Australian Curriculum: Mathematics', in B Atweh, M Goos, R Jorgensen & D Siemon (eds), *Engaging the Australian Curriculum: Mathematics – perspectives from the field* [e-book], Sydney: MERGA, pp. 19–45, <http://www.merga.net.au/node/223>.

Steinle, V & Stacey, K 2001, 'Visible and invisible zeros: Sources of confusion in decimal notation', in M Mitchelmore, B Perry & J Bobis (eds), *Numeracy and beyond* (Proceedings of the 24th Annual Conference of the Mathematics Education Research Group of Australasia), Sydney: MERGA, pp. 434–41.

CHAPTER 7

Exploring patterns and algebra

LEARNING OUTCOMES

By the end of this chapter, you will:

- understand the central importance of patterns in early childhood and primary school mathematics
- understand the importance of mathematical structure and its relevance to children's learning of mathematics
- use sequences effectively to find and justify rules, and to explain
- represent and resolve number sentences, equivalence and equations
- be able to describe relationships between variables
- be able to use technology effectively to explore algebraic situations.

KEY TERMS

- **Algebraic thinking:** Thinking that considers the general relationships between numbers, rather than the manipulation of numbers
- **Equation:** A number sentence that states two quantities are equal
- **Equivalence:** A state of being equal or equivalent; balanced
- **Generalisation:** The formalisation of general concepts; general statements that can be used to form conclusions
- **Pattern:** A set of objects or numbers in which each is related to the others according to a particular regularity or rule
- **Variable:** A value that can change within a problem or set of operations

Until fairly recently, algebra was regarded as the domain of the secondary school years in most countries. In addition, it was often regarded in quite narrow ways by non-mathematics teachers, parents and students themselves as being concerned with the manipulation of symbols according to tightly prescribed rules. Recent attention to algebra in the primary school has not regarded it as appropriate that such a narrow view of algebra be taken, leading to the use of terms such as 'pre-algebra' or 'early algebra' to describe the mathematics involved.

In this chapter, it is recognised that students' understanding of algebra in the secondary school rests on foundations that are laid in the primary school, as reflected in the Australian Curriculum: Mathematics and in the New Zealand Mathematics Curriculum. These foundations are concerned with key algebraic ideas about patterns and **generalisations**, rather than with symbolic representations of these using xs and ys. This chapter explores developmental models associated with patterns and algebraic concepts, with a focus on how ICT and other resources can be used to develop **algebraic thinking**.

Linking with curriculum

In the Australian Curriculum: Mathematics, Patterns and Algebra is included as a sub-strand within Number and Algebra. The Foundation and Year 1 descriptions both refer to patterning with objects:

- *Foundation:* Sort and classify familiar objects and explain the basis for these classifications. Copy, continue and create patterns with objects and drawings (ACMNA005).
- *Year 1:* Investigate and describe number patterns formed by skip counting and patterns with objects (ACMNA018).

In the later years, the descriptors focus more on number patterns, identifying number properties, sequencing and the formulation of rules. In addition, algebra is critically related to the development of some of the proficiencies of problem-solving and reasoning. In particular, when students identify a **pattern** and look to generalise it, inductive thinking is involved, and conjectures need to be made: will this pattern continue in a certain way? Conjectures need to be tested (was the prediction correct?) and, if found to apply consistently, reasons need to be found to explain this. The kind of reasoning that explains the origins of a pattern and justifies why it must always be true is endemic to mathematical thinking, and lies at the heart of the Reasoning Proficiency strand of the Australian Curriculum: Mathematics.

The New Zealand Curriculum is structured around five developmental stages that follow the same sequence. The five stages are:

- Copy a pattern and create the next element.
- Predict relationship values by continuing the pattern with systematic counting.
- Predict relationship values using recursive methods – for example, table of values, numeric expression.
- Predict relationship values using direct rules.
- Express a relationship using algebraic symbols with structural understanding.

Pattern and structure

Virtually all mathematics is based on pattern and structure, and there are many indications that an understanding of pattern and structure is very important in early mathematics learning (Mulligan & Mitchelmore 2009). The idea of a pattern is central to thinking about algebra, so it is unsurprising that the sub-strand of the Australian Curriculum: Mathematics dealing with algebra in the primary years is called Patterns and Algebra. Although patterns occur throughout mathematics (including geometric, statistical and probabilistic patterns), the focus here is on patterns related to numbers, because of their location within the Number and Algebra strand. The integration of Number and Algebra is a design feature of the New Zealand curriculum that assists teachers to explore pre-algebra ideas with young children through number investigations.

In its broadest sense, a pattern is a regularity of some sort, and may range in sophistication and familiarity from the observation that odd numbers and even numbers alternate:

odd, even, odd, even, odd, even, odd, even … etc.

to rather more surprising observations, such as that if three numbers are in a line on a calendar, the sum of the end two numbers is twice that of the middle number:

For millennia, people have been fascinated by surprising patterns and relationships, such as that adding up successive odd numbers produces square numbers:

$$1 = 1^2$$
$$1 + 3 = 4 = 2^2$$
$$1 + 3 + 5 = 9 = 3^2$$
$$1 + 3 + 5 + 7 = 16 = 4^2, \text{ etc.}$$

Children need opportunities during the primary years to engage with a variety of patterns, to make sense of them, to describe them clearly, to make suitable use of them, to understand where they come from and to be able to confidently explain why they will continue.

Patterning in the early years

Many young children's early experiences of patterning occur through activities that require them to use informal materials to identify, make, compare and extend repeating patterns. A repeating pattern has a core that contains the shortest string of elements that repeat. For example, in Figure 7.1, the core is the oval and the rectangle.

Figure 7.1 Oval and rectangle

Early childhood educators advocate that the most meaningful patterning activities involve the use of physical materials that children can manipulate to create and extend patterns, and use trial and error. Activities that are presented in a static form, such as colouring patterns on a worksheet, tend to restrict children to a given number of elements and lack the creative aspect of physically creating one's own patterns. Virtual patterning experiences are readily accessible to children, and – like physical materials – can easily be manipulated to develop children's abilities to search for, and extend, visual and even auditory patterns.

Scootle, <http://www.scootle.edu.au/ec/p/home>, and the New Zealand Digital learning objects, <http://nzmaths.co.nz>, have a number of interactive activities that focus on repeating patterns. For example, the learning object Musical Number Patterns (see Figure 7.2) asks students to:

> Test your understanding of counting rules by building up rhythms for four musical instruments. Make counting patterns by following the given rules, or use the musical patterns to work out what the counting rules are. Then, make your own music by creating rules to make the counting patterns. View and print a report of your results.

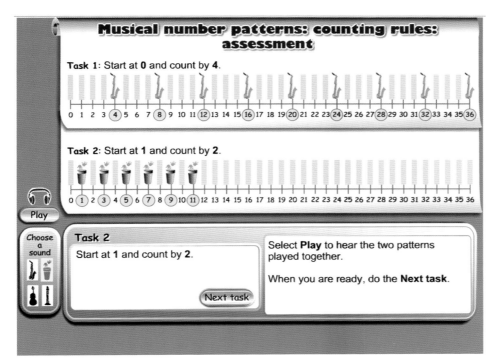

Figure 7.2 Screenshot from the Musical Number Patterns activity

Figure 7.3 Screenshot of the Learning Federation's Monster Choir activity

Similarly, on the same websites, Monster Choir focuses on matching oral patterns with visual patterns, using symbols and sounds (see Figure 7.3).

ACTIVITY 7.1

Access Musical Number Patterns and Monster Choir.

- Look at the learning objectives for each of them. Do you think the learning objectives are achieved?
- What questions would you ask the students either before, during or after doing the activities?
- How would you extend students' skills beyond creating repeating patterns?

There are also a number of apps that focus on repeating patterns. For example, the Bugs and Buttons iPad app contains a number of repeating pattern activities that can be used to reinforce the concept of patterning and looking for regularity (see Figure 7.4).

Figure 7.4 Screenshot from the Bugs and Buttons app

Growing patterns

It is important for young children to be exposed to growing patterns, and to understand the difference between a repeating pattern and a growing pattern. Experiences with repeating patterns lead to the development of multiplicative thinking, whereas experiences with growing patterns lead to functional thinking (Siemon, Bleckly & Neal 2012). When children work with repeating patterns, the focus is on identifying regularity and extending the pattern. With growing patterns, however, the aim is not to extend the pattern, but rather to predict what the next element will be, then predict any missing term. Figure 7.5 provides an example of a growing pattern.

Figure 7.5 Growing pattern

Although there are many interactive sites that focus on repeating patterns, some are particularly are suitable for younger children; these involve exploring growing patterns. For example, the TEAMS Educational Resources website, <http://teams.lacoe.edu/documentation/classrooms/linda/algebra/activities>, contains a number of activities that focus on completing the next term in the sequence, and move beyond repeating to growing patterns (see Figure 7.6).

Figure 7.6 Screenshot showing a focus on growing patterns

Patterning with number

In order to develop algebraic thinking, children need experiences that will enable them to see regularities in the ways that numbers work. According to MacGregor and Stacey (1999), students find algebra difficult to learn unless they have a good knowledge of number and basic operations. These authors consider that there are five aspects of number knowledge that are essential for number learning:

- understanding equality
- recognising the operations
- using a wide range of numbers
- understanding important properties of numbers
- describing patterns and functions.

The hundreds chart is often used as a tool in early childhood classrooms to assist students with counting and place value, and to identify patterns in the number sequence. However, it appears that this tool is only useful if students can understand the structure of the chart, and according to Thomas and Mulligan (1994), many children tend to represent the numbers 1–100 in unconventional ways.

ACTIVITY 7.2

In their article, 'Dynamic imagery in children's representation of number', Thomas and Mulligan (1994) include many examples of how children have constructed unconventional images of the 1–100 chart. Figure 7.7 shows one example.

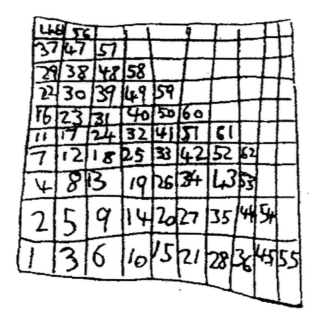

Colin (Grade 5)

Figure 7.7 Image of the 1–100 chart from Colin (Grade 5)

Chapter 7: Exploring patterns and algebra

1. Can you provide an explanation for how Colin has constructed the chart? How would you help him to form a more accurate image of the way the chart should be constructed?
2. Revisit the sites in Chapter 2 that made use of the 100 chart. How could you use these sites to help Colin appreciate the structure of the chart?

ACTIVITY 7.3

In order to further demonstrate how pattern and structure may not be obvious for all students, visit the nrich website at <http://nrich.maths.org> and access the 'Coded 100 square' (Figure 7.8). Print out the pieces of the puzzle, cut them out individually and try to piece the puzzle together.
- What strategies did you use to try to reconstruct the chart?
- What clues did you look for?
- Did you find it difficult to identify a pattern to assist you?

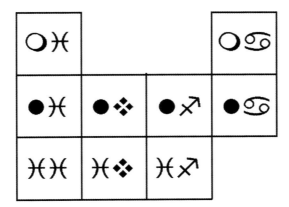

Figure 7.8 Coded 100 square

ACTIVITY 7.4

Make a copy of the 100 chart and project it onto the IWB. Use features such as screens or curtains to cover numbers, and have children name the hidden numbers. Question children about how they know what numbers are hidden, encouraging them to focus on the structure of the board (e.g. numbers 44 and 54 because the rows go down in tens). Hundred grids (interactive and static) can be found using the search field on the HOTmaths website. Figure 7.9 shows an example of a Hundred Grid widget.

Figure 7.9 Screenshot of HOTmaths Hundred Grid widget

Developing an understanding of relationships

The NZmaths website, <http://www.nzmaths.co.nz>, is a useful teacher resource that contains a number of unit plans, lesson plans and resources for teaching a range of mathematical strands, including number and algebra. A feature of the site is the inclusion of specific learning outcomes and descriptions of mathematics, such as the following in relation to the unit of work 'Ten in the Bed' which is aimed at Level One students:

Specific learning outcomes are:

- to see what a number pattern is
- to be able to guess and check the next number in a pattern
- to skip count in twos, fives and tens.

Description of mathematics:

- A pattern involves a continual repetition in some way.
- The next term in a pattern can be guessed.
- This guess should be checked.

Each activity provides details on how the lessons can be implemented, what resources are required, links with home and similar activities. In this particular sequence, the book *Ten in the bed* by Penny Dale (2007) is used to explore the relationship between the number of people and the number of eyes. Classroom snapshot 7.1 describes how a teacher used a similar activity to explore patterns in numbers.

CLASSROOM SNAPSHOT 7.1

Using an IWB, the teacher shows a class of Year 2 students the following YouTube clip: <http://www.youtube.com/watch?v=18nthFexzBc>.

In the clip, eight spiders are eventually reduced to zero due to a set of unfortunate circumstances. At the beginning of the clip, the link is made between the number of spiders and the number of legs.

After watching the clip, the teacher asks the children to recall how many legs a spider has. She then draws up a table and asks the children to help complete it to show the relationship between the number of spiders and legs.

1	2	3	4	5	6	7	8
8	16	24	32	40	48	56	64

The teacher then asks the students to describe the patterns in the table. The following responses are received:

JAMES: The top numbers go up in ones.
ROBERT: You just add 8 each time to the second row.
SUSAN: You times the top number by 8 to get the bottom number.

PAUSE AND REFLECT

After reading Classroom snapshot 7.1, consider what each student's answer revealed about their thinking. Which students showed evidence of algebraic thinking? Which student's answer would be most useful for determining how many legs 100 spiders have?

Classroom snapshot 7.1 demonstrates an activity that can be used to encourage students to identify mathematical relationships and to specify a functional relationship as a general rule. A function 'machine' can be used with younger children to look at the relationship between two **variables** instead of just looking at changes in one variable.

ACTIVITY 7.5

Professor Mad Maths created a machine called the 'Number Cruncher'. If he sets a rule on the machine and feeds numbers in at one end, different numbers come out at the other end.

- What is happening to the numbers as they go in, if they come out like those in the following table?
- Can you identify the relationship between the 'in' and 'out' numbers?

In	Out
4	17
7	26
2	11

An interactive function machine can be accessed at Math Playground, <http://www.mathplayground.com/functionmachine.html>. It provides different levels of difficulty and allows students to input numbers and then identify the function (see Figure 7.10).

Figure 7.10 Screenshot showing interactive number function machine

Equals and equivalence

The concept of equality is one of the five aspects of number knowledge essential for algebra learning as identified by MacGregor and Stacey (1999). Students, however, often literally interpret the equals sign as 'the answer is' (Assessment Resource Banks 2011) or as a command to take an action rather than as a representation of a relationship. This could partly be attributable to teaching practices that tend to present number sentences in which the answer goes at the end.

Consider Classroom snapshot 7.2, which comes from teaching resources provided by the Australian Mathematics Sciences Institute, <http://www.amsi.org.au>, which supports the enhancement of mathematical capacity and capability through a range of mediums. One of these is online support materials. An example of these is a document titled 'Addition and subtraction: the learning and teaching directory', prepared by Janine McIntosh and Michael O'Connor. It includes a student transcript provided by Dr Max Stephens from the University of Melbourne. This transcript is the stimulus for the following snapshot.

CLASSROOM SNAPSHOT 7.2

Students were asked to find the missing number in the open number sentence

$$7 + 6 = ? + 5.$$

The following responses were given by students:

Luke:	7 + 6 = 13 + 5.
Teacher:	Luke, what number did you put in the box?
Luke:	13.
Teacher:	How did you decide?
Luke:	7 and 6 are 13.
Teacher:	What about the 5?
Luke:	It doesn't matter. The answer to 7 + 6 is 13.
Teacher:	What is the 5 doing then?
Luke:	It's just there.
Teacher:	Cameron, what number did you put in the box?
Cameron:	18.
Teacher:	How did you decide?
Cameron:	7 and 6 are 13 and 5 more is 18.
Teacher:	Does 7 plus 6 equal 18 plus 5?
Cameron:	7 + 6 is 13 and 5 more is 18.
Teacher:	Fiona, what number did you put in the box?
Fiona:	8.
Teacher:	How did you decide?
Fiona:	7 and 6 gives 13 and I thought what number goes with 5 to give 13. 7 + 6 is 13 and 5 + 8 is 13.
Teacher:	Chris, what number did you put in the box?
Chris:	8.
Teacher:	How did you decide?
Chris:	(points to numbers) 7 + 6 = 8 + 5. 5 is one less than 6, so you need a number that is one more than 7 to go in the box so it all balances.

1. What do you notice about the responses given?
2. How do Luke's and Cameron's responses contrast with those of Fiona and Chris?
3. What would you do to assist Luke and Cameron?

In order to encourage students to develop a more algebraic view of equality, they should be given lots of exposure to various number sentences that do not include the equals sign at the end (for example, 8 = ? + 5). In addition, the concept of balance can also be used to reinforce the idea of equality. A set of simple balance scales can be used with different-coloured blocks to demonstrate visually the equality of relationships. Alternatively, a coat hanger suspended containing string bags containing different objects can be used to visually represent the concept (see Figure 7.11).

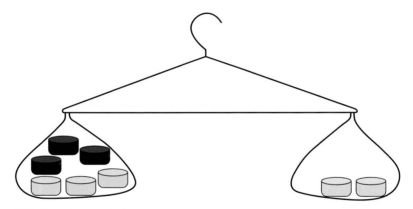

Figure 7.11 Coat hanger balance
Source: Assessment Resource Banks (2011).

While many of the interactive websites that use balance scales to model **equations** are more suitable for older students, the Learning Today website, <http://www.learningtoday.com/corporate/files/games>, uses simpler equations to demonstrate equality (see Figure 7.12).

Patterns can be made in a variety of ways, including by applying geometrical transformations such as rotation or reflection. Children in the middle years of primary school should explore these options. We use patterns of this type for aesthetic purposes, and this presents an ideal opportunity to link mathematics with, for example, art. The Patch tool on the Illuminations website, <http://illuminations.nctm.org>, provides an opportunity for children to design their own square patch and see whether the design will repeat. The

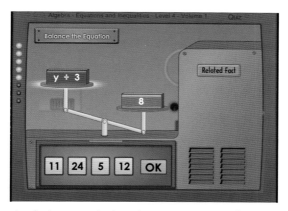

Figure 7.12 Interactive balance scale that demonstrates equality

HOTmaths Moving Shapes widget also provides an opportunity to explore creating patterns through geometric transformation. You can access this widget by logging into HOTmaths and entering its name in the search field.

Older children could benefit from exploring design ideas from other cultures, and from looking at repeating patterns in friezes, railings, floorings and fabric designs.

In the middle primary years, the curriculum emphasis on patterns shifts from recognising and making patterns with objects to recognising patterns in numbers. There are strong links here with developing computational skills, and children should be encouraged to use patterns as a way of solving computational problems. In Year 3 of the Australian Curriculum: Mathematics, children are expected to 'Describe, continue, and create number patterns resulting from performing addition or subtraction' (ACMNA060); by Year 4, children should use multiplication patterns (ACMNA081), solve word problems by drawing on patterns in number sentences (ACMNA082) and recognise equivalent expressions using addition and subtraction (ACMNA083). Activities like looking for patterns in addition problems such as 7 + 5, 17 + 5, 27 + 5 and so on are important experiences for the middle years of primary school. This is consistent with Levels 2 and 3 of the New Zealand Curriculum, where students work with number patterns from those involving addition and subtraction to those involving multiplication. The Supermarket Displays learning sequence, which targets finding rules for the next member in a sequential pattern (NA2-8), is an example of a contextualised open-ended set of student tasks. The students are familiar with the context of supermarket displays, providing an opportunity for them to display insight and formulate new ideas at individual entry points.

As previously mentioned, using patterns in the hundred chart is a useful way to help children see the uses of patterns and to develop ideas about multiples and factors.

ACTIVITY 7.6

Log into HOTmaths and access the Hundred Grid widget (see Figure 7.9) by entering its name in the search field.
- Colour all the multiples of 3 in green.
- Write a few sentences to describe the pattern that you see. There may be different ways of looking at the pattern.
- Now colour all the multiples of 4 in red.
- Write down all the numbers that are coloured green and red. A new number pattern has been created. Describe a rule for the new number pattern. What would be the next number after 100?
- Activities like this one can be used in a variety of ways. The new pattern is that of multiples of 12. Why is this pattern created from multiples of 3 and 4?
- Try some other multiples. What can you say about all multiples of 5? What digits do they end with? Using that information, what might you say about the number 375? Or 4980?

Going one step further, such as asking children what they now know that could tell them something about a new number, allows children to solve a problem and demonstrate their understanding – both proficiencies that the Australian and New Zealand curriculum documents have indicated are important for children to develop. While the Australian Curriculum: Mathematics describes this as a Year 3 outcome, not all children will be able to work from the rule they have found – such as all multiples of 5 end in 5 or 0 – to draw an inference about new numbers; however, it is important to give children the opportunity to develop this sense of working like a mathematician.

CLASSROOM SNAPSHOT 7.3

Children in a Year 3 classroom were exploring patterns using their calculators. The teacher asked them to start at any number and count by nines. They recorded each new number on a long strip of paper. The teacher asked them to record as many numbers as they wished and then to write two interesting facts about the patterns they saw.

Jay started at 1 and recorded 10, 19, 28, 37 … He wrote, 'The pattern goes down by 1 each time.'

Sam started at 714 and recorded 723, 732, 741, 750, 759, 768 … She wrote, 'The last number goes down by 1s to 0, and then it goes to 9 and starts again. The middle number goes up by 1 each time to 5 when it stays at 5 and then goes up next time.'

Asking children to develop rules for patterns is a staging post on the way to generalisation and the beginnings of the mathematical concept of a function. Jay is looking only at the last digit of each number. He is not working with the number as a whole, and needs to be encouraged to consider the place value inherent in the numbers he is generating. Sam is somewhat more sophisticated in her number sense. She chooses to start with a larger, three-digit number and is looking at the tens and units places. Her pattern description is still focusing on separate digits, however. She could be encouraged to consider the pattern more holistically by focusing on the idea that when she adds on 9 it is the same as adding 10 and subtracting 1.

The notion of using a rule to describe a pattern is one that becomes increasingly important as children move into the more abstract thinking required for algebraic manipulations in secondary school. In the middle years of primary school, rules will generally look at the growth or contraction of a number sequence rather than at a relationship between an input and an output number. Nevertheless, activities such as 'Think of a number' are useful 'fillers' – educationally sound activities that stand alone as experiences and provide a basis for the development of more complex ideas when children have the necessary skills and knowledge to progress.

For example, a starter question on the IWB as children arrive might be, '[Name of child] in the class is thinking of a number. If he adds 7 to the number he gets 13. What is [name's] number?' Such activities can be varied according to the level of understanding of the class.

There are also a number of interactive activities that could be used to generate questions around the ideas of patterns and algebra, such as the TES interactive 'Thinking of a Number', <http://www.iboard.co.uk/iwb/Im-Thinking-of-a-Number-Visual-616>, which has options such as doubling and halving and counting by tens. This activity could be helpful early in Year 3.

The HOTmaths Dot-to-Dot widget could be used in the same way. Log into HOTmaths and access the widget by entering its name in the search field. Each child completes the next dot in the pattern as they come into the classroom. This widget allows increase or decrease by the numbers 2–9.

Both activities apply a consistent rule to a starting number.

Another way of developing relationship ideas through the application of rules is to illustrate 'Think of a Number' activities with diagrams. Technology can be very helpful. For example:

I am thinking of a number.	■
I double that number.	■■
I add 6.	■■☆☆☆☆☆☆
I divide by 2.	■☆☆☆
I take away the number I first thought of.	☆☆☆
My answer is 3.	

ACTIVITY 7.7

Use a word processing program or a stamp tool to create some 'Think of a Number' diagrams of your own. What might you need to consider when generating these diagrams for children? What happens, for example, if you add an odd number and then choose to divide by 2?

Classroom activities with which the children may be familiar, such as 'Today's number is …', in which they make the target number in as many creative and different ways as they can, are useful to develop ideas about patterns and relate these to computation.

CLASSROOM SNAPSHOT 7.4

Ian's Year 4 class is working on 'Today's number is …' Ian has placed 11 as today's number on the IWB. He asks the class to find addition patterns such as 10 + 1. Chloe says, '5 + 6'. Ian puts this on the board below the 11, leaving space above for other suggestions. Tim contributes, '9 + 2' and Ian writes this immediately under 10 + 1. He then asks for subtraction suggestions. Ben says '12 − 1' and Ian writes this immediately above the 11. Jo then says, '13 − 2, 14 − 3 …' and Ian puts these two suggestions above 12 − 1.

With no further comment, Ian then asks the children to find as many combinations as they can in 10 minutes. Mark quickly completes all the additions to 11 + 0 in a sequential fashion and then continues with −1 + 12, −2 + 13, −3 + 14 … When Ian asks him to explain his answers, Mark says, 'Look, it's a pattern. You just put a minus sign in front. It's the same as 12 − 1 but the other way round.' Ian congratulates Mark on his interesting finding and suggests that he now looks for other patterns that make 11.

Sally is struggling with the task. She has written 7 + 4, 1 + 10 and nothing more. Ian draws her attention to the IWB and says, 'We have 10 + 1, 9 + 2. You have found 7 + 4. Is there something that we could put in there between 7 + 4 and 9 + 2? Sally still looks confused, so Ian asks, 'What number comes between 7 and 9 when we count?' After some time, Sally says '8'. Gradually, Ian leads Sally to writing 8 + 3 and points out the pattern.

After the 10 minutes is up, Ian brings the class together again. Starting with the subtraction patterns, he gets individual children to contribute their ideas, gradually building the patterns. He then does the same thing with the addition patterns, including Sally, who adds 1 + 10, and Mark, who offers −1 + 12.

Ian then leads a discussion about the links between the addition and subtraction number patterns, but focuses on the patterns developed. The lesson ends with Ian asking the children to write two interesting things that they have learnt during the lesson in their maths journals. While they complete that task, he saves the IWB page that has been created as a starting point for the next day's lesson that will focus on equivalent expressions.

There is a lot going on in this classroom, and other children would also show diverse understandings. Mark has a strong sense of pattern and extends this logically, even though he has not been taught negative numbers. Although he was not expecting Mark's pattern, Ian accepts the understanding that Mark is displaying. Sally has some specific difficulties, and Mark uses some questioning as well as explicit teaching to help Sally begin to recognise the number patterns in the task he has set.

The lesson finishes with a session that pulls together the ideas that Ian wanted to develop about patterns and using patterns to help with computation, and offers an opportunity for children to articulate their personal understanding through their journals.

Ian uses technology to collect children's ideas in a systematic way and then saves the page created as a starting point for the next lesson.

Ian is demonstrating good pedagogical understanding in the ways in which he interacts with the individual children, asking questions at their level and being explicit when needed, in his organisation of the information he elicits from the class in a systematic way so that the children can see the developing patterns, and in his intention to use the information the next day to develop the concept of **equivalence**. The IWB technology makes it easier for him to use the class-generated material as a starting point.

> ## PAUSE AND REFLECT
>
> Ian's students have maths journals. Whenever they use the journals, Ian provides a prompt or scaffold for the day's entry. Sometimes he asks the children to write down one problem they can do now that they couldn't do before. At other times he might ask them to indicate something in mathematics with which they would like more help or, as in the lesson described in Classroom snapshot 7.4, to reflect on something they have done.
>
> What do you think of the idea of a mathematics journal? How does the use of a journal help children to develop mathematical understanding? What might be the advantages or disadvantages of this tool?

By Year 4, children are expected to use the ideas of equivalence (ACMNA083). The HOTmaths Equal Amounts widget demonstrates equivalence in a visual way. You can access this widget by logging into HOTmaths and entering its name in the search field.

Children need to begin to recognise that there are different ways to express numbers, and that different numbers can be used to make equivalent expressions – for example, asking children to find numbers that make equations true, such as:

$$\heartsuit + 5 = \square - 4$$

Using a guess-and-check approach and working in small groups, children in Years 3 and 4 are well able to solve these kinds of problems, drawing on their problem-solving proficiency. Developing these ideas to explore unknown numbers also provides powerful opportunities for learning, such as:

If $\star + \blacksquare = 12$, and $\blacksquare$ is less than $\star$, what can you find out about $\star$ and $\blacksquare$?

These kinds of activities lay the foundations for the notion of a **variable**, which becomes important in the upper years of primary school.

Generalisation in upper primary

Generalising requires an idea of a variable, now recognised as a subtle and difficult concept that cannot be taken for granted. In secondary school, variables generally are represented by letters, although the available research makes it clear that many students do not interpret the letters in this way. In the primary school, the use of letters to represent variables develops only slowly, although children need to experience the idea of variables being a key to understanding a general property of numbers or to representing a relationship succinctly.

As well as understanding patterns and their representations, primary children need to acquire a good grasp of the key mathematical idea of equivalence, referring to two quantities being the same in some sense, and its representation in mathematics. In the early years, this idea manifests in seeing that there are several different, although equivalent, ways of pairs of numbers adding to 10 (the base of the decimal number system).

This in itself involves a subtle shift in meaning for an expression such as 7 + 3 from being an *instruction* (add 3 to 7) to being an *object* (the result of adding 3 to 7, which is 10). While sophisticated learners in secondary school may have mastered the idea that equivalence is represented using an equals sign, as previously mentioned, young children may at first interpret an equals sign as a reference to the 'answer'. That is, 7 + 3 = 10 is often interpreted (almost literally) as 7 + 3 *makes* 10, although it is perfectly sensible to also write 10 = 7 + 3, indicating that 10 and 7 + 3 are equal to each other. Further, representations such as 7 + 3 = 6 + 4 also represent an equivalence correctly, although young children who are interpreting an equals sign as 'makes' will not find this at all sensible.

The ideas of equivalence lead naturally into the idea of an **equation**, which requires children to find which values (if any) make a number sentence true. Again, in the early years this may be described quite informally (such as 3 + ? = 12), while in the later years more sophisticated notations and procedures are needed.

It is interesting to note that the Algebra sub-strand lends itself to targeting multiple outcomes during the same task. Often these tasks include data collected from physical or concrete manipulations related to real experiences. The matchstick activity in Classroom snapshot 7.5 demonstrates how the same activity can target each of the outcomes listed below:

- Explore the use of brackets and order of operations to write number sentences (ACMNA134).
- Introduce the concept of variables as a way of representing numbers using letters (ACMNA175).
- Create algebraic expressions and evaluate them by substituting a given value for each variable (ACMNA176).
- Extend and apply the laws and properties of arithmetic to algebraic terms and expressions (ACMNA177).
- Given coordinates, plot points on the Cartesian plane and find coordinates for a given point (ACMNA178).
- Solve simple linear equations (ACMNA179).
- Investigate, interpret and analyse graphs from authentic data (ACMNA180).

The formation of generalisation can be investigated in numerous settings. One of the simplest settings to get your 'head around the idea' of maintaining ownership of the idea with the students is when using headless matchstick patterns. A sample sequence is provided below in Classroom snapshot 7.5.

CLASSROOM SNAPSHOT 7.5

Sample headless matchstick activity sequence

- Brainstorm patterns noticed in daily life.
- Create and continue the matchstick pattern for five terms:

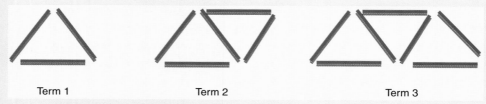

Term 1   Term 2   Term 3

Figure 7.13 Matchstick patterns

- Represent pattern in a drawing (similar to above).
- Tabulate data.

Number of triangles	Number of matchsticks
1	3
2	5
3	7
4	9
5	11

- Ask students to describe a rule for the identified pattern in words. You will notice a developmental structure in the quality of the responses. Common responses will include:
  - 'It is going up by 1s'.
  - 'In that column you just add 2 each time.'
  - 'To get the next one you just double it and add 1.'

- 'The number of matchsticks is equal to two times the number of triangles plus 1.'
- Write the pattern in algebraic notation (often described as shorthand for primary students). The students may start with:

$$m = 2 \times t + 1$$

- The students will find it interesting that we do not need to use the multiplication sign when dealing with pronumerals, and instead they may write:

$$m = 2t + 1$$

- Draw a graph to represent the pattern by plotting the points. This can be done with pen and paper or with software such as GeoGebra, as in Figure 7.14.

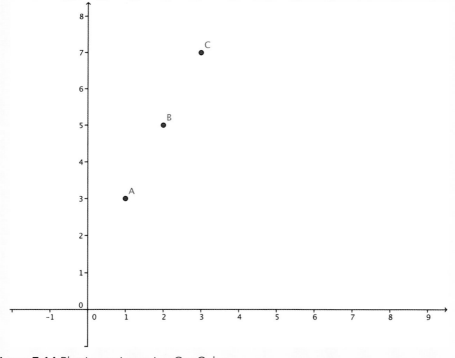

Figure 7.14 Plotting points using GeoGebra

Students may then graph the relationship in GeoGebra and finally superimpose it over the plotted points as a final step. Graphed function (m=) appears in Figure 7.15).

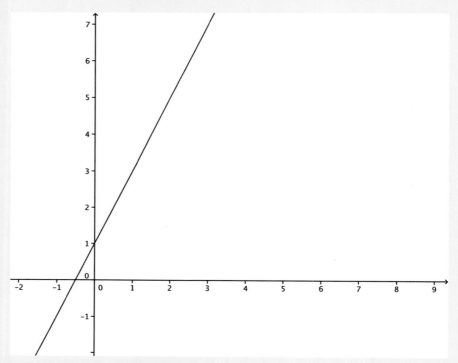

Figure 7.15 Graphing a function using GeoGebra

A patterns lesson can be found in HOTmaths that includes a link to a HOTsheet named Toothpick Patterns, which will provide similar stimuli for a lesson such the one above. Log into HOTmaths and select 'Cambridge Primary Maths' under Course list, then select 'Aus Curric 6' as the Course. Then choose 'Describing patterns' as your Topic, and finally 'Numbers in designs' as your Lesson.

ACTIVITY 7.8

Design a similar sequence using the pattern in Figure 7.16. How many different matchstick pattern ideas can you find that will be suitable for upper primary patterns investigations?

Complete the HOTsheet shown in Figure 7.16, which can be found at the above location.

Chapter 7: Exploring patterns and algebra

TOOTHPICK PATTERNS

You will need toothpicks or headless matchsticks for these tasks.

TASK 1 Describe a pattern

1 Complete the table for this pattern which was created by repeating this shape:

Number of shapes	1	2	3	4
Number of toothpicks used				

2 Use toothpicks to build the next part of the pattern. Add a column to the table for your new design.

3 Describe the link between the number of shapes and the number of toothpicks used.

_____

_____

_____

4 Use your description to predict how many toothpicks would be needed to complete the design with:

   a 6 shapes _____   b 20 shapes _____   c 100 shapes _____

5 How many shapes could you build with:

   a 40 toothpicks _____   b 100 toothpicks _____   c 48 toothpicks _____

Figure 7.16 HOTmaths Toothpick Patterns HOTsheet

The patterns and algebra explorations cannot be viewed in isolation from the other content strands, no matter what syllabus structure is being utilised. While algebraic concepts are often structurally arranged in syllabus documentation alongside number concepts, algebraic concepts are an integral component of all strands, including measurement, geometry, statistics and probability. Classroom snapshot 7.6 demonstrates this in the area of measurement.

CLASSROOM SNAPSHOT 7.6

Mrs Kirby, a Year 5 teacher, was teaching a unit on area. The children had explored area 'as a measurement of a surface', had demonstrated finding areas using informal units and formal units, and were ready to explore a formula for finding the area of any rectangle.

Mrs Kirby was considering the design of the lesson for the next day and thought she might begin with area revision questions, then give the children the formula and show them how to use it. The students could then complete a series of questions using the given formula to find the area of various rectangles. She reflected upon what had happened in previous years when she used this approach. She remembered the students often confused the perimeter formula with the area formula. Many students found it difficult to identify the 'base' and 'height' of the rectangles they needed to substitute into the formula.

On the basis of these past experiences, Mrs Kirby decided to try something different this year. She decided to use technology to facilitate student ownership of the mathematical ideas involved with the formula for finding the area of a rectangle. The lesson went as follows.

Introduction

Students were supplied with 1 cm^2 grid paper on which they were asked to draw six different rectangles using the grid lines as sides of the rectangles. They were then asked to label the dimensions of the rectangles and record the area of the rectangle on each one. (Keep in mind that they have already covered repeated units and formal units of area by counting the number of units covered.)

Development of the idea

Students then created a rectangle using GeoGebra, and used a table to record the area of the rectangle as the dimensions were changed. The students were asked to find any patterns and relationships, and to express them in words. They came up with responses such as:

> 'When the two different sides are larger, the area is larger.'
>
> 'To get the area I always multiply the two different sides.'
>
> 'You can multiply the sides that are at right angles to each other to get the area each time.'

In a class discussion, the students shared their findings and the teacher recorded them on the IWB. The students were prompted to think about a way of expressing how to find the area of any rectangle using symbols. They started with:

> Area = one side length times the different side length.

The teacher asked them to think about the square, and whether it also worked for that shape (a few students had an understanding of the square as a rectangle with additional properties). The students decided that 'different sides' wasn't needed, but instead it was always the sides at right angles to each other. This was a really important phase of the lesson, and Mrs Kirby was glad that the idea came from the students. A discussion followed, in which Mrs Kirby introduced the convention of using the terms 'base' and 'height', and 'length' and 'breadth'. They decided as a class to use the terms 'base' and 'height' and came up with the formula:

> Area = base × height, or
>
> $A = b \times h$, or
>
> $A = bh$

Application

The students drew two rectangles on grid paper and found the area using the formula they had derived. Mrs Kirby then drew two more rectangles on the board in different orientations and asked the children to find the area using the strategy they had discovered that day.

Conclusion

The students summarised in a diagram, in their own words and in symbols, how to find the area of any rectangle.

Follow-up lesson ideas

- Design a spreadsheet for finding the area of an area. Creating a formula in the formula bar is another excellent strategy for enhancing algebraic concepts. These concepts have been around for a long time, and their affordances are rarely realised in the primary mathematics classroom.
- Students could find the area of larger rectangles outside by working with a trundle wheel to measure the dimensions of the rectangles, then using estimation and the student-designed spreadsheet.

ACTIVITY 7.9

Explore the range of contextual stimuli that can be incorporated into a similar lesson design as in the matchstick pattern example in Classroom snapshot 7.5. Some ideas to explore are:

- the Handshake Problem, <http://www.youtube.com/watch?v=nuOgQQd8iVw> – a YouTube video demonstrating solutions with unifix cubes and a calculator
- NCTM's Illuminations website, <http://illuminations.nctm.org/LessonDetail.aspx?ID=U168>, which provides complete lesson plans that assist the students in moving from the physical simulation, tabulating the data and using technology to explore the relationships. The unit is called the Supreme Court Welcome.
- Garden Beds, <http://www.maths300.com>, an excellent Maths300 lesson
- the Chairs Around the Table activity, an exploration based on the context of a school celebration where students need to consider possible table configurations. The students can tabulate, graph and describe the relationships identified according to the different arrangements. See Illuminations, <http://illuminations.nctm.org>.
- an NZmaths unit titled Pede Patterns, which is concerned with generating number patterns within the context of insects from a mythical planet.
- Tower of Hanoi, an oldie but a goodie, in which patterns arise from the exploration of an ancient game of moving different-sized disks on three spikes, always having a smaller disk on top of a larger disk. The students will develop a rule that will enable them to find out the number of moves required to move to another spike with n number of disks. Complete a search for the Tower of Hanoi. Find interactive software for whole-class investigation on the IWB, lesson plan suggestions and solutions to the problem.
- HOTmaths Mystery Machine widget (see Figure 7.17), accessed by logging into HOTmaths and entering its name in the search field.

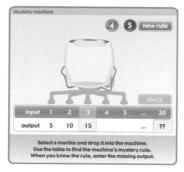

Figure 7.17 Screenshot of HOTmaths Mystery Machine widget

Chapter 7: Exploring patterns and algebra

In each of the contexts described above, not only do the student tasks lend themselves to the development of the notion of a variable and relationships; they also lend themselves to the early investigation of equations. For instance, in the Garden Beds example, the students may develop the relationship of, 'The number of pavers is four times the length of the garden bed plus four.' The power of equations is developed when students begin exploring questions such as, 'If I have 424 pavers, what is the length of the largest square garden I could pave?' Another example of working with equations is provided in the HOTmaths Symbols and Rules HOTsheet (see Figure 7.18), which can be accessed by logging into HOTmaths and entering its name in the search field.

Global

HOTsheet

SYMBOLS AND RULES

TASK 1 Complete the patterns

Use the rule that links the top and bottom numbers to complete the table.

1 Bottom number is top number plus 13.

Top		16		33		46
Bottom	15		42		54	59

2 Bottom number is top number divided by 7.

Top	14		42	49	
Bottom		3		7	9

3 △ = ✖ × 3 + 10

✖		2	3	5		8	
△		16		25	31		40

4 □ = (△ − 8) ÷ 6

△		14	20		62		80
□			2	6		10	

Figure 7.18 HOTmaths Symbols and Rules HOTsheet

Reflection

This chapter has looked at the breadth of algebraic concepts explored in the primary classroom according to many curriculum documents. Over the past few years, we have seen a growing emphasis on the development of algebraic concepts in the early years of schooling, leading to many students being able to formulate algebraic generalisations in the upper primary years. The most exciting development for today's primary teachers is the number of ICT tools that integrate seamlessly with other concrete materials used in the classroom. As shown in the examples provided in this chapter, it is definitely a case of 'not throwing the baby out with the bathwater', and instead enjoying the timeless nature of hands-on algebraic tasks within a technological environment.

Websites for exploration

Illuminations: <http://illuminations.nctm.org>
Math Playground: <http://www.mathplayground.com/functionmachine.html>
Nrich: <http://nrich.maths.org>
Scootle: <http://www.scootle.edu.au/ec/p/home>

CHAPTER 8

Exploring data and statistics

LEARNING OUTCOMES

By the end of this chapter, you will:

- be able to choose suitable questions for investigation for children of different ages
- understand the importance of variation in data and different types of variable
- understand the difference between a population and a sample
- recognise different ways of displaying data to 'tell a story'
- understand the importance of drawing inferences from data, and the uncertainty associated with these inferences
- be able to draw on technology to support the development of statistical understanding.

KEY TERMS

There are many specialist words used in statistics. The ones listed here are the most commonly encountered in primary mathematics. The explanations provided are not definitions as such. For detailed mathematical definitions, go to your favourite mathematics dictionary.

- **Column or bar graph:** A way of displaying categorical data using vertical columns or horizontal bars and frequency counts
- **Data:** Information collected in a systematic way
- **Distribution:** The look or shape of the data when displayed systematically on an axis

- **Interquartile range:** Contains the 'middle half' of the data set
- **Mean:** A balancing point that provides a summary of the data
- **Median:** The middle value of an ordered set of data
- **Mode:** The most frequently occurring value
- **Pictograph:** A column graph that uses pictures or symbols to show the data
- **Pie chart:** A sector graph based on a circle, with each sector in proportion to the percentage of the whole
- **Population:** The entire group of interest – for example, all Year 3 students in the school
- **Range:** A measure of the spread of the data
- **Sample:** A sub-group of the population, often intended to be representative of the whole group – for example, Ms Pitt's Year 3 class is a sample of the population of Year 3 children in the school
- **Scootle:** An online repository of digital resources for Australian teachers
- **Standard deviation:** A measure of the variability or spread of a set of data
- **Stem-and-leaf plot:** A display based on splitting each data value into a 'stem' and 'leaves' – particularly useful for comparing two groups on the same attribute
- **Variation:** The concept that underpins statistics

The importance of understanding statistics

We experience uses of statistics every day. Some of these uses are obvious – in sports reporting or finance news, for example. Others are more hidden – for instance, supermarkets decide which items to stock and where to place these on shelves as a result of **data** gathered at checkouts. Our society has been described as 'data-drenched' (Steen 1999). It is becoming increasingly difficult to make informed decisions without understanding statistical information. This reality has been recognised in the Australian Curriculum: Mathematics by having one strand focused on statistics and probability. In New Zealand, the curriculum is named Mathematics and Statistics, emphasising the growing importance of data in our society.

The use of technology is essential for developing an understanding of statistics. Spreadsheets allow us to manipulate data in ways that were tedious and difficult before ICT became so widespread. In this chapter, the focus is on developing statistical understanding through the use of technological tools.

Underpinning any understanding of statistics is the notion of **variation**. People are different, events are different, objects differ, and without these differences there would be no need for statistics. Young children intuitively understand differences when they make comparisons – for example, 'I am bigger than Walter'; 'This red brick is bigger than the blue brick'. The notion of variation is not difficult for children to understand. When used in a statistical way, however, it can seem very complicated. In a class of 25 children, there will be a variety of heights, eye colours and shoe sizes. Often we want to know 'what is typical' of the class, year group, school or **population**. We have developed a variety of techniques to summarise data, including the **mean**, **median**, **mode** and **range**, as well as more complex measures such as the **interquartile range** and **standard deviation**. Often, applying some common sense and intuition will tell you which of these is most appropriate.

Statistics is about problem-solving. Children need to recognise, or set up, a problem that they wish to solve, which requires data to be collected in order to solve it. This problem may be simple, such as 'How do children get to school?', or more complex, such as 'Do people who play sport have better balance than those who don't play sport?' Once the problem is defined, children must plan to collect the data. They need to develop an understanding of the ideas behind sampling, and of the importance of being systematic and collecting the data in consistent ways.

There are practical considerations as well. How will the data be collected, or how were the data collected if they are using an archived data set? What kinds of analysis will be undertaken? How will the data be displayed to tell the story?

Finally, the findings must be interpreted and communicated to others. This cycle of problem, plan, data, analysis and conclusions has become known as the PPDAC cycle (Wild & Pfannkuch 1999). PPDAC provides a framework for statistical investigations, an area that is central to the New Zealand curriculum and that also fits well with the focus of the Australian Curriculum: Mathematics.

The statistics strand of the mathematics curriculum in the primary years focuses on making meaning from statistics rather than the mechanics of calculating particular statistics. In Australia, the key ideas are related to posing questions, data collection and interpretation, and these ideas appear at every year level. In New Zealand, from the earliest years of primary school children are expected to carry out investigations using the PPDAC cycle. This strand of the mathematics curriculum links to many other subjects, including science, geography, and health and physical education. These other curriculum areas provide opportunities to collect and explore data of interest to children, and create a platform upon which statistical understanding can be built. Children respond to questions of interest to themselves, and about themselves. By the

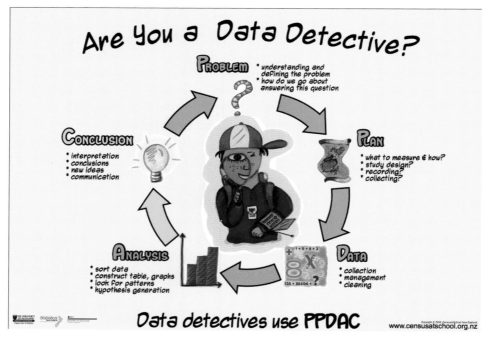

Figure 8.1 PPDAC cycle

upper years of primary school, they also need to engage with and critique data presented by others – in the media, for example.

In the early years of primary school, the skills developed in measurement, such as choosing the appropriate attribute and developing an understanding of standard units, are those that will be used when dealing with statistical tasks. By the middle primary years, children are beginning to develop a sense of collecting data systematically, and displaying this in ways that 'tell the story' to someone else. By the upper primary years, children should be developing a critical sense about data. They are starting to summarise data in appropriate ways, and to analyse the data they have collected, as well as data from elsewhere. At all levels of schooling, children should be asked to explain their thinking and justify their views, demonstrating the earliest ideas of statistical inference.

The role of language and communications is critical. Children must be able to explain and justify their conclusions. Even very young children can have opinions and ideas relating to cause and effect. All children must be given opportunities to explore, analyse and present data in ways that make sense to them, and to explain why they have drawn their conclusions by referring back to the data they have used. This chapter is presented around the PPDAC cycle to illustrate how technology can be used to support statistical investigations.

Asking questions (problem)

The question posed is of key importance because it determines the nature of the data to be collected. A variety of data displays and manipulations are used with different kinds of data, and understanding the opportunities and limitations of ways of dealing with data is important in order to identify misrepresentations and interpret data presented by others. Even young children can ask good questions, and children in all years should be encouraged to pose questions that they can answer by collecting data.

Initially, these questions will be straightforward, such as 'What pets do our class own?' or 'What is the favourite colour of our Year 2 class?' These questions ask about categories – different types of pets, or a variety of colours. Later, children will be able to pose questions about numerical variables, such as height or time spent on different activities. As their questions become more complex and sophisticated, children's natural curiosity will lead to questions comparing two groups, such as boys and girls, or questions about what is typical for the class or the year level. They also want to look at associations, such as whether blue-eyed people sleep longer. These questions may seem trivial from both a statistical and a substantive perspective – that is, the questions children pose are not necessarily those that statisticians would ask or that adults would find interesting to investigate, but exploring how to answer questions of interest to themselves is likely to lead to more learning for children, and to develop an increased ability to develop worthwhile questions of interest.

It is also important that children learn to ask sensible questions in a survey that will allow them to answer their question of interest. There are experiences from other learning areas that are useful. Health and science, for example, provide rich opportunities for thinking about data and recognising different types of data.

ACTIVITY 8.1

To undertake this activity, you need to register with **Scootle**, <http://www.scootle.edu.au>, a resource that is available to all Australian teachers and pre-service teachers in Australian institutions. In New Zealand, the same material is available from the Digital Learning Objects resource, <http://www.nzmaths.co.nz/digital-learning-objects>. To access these websites, you will need to use your Education Sector login.

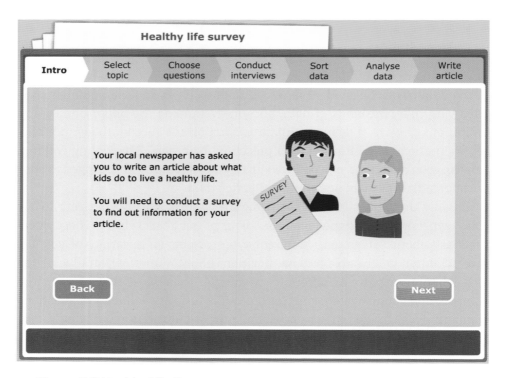

Figure 8.2 Healthy Life Survey

Find resource L3158, Healthy Life Survey (in New Zealand this can be found at <http://www.nzmaths.co.nz/node/1650>). In this activity, children ask questions to enable them to write an article for the local paper. Choose a topic that interests you, then work through the questions that are posed. Resources of this type provide rich opportunities to discuss the process of interpreting data to answer questions. Another resource of the same type is L3151, Home Internet Survey.

After exploring these activities, make up one of your own, using a relevant context such as 'How do children spend their pocket money?' or 'What do Year 6 children want for their birthdays?'

Develop a series of questions similar to those posed by the learning object. What technological tools might help you to find the answers to your questions?

Collecting and recording data (plan, data)

Once children have posed a question, they need to collect and record data. It is important to emphasise the systematic nature of the collection process, and to ensure that the information collected is not ambiguous. For example, children have to learn that they may need to choose only one item from a list, rather than saying they want two or three choices. This requirement has social benefits as well, teaching children about making choices. Consider the teaching that is happening in Classroom snapshot 8.1.

CLASSROOM SNAPSHOT 8.1

Lee's class of 5-year-old children is sitting on the mat. He asks the children, 'What question would you like to answer today?' Several children say they would like to know about favourite animals. Lee asks the children to suggest some animals. As they say some animal names, Lee writes these on the IWB and suggests the snake as his animal. Finally the class settles on four animals: dog, cat, snake and bird. Lee goes to the HOTmaths Data Tables widget that he has previously opened but hidden from the children. He creates a table with the four categories, and asks a child to drag the picture on to the relevant tile each time. Then Lee asks for children to vote for their favourite animal by raising their hands. He makes it clear that they can only vote for one animal.

As they vote, two children who are not voting for that animal count the number of hands raised. Another child enters the number of the count in the table. After all the votes are taken, snake has no votes. The class has a discussion about the most popular animal and the importance of recording that there were no votes for snake. The children return to their seats to draw a picture of the most popular animal in the class, and their own personal favourite animal, as a way of reflecting on the lesson. Lee saves the page created by taking a screenshot so that he can remind the children of their discussion later. He also prints the page using the 'Print' button in the widget.

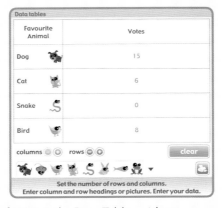

Figure 8.3 Screenshot of HOTmaths Data Tables widget

Global

To access the HOTmaths Data Tables widget for yourself, log into HOTmaths and enter its name in the search field.

Lee was using technology effectively to undertake an activity that he might previously have done using an ordinary whiteboard. He planned for the lesson by having the widget open but hidden, so that when the appropriate moment came in the lesson he could access this quickly. Children were actively involved in recording the information by counting and writing on the IWB. He allowed the children to suggest their own question, but also made sure that the animal categories were some of those available in the widget.

He deliberately suggested snake, anticipating that no child would choose this as a favourite, so that he could start the discussion about recording empty categories. Young children often leave out categories where there are no data, because for them zero means it does not exist. They need to learn about the importance of empty groups, and that these also tell us something. Finally, he recorded the page for possible future use. This ability to be able to recall what happened earlier is a powerful use of technology.

Global

One of the most common ways of recording data by hand is by using tally marks. The use of tallies for counting data links to the Number and Algebra strand and counting by fives. A variety of resources are available to develop the notion of tally marks. In HOTmaths, the Capsule Toy Machine widget (Figure 8.4) provides a visual demonstration of using tallies to collect data. This widget can be accessed by logging into HOTmaths and entering its name in the search field. This widget could be used as a whole-class activity, initially to demonstrate how tallies work and then by having the class do their own recording as the toy is produced, which is then checked and reinforced by the widget use. A similar example is available from Kids Math Games Online at <http://www.kidsmathgamesonline.com/numbers/mathdata.html>.

Often, data are collected by hand and then entered into some form of technology to transform them into useful information, using tables, graphs or other graphic organisers. Initially, data will be recorded on an individual basis and then the individual information will be combined to create a data set. Consider, for example, a series of simple surveys collecting categorical data about favourites: food, activity, colours – the possibilities are endless. Create a sheet with a limited selection of favourites on it – usually five or six favourites with up to five categories in each. These can be organised so that

Figure 8.4 Screenshot of HOTmaths Capsule Toy Machine widget

each child has a strip with the surveys on it (see Table 8.1). Every child ticks their own favourites and then cuts the strip into the separate parts and places each part in the appropriately labelled container. This activity could be undertaken with children from the middle primary years upwards.

Table 8.1 Suggested categories for a 'favourites' investigation

Favourite sport	Favourite soft drink	Favourite colour	Favourite food	Favourite activity
Football	Lemonade	Pink	Bolognaise	Reading
Cricket	Cola	Blue	Fish and chips	Dancing
Hockey	Water	Purple	Burger	Playing sport
Swimming	Orange juice	Orange	Sausages	Computer games
Basketball	Ginger beer	Red	Fried chicken	Hanging out with friends

In small groups, children then summarise the data by creating a spreadsheet in Excel or a similar program. A New Zealand resource called Which Graph? With Excel, <http://nzmaths.co.nz/resource/which-graph-excel>, has useful master sheets with clear instructions for using Excel to create graphs. Another possibility is to use the online Create a Graph tool from the National Center for Education Statistics (NCES) Kids' Zone, <http://nces.ed.gov/nceskids/createagraph/default.aspx>. The children prepare a poster or a report about their findings. Discussion with the class should focus on the appropriateness of the display to represent the data and answer the question posed about 'favourites in our class'. Because the data are frequency counts of categories, the best representations are **pie charts**, or **column or bar graphs**.

When collecting data, it is worth having a discussion about the conditions under which the data should be collected. This is an aspect of the work of statisticians that frequently is overlooked. Consider the discussion in Classroom snapshot 8.2.

CLASSROOM SNAPSHOT 8.2

Teacher: We've decided to investigate how well students in our Year 6 class balance on one leg with their eyes closed. Okay. What data do we need to collect?

Linh: How long each student balances on one leg with their eyes closed.

Teacher: Yes, but is there anything else we need to think about?

Mark: Which leg?

Neema: Yes – it should be the strongest leg.

Teacher: Would that be the same for everyone?

Adib: No – some people kick with their left leg and some with their right and some can kick with both. I think it should be both legs. That way it's fair. [General agreement from the class]

Teacher: (records 'both legs' on whiteboard): Okay. What does it mean to balance on one leg?

Aiden: You take your other foot off the floor.

Grace: But how high off the floor does it have to be? (Grace demonstrates lifting her foot just off the floor and then high off the floor)

The discussion continues in this manner for some time, as the children try to come up with conditions that would be fair to all people and could sensibly be enforced. They consider whether arms can be used for balance, what constitutes the end of timing, whether hops are allowed, whether shoes should be off or on and other factors that might affect the outcome, such as being right- or left-handed, what sport is played and so on. This time is not wasted. Rather, it is developing a sense in the children of how difficult it is to collect data under consistent conditions, and being sure to consider factors that might impact on the attribute to be measured. Once they have decided on all the conditions, the children collect the data in threes – one balancing, one timing and the third acting as a referee to see that all the conditions are complied with. As the members of each group finish collecting and recording their individual data, the students enter their own data into a spreadsheet displayed on the IWB. This is saved in the class folder so that all children can access it later for analysis. Having each child or group responsible for their own data entry develops a sense of ownership, and is easily managed with either a data projector or the IWB. The electronic data set can then be used at a time convenient to the teacher.

One quick and efficient way to collect data is by using a data card. An example of a data card used to collect information about litter at a school is shown in Figure 8.5. Each child in the class completes one or two cards, and enters the data into a pre-prepared spreadsheet. Data cards have been used to collect data from a variety of situations (Callingham 1993; Watson & Callingham 1997). Using tools such as this together with appropriate technology allows for the best of both concrete experiences and the efficiencies technology can bring. The nature of the information can be adjusted for the age group or interests of the class.

Litter Survey

Day: Wednesday

Area of school: Top Oval

Kind of litter	Number of pieces
Icy pole sticks	7
Chocolate wrappers	4
Milk cartons	2
Chip packets	6
Hamburger wrappers	0
Fruit box straws	9

Figure 8.5 Data card for litter collection

Analysing and representing data (analyse)

Data alone do not tell a story. Having been collected, the data have to be analysed and represented to make sense to other people.

Even young children have quite sophisticated understandings of how to display information. In the two pictures shown in Figure 8.6, Year 1 children

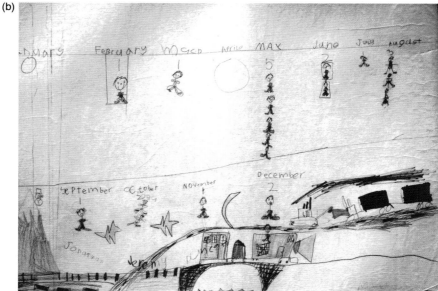

Figure 8.6 Year 1 children's informal recording of data

were recording the birthdays in their class in a way that would make sense to their parents or someone who was not in their class. They worked in small groups to record the information and were free to choose whatever representation they wished. They had not previously had any formal teaching about graphs, but were able to show the information in lists, tables and **pictographs**.

If you look closely at the pictures, you will see that they don't have the same information. This is because the children were collecting the information from birthday cards that they had made for themselves and hung up across the room. They removed the cards each time they wanted to record the information, and as a result some were missed out and others were counted more than once. This situation led to a discussion in the class about the importance of being accurate and systematic about recording data. The freedom that the teacher gave the children led to a greater range of types of recording, and brought up issues that had not been anticipated. Often we constrain children to recording in a particular way. Although it is important to develop conventional formats for displaying data, if we never give children opportunities to make decisions for themselves, we reduce opportunities for them to display what they know.

PAUSE AND REFLECT

In the birthday activity, children first made birthday cards, and these were displayed on a string hung across the room, grouped by months. How important do you consider this concrete activity to be? How would the learning experience change if the teacher had decided to use technology to collect the information?

There are often good reasons for doing an activity with concrete materials, like the birthday activity, before moving to a technological approach. Sometimes with technology use, children lose sight of the aim of the activity and become caught up in the technology itself. Collecting data from the class and then transferring the information to some form of technology can help to develop notions about the appropriateness of the display. Using an interactive graph plotter, for example, is useful to show children that the same data may look different even though they tell the same story. Recognising different representations of the same data relates to the Australian Curriculum: Mathematics content descriptors in Year 2 (ACMSP050), Year 3 (ACMSP069, ACMSP070) and Year 4 (ACMSP096, ACMSP097), and the New Zealand National Standards for Statistics at Level 2, Level 3 and Level 4. This skill is critical for developing understanding for later, more complex data

Global

displays. For example, the two representations shown in Figure 8.7 are of favourite fruit in a class, created using HOTmaths widgets. They show the same data in different ways – as a table and as a column graph. To access these HOTmaths widgets for yourself, log into HOTmaths and enter their respective names into the search field.

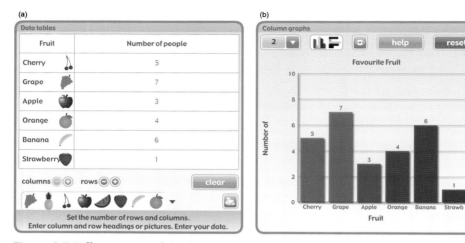

Figure 8.7 Different ways of displaying the same data using HOTmaths widgets

Older children might create a pie chart of the same data, as shown in Figure 8.8. The advantage of using technology for this purpose is that complicated calculations are not needed. Because the purpose of the activity is to compare data displays, focusing only on the display and not on the computation behind it allows an appropriate discussion. Later, children will need to understand how to create such displays, and be able to articulate how the pie chart segments relate to a circle, but in the primary years familiarity with different types of representation, without the distraction of calculating the necessary angles, meets curriculum goals. The graph shown here was created using a free chart-maker program, at <http://www.meta-chart.com/pie>.

Asking different groups of children to create these representations and showing them together to the class – which is possible using IWB tools – provides an opportunity to discuss which presentation might be appropriate for a particular purpose. For example, for a fruit shop it is easier to plan orders using numbers, so a table might be more helpful, whereas a report to fruit growers might want to emphasise the visual image of strawberries being least popular.

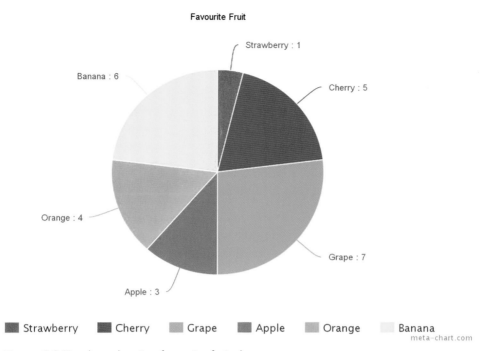

Figure 8.8 Pie chart showing favourite fruit data

Children from the middle years of primary school upwards are able to discuss these kinds of ideas, and to understand the power of different displays of data. They could be asked to produce a piece of persuasive text targeting a particular audience and use data representations to get their message across, hence developing cross-curriculum competence in literacy through mathematics.

Read Activity 8.2 for a creative use of different types of display, built into a classroom routine.

ACTIVITY 8.2

Ang has a mixed Year 3 and Year 4 class. All the children have a personal icon that they have created using their initials by using a drawing program. These are stored in a special folder accessible from the IWB. Every Tuesday morning, the children come into the class to find a question on the IWB and the beginnings of a data display, such as labelled axes or two overlapping circles. The children decide where their own icon should go and position this in the appropriate place. The display is saved and used as a basis for discussion later. The discussion doesn't always focus on the display. Sometimes the aim is to discuss the topic of the question, such as the local show. At other times, the discussion is used to focus on aspects of the data display, such as the meaning of the overlapping section of the two circles.

1. What features of this routine make it educationally useful?
2. What kinds of questions might Ang have used?
3. What preparation is needed to make this activity successful?
4. Why doesn't Ang do this on Monday mornings?
5. How does technology support this activity?

This is a productive routine that Ang uses to create discussion starters for whatever topic she has in mind for the week, or simply to get to know the class better. The technology use is efficient, but it also allows the children to have control over their own information, and leads to later discussion about statistical concepts.

It is particularly important that concept development – such as ideas about the middle of the data, which lead to statistical notions about summarising data using mean, median and mode – happens slowly so that children develop a deep understanding of the concept and don't simply use technology unthinkingly for computation of statistics that are not well understood. The emphasis should be on developing the concepts of summarising data in meaningful ways to answer questions posed in an investigation rather than calculating averages. In New Zealand, there is an emphasis across the primary years on children developing gradual understanding of variation, patterns, relationships and trends within data, and on evaluating statements made by others about data. Through the Australian Curriculum: Mathematics, children in Year 7 not only calculate the different summary statistics – mean, median and mode – but also have to describe what happens to these measures in relation to outliers, for example, or how changing the data set slightly affects the measures. Many children are ready to understand the implications of outliers earlier than Year 7, and exposure to these ideas can be useful earlier (see also Chapter 11). The key idea is that all children need time to develop deep understanding of the ways in which data can be collected, summarised and displayed, and opportunities to experiment and 'play' with data are not time wasted.

AC

There are several widgets in HOTmaths (listed below) that allow children to consider what happens to the mean, median, mode and range of a data set as one value is changed. They can be accessed by logging into HOTmaths and typing their respective names in the search field. The Moving the Mean widget allows children to develop an understanding of the sensitivity of the mean to extreme values by moving dots on a dot plot (Figure 8.9). The Changing Data widget demonstrates the impact changing a single value has on the mode, mean, median and range (Figure 8.10). This widget also could be used to encourage children to predict changes when one value is changed, or to challenge them to change a given data set so that, for example, the mean and median are the same. The display shows the result of four changes only, so children can be challenged to make the changes to the median and mean by altering one value at a time in four moves or less. This kind of activity can become a productive 'filler' for a short time – before lunch, for example.

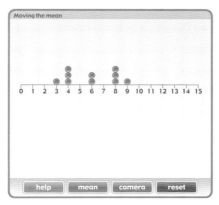

Figure 8.9 Screenshot of HOTmaths Moving the Mean widget

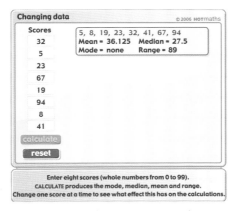

Figure 8.10 Screenshot of HOTmaths Changing Data widget

The Exploring Dot Plots widget could be used in a similar way, and the use of a different form of data presentation is valuable (Figure 8.11). Children need to recognise that the same processes can be used to summarise data when they are presented as numbers or in graphical form. They need to experience a

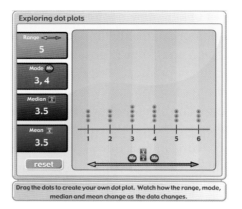

Figure 8.11 Screenshot of HOTmaths Exploring Dot Plots widget

variety of activities in which they learn about summarising data. Another tool that could be used to support concept development is the Mean and Median applet, available via the search field at the Illuminations website, <http://illuminations.nctm.org>.

Using technology tools such as these provides a way of playing with data that is difficult to do with paper-based methods. Dealing with middle measures, such as means, medians and modes, requires considerable computation, and children get caught up with the process of calculating the different measures rather than understanding what they are doing. Using technological tools can remove the computational burden, but the technology must be accompanied by discussion and informal recording of the ideas to help children consolidate their learning.

Data can be represented in many different ways, and technology makes this easy to do. Not every representation of data is helpful or correct, however, and technology tools will simply do what they are told to do. Activity 8.3 suggests ways of changing representations as part of exploring a very small data set.

ACTIVITY 8.3

Table 8.2 shows a **sample** of 10 students in Years 5 to 8, with their sex and height recorded.

- Copy the table into a spreadsheet, using a software program such as Excel. In table form, these data don't provide much information. It is difficult even to see how many males or females are included in the sample.

Table 8.2 Students' heights by sex

Sex	Height (cm)
Male	150
Male	151
Female	152
Female	153
Female	160
Male	161
Female	165
Male	165
Male	165
Male	173

- Use your software program to create a column graph of the data. It should look something like the graph shown in Figure 8.12. This shows a bit more information, but is still fairly uninformative.

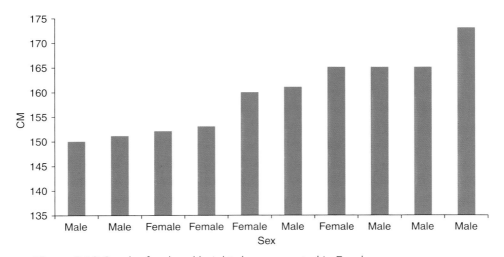

Figure 8.12 Graph of ordered heights by sex created in Excel

- We need to consider what we want to know about the data – what information do we want? For example, we may want to find out whether, in this sample, girls are taller than boys. We may want to find out what height is typical of school students in Years 5 to 8. We may want to consider how the data are distributed. Each of these questions suggests a different data display.

- In Excel, sort the data by sex. How does this change the graph? Is the display more helpful than the first column graph? Can we confidently answer any of the questions we have posed?

A better way to look at the comparison between males and females might be a **stem-and-leaf plot**. In this instance, we will make a back-to-back stem-and-leaf plot that will show males on one side of the stem and females on the other side of the stem. Excel will not create this representation automatically, so it has to be done manually. The stem is created by looking at the data set and choosing an appropriate value that will form the backbone of the plot, the first digit if two-digit numbers or, in this instance, the first two

		M	Stem	F	
	1	0	15	2	3
5	5	1	16	0	5
		3	17		

Figure 8.13 Stem-and-leaf plot of students' heights by sex

digits representing the hundreds and tens values: 15, 16, 17. The other digits then form the leaves. If you place male values on the left and female values on the right, the stem-and-leaf plot should look like Figure 8.13.

Although the data set is too small to be able to draw firm conclusions, the **distribution** of the data suggests that boys are a little taller than girls. This form of representation is more helpful than the column graph for answering the comparison question. Although in this instance means could have been used to compare the data sets, sometimes the means are very similar and looking at the distribution using a stem-and-leaf plot or similar representation will provide better information. For an example of an activity that could be used to emphasise the importance of the distribution, see 'Which List is Which?' on the nrich website, <http://nrch.maths.org>.

There are a number of websites that will produce simple stem-and-leaf plots, such as <http://www.shodor.org/interactivate/activities/StemAndLeafPlotter>. Activities such as creating a stem-and-leaf plot can also help to reinforce place value ideas.

Telling a story from the data (conclusions)

Many children persist with representations in which each data point is associated with an individual, but such thinking limits the inferences that could be drawn. Deliberate, careful teaching is needed to move thinking from the representation of individual data to a more abstract conception of data that can lead to inferences about a situation. This is what Mike is trying to achieve in Classroom snapshot 8.3.

CLASSROOM SNAPSHOT 8.3

Mike's Year 6 class was discussing the amount of sleep needed each night as part of a health unit. First Mike asked each child to write down on a sticky note how many hours' sleep they got for a typical school night. He then created an informal graph by asking each child to place the sticky note on to an axis drawn on a whiteboard. He posed the question 'How long is a typical night's sleep for our class?'

Through careful, structured discussion, he led the children to the idea of looking at the concentration of data in the middle, and used the 'middle half' as the boundaries. There were 24 children in the class, so he counted up six data points from the lowest value and drew a line. He then counted six down from the highest value and drew another line. He showed the children that there were 12 data points in the middle – the middle half of the data – and they agreed that the typical value lay somewhere within this range.

Mike left the concrete representation with sticky notes on the whiteboard for the children to refer to, and photographed it for future reference using his phone, transmitting the image to a folder on his laptop.

The children were then asked to explore similar information on their laptops. Mike previously had downloaded a sample of 50 responses from the Australian Census at School site, <http://www.abs.gov.au/censusatschool>, choosing questions that included one about the number of hours of sleep the respondent had on a school night. The children used the data exploration program TinkerPlots, <http://www.tinkerplots.com>, to explore the data. Kavitha investigated whether boys or girls had more sleep. Her TinkerPlots report is shown in Figure 8.14.

Kavitha has used a representation of a 'hat' with a crown, available from within TinkerPlots, that covers the middle 50 per cent of the numerical data to illustrate what is typical of the two categories of male and female. This representation provides an introduction to box-and-whisker plots in an intuitive way. It does not

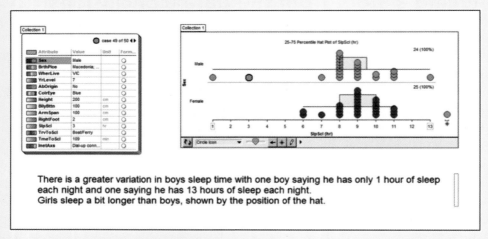

Figure 8.14 Kavitha's TinkerPlots report

require complex explanations of the interquartile range – the key idea is that the hat represents the middle half of the data and the brim shows the range from lowest to highest value (Watson et al. 2008). Using hat plots is one way by which technology use – in this instance, a commercial specialist program – can support the curriculum. Hat plots are not included as a formal representation in the curriculum, but are useful for providing a stepping stone to understanding. They give a visual display that tells a story and helps children to build on intuitive ideas about summarising data. They also provide a way of introducing the notion of the median as 'the middle of the middle' in a conceptual rather than a procedural way. Kavitha's simple report shows the beginnings of statistical inference. There is a reference to the summary of the data, but the interpretation is still very individualistic, shown by the focus on the boys' extreme values.

TinkerPlots has many features that cannot easily be replicated using other programs, including intuitive drag-and-drop displays. It is particularly useful for exploring the distribution of data. There is a free version on the TinkerPlots website that can be downloaded. Try some of the sample data sets provided with the trial version.

Developing the relationship between a real situation and a graphical representation is important, and this was the approach that Mike took by creating a concrete representation before moving into a technological solution.

ACTIVITY 8.4

Activities such as You Tell the Story from nrich, <http://nrich.maths.org/4802>, are useful ways of developing an understanding of what a graph represents. Go to this site and try the various story activities.

In the You Tell the Story activity, showing the distance of a man from a sheep, what does the graph show? Write down your thoughts about the activities. How might you use these kinds of activities in the classroom?

Many children have difficulty understanding that a graph of distance against time shows the distance from the starting point, not the direction. Playing with dynamic graphical interfaces can help children understand what the graph is showing them. The interactive nature of this kind of activity makes it suitable for individual use – for example, as part of a series of work stations about data and graphs.

Graphing Stories, <http://www.graphingstories.com>, is a useful website that contains videos of a variety of situations, which can be used to create graphs that tell the story of what the video is about. Many of these would be suitable for upper primary students. Try creating your own videos of different situations, such as filling bottles of different shapes at a constant rate, graphing height against time or running up and down stairs and graphing the number of steps away from the ground against time. Videos can easily be captured using mobile phones and then saved to a folder on a computer for later use. Children could work in groups to produce their video and a related electronic report. The conclusion step in the PPDAC cycle is crucial, and requires time to develop the necessary language and conceptual understanding (Pfannkuch et al. 2010).

Children need considerable experience to develop the skills of writing reports that explain what the data are telling them, and the language focus may be unusual in a mathematics lesson.

There are several relevant HOTmaths widgets and HOTsheets, including Graphs of Filling/Emptying Containers, Graphing the Burger Run (Figure 8.15), Replicating the Journey and Dunking Machine widgets. Although too complex for most primary students, they provide useful background and practice for the primary teacher, and can be accessed by logging into HOTmaths and entering their respective names in the search field.

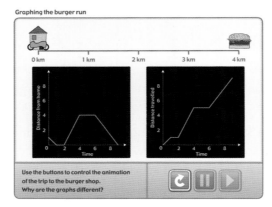

Figure 8.15 Screenshot of HOTmaths Graphing the Burger Run widget

The NZmaths site (<http://nzmaths.co.nz/statistical-investigations-units-work>) includes many useful units of work to support statistical investigations appropriate for every level of schooling. It includes examples of children's work, ideas for investigations and links to appropriate learning objects, and can be accessed from outside New Zealand.

Using data in social contexts

Data are used to problem-solve in many situations. A doctor making a diagnosis may use data of different kinds from a variety of sources, including the patient's description, the results of tests (which may include imagery and/or numbers) and a physical examination of the patient. We rarely see this as a data-driven activity, but it involves using different types of data and putting them together to answer a question – in the doctor example, 'What is wrong with this patient?'

In mathematics classes, we are trying to develop understanding of the processes involved with quantitative data – that is, data expressed as numbers – while in other subject areas we may be using a variety of other forms of data as well as numeric data. Other learning areas provide opportunities to reinforce the statistical ideas in social contexts. Scootle, <http://www.scootle.edu.au> and Digistore, <http://digistore.tki.org.nz>, have a number of activities that can be used to reinforce understanding of statistical ideas in different contexts. For example, L2631 Media Report: Water Usage could be used as part of a unit about water that combines studies of society, English and mathematics, and also addresses the cross-curriculum priority of sustainability. Using a data set from the First Fleet or the *Titanic*, for example, links to history and English. The numerical data must be presented in ways that are useful to support the reasoning about the social context.

Media articles are also a source of statistical ideas. A collection of archived media articles, together with teaching ideas that could be used as part of a social studies unit as well, is available from the Numeracy in the News site, <http://www.mercurynie.com.au/mathguys/mercury.htm>. MathsPig, <http://mathspig.wordpress.com/category/lists/10-big-media-maths-errors>, is another source of misleading articles.

Children as young as those in Year 3 or 4 can draw sensible conclusions from data presented by others if the social context is suitable and the questions posed are straightforward. The text in the media may be difficult, but understanding ideas such as 'the top 10 toys for Christmas' is not. This context, for example, could provide a starting point for a survey about toys, a discussion about toy safety and a design challenge to create a new toy, all of which can be undertaken by middle primary students. Technology could be used to enhance these activities.

The National Geographic website, <http://education.nationalgeographic.com>, has a novel idea for creating a map of the school for children in Years 1 and 2, which allows them to represent concretely the usage of areas in the school by sticking pasta on to a map of the school. You can access this map by logging into the site and typing its name into the search field. This data representation builds understanding about population density, for example, which is a concept that children will meet later in their school career, and links to sophisticated geographic information systems that are widely used in environmental science.

Reflection

In this chapter, several ideas about teaching and learning about statistics have been presented. Technology provides ways of collecting, presenting and analysing data. The use of data is likely to increase, placing greater demands on educators to develop sound statistical sense in their students. From reading this chapter, and engaging in the activities, you should have begun to develop an understanding of some of the key ideas in teaching statistics and the ways in which technology can enhance this endeavour. Statistics is also intrinsically linked to uncertainty and probability, and forms the focus of Chapter 9.

Websites for exploration

BBC: Interpreting Data: <http://www.bbc.co.uk/bitesize/ks2/maths/data/interpreting_data/play>

Census at School: <http://www.abs.gov.au/censusatschool>

Create a Graph: <http://nces.ed.gov/nceskids/createagraph/default.aspx>

Digistore: <http://digistore.tki.org.nz>

Illuminations: <http://illuminations.nctm.org>

Kids Maths Games Online: <http://www.kidsmathgamesonline.com/numbers/mathdata.html>

Meta-chart: Create a Pie Chart: <https://www.meta-chart.com/pie>

National Geographic: <http://education.nationalgeographic.com>

NZmaths: Digital Learning Objects: <http://www.nzmaths.co.nz/digital-learning-objects>

NZmaths: Healthy Life Survey: <http://www.nzmaths.co.nz/node/1650>

NZmaths: Which Graph?: <http://nzmaths.co.nz/resource/which-graph-excel>

Scootle: <http://www.scootle.edu.au>

Tinkerplots: <http://www.tinkerplots.com>

CHAPTER 9

Exploring chance and probability

LEARNING OUTCOMES

By the end of this chapter, you will:

- understand the difference between objective and subjective views of probability
- be able to use a range of random generators to determine probabilities
- recognise the applications of probability in daily life
- be confident in using technology effectively to develop ideas about uncertainty.

KEY TERMS

- **Independent events:** Two or more events that have no dependence on each other – for example, the chance of getting heads when tossing a fair coin is always 50 per cent, regardless of any previous tosses
- **Odds:** A way of expressing probability as a ratio, usually expressed as the probability of an event not being likely to occur
- **Outcomes:** The set of possible predictable events in a given situation, such as heads and tails when tossing a single coin
- **Probability:** A quantification of the chance of an event based on the possible outcomes
- **Randomness:** A situation where any particular outcome is uncertain
- **Relative frequency:** The number of times a given outcome occurs relative to other possible outcomes. This is the basis for classical probability theory.
- **Simulation:** Any created activity that aims to represent a 'real-life' situation.

Why is probability important?

Probability is concerned with how likely it is that something will happen, or that a particular outcome will be achieved. As such, it can be both objective – as in calculating the probability of tossing a 6 on a fair die – or subjective – as in assigning a probability to the chance of rain. In both examples, there is some uncertainty attached, and probability aims to quantify that uncertainty. Both objective and subjective views of probability are met in the primary mathematics classroom. In this chapter, the use of technology to enhance understanding of these different interpretations of uncertainty is considered.

As a teacher, it is important to understand the ideas of probability, and for many people these are difficult to grasp. The formal ideas challenge our intuitions, even for experienced and knowledgeable people. There is a famous probability problem based on a game show, called Monty Hall's problem. When this was first presented, it caused great controversy – including among mathematicians – because the formal solution is so counter-intuitive (see <http://en.wikipedia.org/wiki/Monty_Hall_problem> for an account).

ACTIVITY 9.1

There are many websites devoted to social aspects of probability, such as gambling, weather forecasting or determining risk. Search for some of these websites; some suggestions are provided below.

- Understanding Certainty, <http://understandinguncertainty.org>: This website includes videos, a blog and links to numerous articles about uncertainty.
- Understand Risk, <https://www.understandrisk.org>: Understanding risk is a worldwide forum that runs a series of conferences with different foci. This website provides information on risk assessment.
- Know Your Odds, <http://knowyourodds.net.au>: This is a Tasmanian government website that addresses common myths and beliefs about gambling. A similar New Zealand website is located at <http://choicenotchance.org.nz>. It is worth comparing the material presented on these websites with some of those that promote gambling.

1 Identify the language of probability used on these websites.
2 What is the difference between expressing probability as the chance of an event and as the odds of something happening?

3 How is probability expressed mathematically? You will see uncertainty expressed as odds and chance presented as a percentage, a fraction, a ratio or in words.
4 Think about what makes probability tricky for children (as well as many adults). Discuss your ideas with another person.

These sites indicate the pervasiveness of uncertainty and the complexity associated with expressing it. It is therefore not surprising that this is one of the trickiest areas of the mathematics curriculum. It is, however, also one of the most important because so many everyday situations are expressed in probabilistic terms, such as the risk of dying from a particular disease, insurance and investment risks, or the **odds** of a particular team winning in a sporting contest.

The New Zealand Mathematics and Statistics curriculum explicitly encourages investigative approaches to developing an understanding of probability (see the PPDAC cycle in Chapter 8), beginning to develop the links between statistics and probability. Such an approach is relevant for all children because it allows them to build their knowledge on the basis of experience and develops the sound intuition necessary for understanding probability.

Understanding probability

Bryant and Nunes (2012) suggest that children need to understand four key concepts in probability: **randomness,** sample space, quantifying probability and correlations.

Randomness does not mean disorder. In the mathematical sense, it is about a lack of predictability, and can be thought of as a measure of uncertainty. There is considerable evidence that children – and many adults – do not recognise the independence of events in a random sequence. If you toss a fair coin and get five heads in a row, the chance of getting a head (or a tail) on the next toss is still 50 per cent. Each toss is independent of those that have come earlier.

Sample space is essential for identifying the number of possible **outcomes.** Many children have difficulty identifying all possible outcomes. Although this is fairly easy when tossing a single coin with only two possible outcomes (H/T), tossing two coins has three categories (HH, HT, TT) but four possible outcomes (see the example later in this chapter). Recognising all possible outcomes is critical in being able to identify theoretical probabilities and quantify outcomes.

Quantifying probability is relatively straightforward for a single event based on a concrete example, such as tossing a die or a coin. It does rely on having identified all possible outcomes in the sample space, and this becomes more complex when there is more than one event, such as tossing a coin and rolling a die together. Finally, understanding relationships and associations – correlations – can be quite complex. Although beyond the formal scope of the primary curriculum, this is a key understanding for daily life, and teachers should be aware of potential pitfalls in drawing conclusions from associations. Good examples are provided by Gigerenzer (2002), including breast cancer screening and the OJ Simpson trial.

The research into school children's understanding of probability indicates that they have difficulty coming to terms with its non-deterministic nature. Unlike mathematics, where 3 + 4 always makes 7, every time a probabilistic experiment is conducted, the result may be different. For example, every time a die is rolled, a number from 1 to 6 will appear, but we would be suspicious if only 6 ever appeared. Young children in particular need many opportunities to understand that the chance of any one number appearing when a die is rolled is a random event (Truran 1995). Very young children often ascribe the behaviour of the die to some outside force, or to the die itself, saying, 'The die's naughty', and the phrase 'It's just luck' is used by all ages as a generic description of uncertainty.

Moving from this intuitive view to an understanding that there may be a set of outcomes for which the chance of occurrence can be determined theoretically is a big step. Watson and Caney (2005) suggest that even upper primary or early high school students struggle with the tension between intuition, expected (theoretical) outcomes and the natural variation that occurs from these expected outcomes. Take, for example, rolling a single, fair, six-sided die 60 times. Many children think that the distribution of numbers that will appear will be something like representation (a) in Figure 9.1 – possibly from many experiences of rolling two dice to create addition problems.

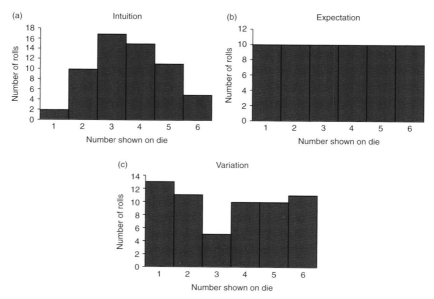

Figure 9.1 Results from rolling a single die 60 times

The theoretical outcomes look like (b) – the expected distribution – but if an experiment is carried out, the result may look like (c) – the variation from expectation. Many children will think there is 'something wrong' with the number 3 in representation (c) (these numbers were generated from an actual experiment using a spreadsheet), but the apparent 'problem' is simply due to natural variation.

Not only do children need to come to terms with variation from expectation; they also have to begin to understand that rolling a die many more than 60 times will lead to observed outcomes that are closer to the expectation. Technology can be extremely useful here because of its capacity to repeat the same action very quickly, setting up a **simulation**. It is important, however, that children have experience in using the concrete materials prior to moving into a virtual world.

The Australian Curriculum: Mathematics expects that, by Year 4, students have early understanding of unconditional probability – that is, **independent events** that occur without the influence any other event, such as the sex of a new baby – and conditional probability – where an event can or cannot occur depending on another event. For example, the walk sign at a crossing (Figure 9.2) cannot show 'walk' and 'stop' at the same time; the 'walk' instruction is dependent on not showing 'stop'.

Figure 9.2 Walk sign

These notions are very complicated for young children, because they rely on statements such as, 'If this happens, then that will happen', requiring children to hold one uncertain idea in their heads and then consider a second uncertain idea alongside it. Everyday experience of such statements often may not lead to the necessary probabilistic understanding. Scratch is a coding site on which there are many examples for children to explore and then create their own conditional statements. Look at 'Learning About If–Then Statements', <https://scratch.mit.edu/projects/97246>, for some simple examples that are accessible to Year 4 students.

Although children have many intuitions about uncertainty and likelihood, often these are not well articulated or are incorrect, and these misunderstandings can persist well into adulthood. The primary years are when many of the ideas are first encountered.

Early primary years

In the early years of schooling, the curriculum emphasis is on developing the language of uncertainty, such as 'possible', 'likely', 'maybe', 'certain', 'impossible' and so on.

CLASSROOM SNAPSHOT 9.1

Helen teaches a Year 1 class in a country school. She wants to introduce the idea of uncertainty through deciding whether an event is *certain* to happen, *might* happen or is *impossible*. For one week, every day the children take a walk to the school car park. They take note of what they see and practise remembering so that they can talk about the visit when they get back to the classroom. In the classroom, Helen asks the children what they saw and they create a collection of observations for each day of the week using Kid Pix™ on the IWB. On Friday, they review the week's observations by showing the relevant pages saved on the IWB. Helen asks, 'What things did we see every day?' The children identify objects like trees, fences, white lines and signs. Then she asks, 'What things did we sometimes see?' This time they suggest objects like a bird, a red car, a caterpillar and the moon in the sky. Finally, she asks, 'What things did we never see?' This question confuses the children initially, so she suggests some ideas, such as a train, a penguin and a volcano. This provides the children with imaginative prompts and they start suggesting objects like monkeys and dinosaurs. One child suggests a horse, and this notion leads to a discussion about whether they would never see a horse or that it would happen only very occasionally, beginning to establish the ideas of chance. Finally, the children return to their tables and create a chart on paper divided into three parts headed NEVER, SOMETIMES and ALWAYS, using pictures cut from magazines. Later, Helen scans each child's work before they take it home, to add to their electronic portfolios (see Figure 9.3).

Figure 9.3 What we saw in the car park

Helen is using technology to record the children's observations efficiently for effective use of the information later. She engages the children in discussion about the language that she is aiming to develop, and reinforces the discussion by referring to the visual imagery that they created during the week. The discussion also supports the reasoning proficiency in the curriculum.

The culminating activity has several purposes. Most importantly, it provides a record of what the children have done and information about how well they understood the language of never, sometimes and always, based on the activity they did during the week. Understanding the language provides a foundation for later work on predicting events. Helen has also created the sheets that the children use to move from NEVER on the left to ALWAYS on the right, replicating the direction of more formal probability number lines that the children will meet later.

The activity also gives the children fine motor skill practice through cutting out and pasting, and encourages language development through the conversations that Helen, the aide and the other children have about the pictures they choose. This activity could be done without technology, but it is more efficient when technology is used.

The language of chance is the focus of a video on the ABC Splash website, <http://splash.abc.net.au/media?id=29637> in which Flynn and Dodly prepare for a day at the beach. Children in Years 1 and 2 could watch the video and discuss the words used to describe uncertainty, such as 'not much chance', 'might', 'likely' and so on. Children could make their own mini-videos about uncertainty, using simple puppets on sticks and shooting the video with a mobile phone. The activity of preparing a storyboard and writing the script also emphasises sequencing, and links mathematics to the arts.

Other technology-based activities that could be used to emphasise the development of understanding are 'The Slushy Sludger' (available in Australia on Scootle and in New Zealand on Digistor), which has several activities for children to use. NZmaths (http://www.nzmaths.co.nz) also has several Probability units of work for young children that help to develop the language of uncertainty.

Children appear to have more difficulty with events that sit at the extremes – impossible or certain events (Fischbein, Nello & Merino 1991). They can accept the inherent uncertainty of events that might happen because of prior experiences. As they grow older, however, they are better able to make judgements and recognise that some events can be classified as certain to occur.

Middle primary years

The language focus is still evident in the middle primary years, but it becomes more focused. Children sequence events using uncertainty language, building on activities such as those in Classroom snapshot 9.1. The HOTmaths What's the Chance? widget provides some practice in deciding how likely an event is, using a five-point scale from impossible to certain, with equal chance providing a reference point (see Figure 9.4).

Activities such as this not only help to develop the language of uncertainty, but also begin to develop the notion of ranking uncertain events. In turn, this idea leads to quantification of uncertainty in the later years of primary school. The HOTmaths Walkthrough and Is That Likely? and Marble Jar Chance HOTsheets from the Chance Language lesson about arranging events in the order of their likelihood are also useful. You can access all of these resources by logging into HOTmaths and selection 'Cambridge Primary Maths' as the Course list and then 'Aus Curric 3' as the Course. Then choose 'Chance & data' as your Topic, and finally 'Chance language' as your Lesson.

Chapter 9: Exploring chance and probability

Figure 9.4 Screenshot of HOTmaths What's the Chance? widget

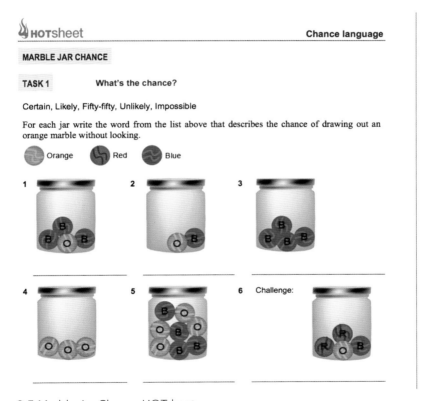

Figure 9.5 Marble Jar Chance HOTsheet

As children move into the middle primary years, identifying all the possible outcomes from a particular event becomes important. The capacity to identify all outcomes underpins objective notions of probability, and it is important to develop this skill. When children played a variety of board games, such as

Snakes and Ladders, they developed a sense of the possible outcomes from rolling a die, but many children now have limited experiences with these ideas. An interactive version of Snakes and Ladders is available from the Virtual Math Fest website, <http://www.counton.org/games/virtualmathfest/snakesladders.html>, and interactive Ludo is available from Funnysgames, <http://www.funnysgames.com/games/Virtual-ludo.html>. These games could be played by two or more players as an educational choice activity.

Developing the skills to recognise possible outcomes involves systematic counting of all possibilities, as illustrated in Classroom snapshot 9.2. This is not probability itself, but an important underpinning capability that children need to develop.

CLASSROOM SNAPSHOT 9.2

Josie, in Year 4, has been to a shop where teddy bears are made. She chose her bear, put a red heart inside it from a choice of either plain or checked red, then chose a blue scarf from a choice of red, blue or green. She tells her story to the class and shows them her bear. The teacher, Neil, decides to capitalise on this by asking, 'How many different bears could Josie have made?'

The next day, he organises the children into pairs, and gives each pair a teddy bear outline, two hearts and three scarves, as in Josie's story. The children are asked to find the answer to the problem posed and to record their working out in any way they like. As they are working, Neil talks to them about their thinking and asks selected groups to record their approach on the IWB. Some groups have chosen to draw all the possible bears, some make lists, some use other kinds of symbols such as coloured lines to represent the hearts and scarves, and one pair develops a form of a tree diagram. All the children correctly answered the question by getting six possible bears.

Neil then decides to take this further and asks, 'What if there were four different scarves or three types of heart?' To explore these extension ideas, Neil uses an interactive activity called Bobbie Bear, from the Illuminations website, <http://illuminations.nctm.org/ActivityDetail.aspx?ID=3>, which uses shorts and t-shirts rather than hearts and scarves. He models the original problem on the IWB with Bobbie Bear to make the link between the two slightly different contexts, then returns to the extension problem.

The children work on this problem and others posed by Neil or by themselves as he monitors each pair's progress. Again he asks selected groups to share their work using the IWB. Finally, Neil returns to the original scenario and asks the class, 'If Josie had just shut her eyes and chosen a heart and a scarf without looking, what would be the chance of her getting the bear with the plain red heart and the blue scarf?' Through discussion, and showing the original solutions that the children had recorded, he leads the class to the idea of one out of six possibilities, but does not go further with this line of thinking at this time.

This activity has a focus on identifying all possible outcomes. Neil uses technology together with concrete materials to enhance the ideas that he wants to develop. He builds on the intriguing story that Josie tells, and hence connects the mathematical ideas directly to the children's experiences. He also understands where the learning will need to go as the children move through school, by introducing the links between counting outcomes and probability. He chooses not to pursue this connection because only a small number of children in the class are really ready to learn this idea, and quantifying probability is not in the curriculum in Year 4.

This snapshot raises a number of pedagogical issues. These include links to real-world situations, and building on these as they occur – such as when Josie told her story. There are questions around formal and informal recording of possibilities, and Neil did not follow this line of thinking although he did capture the various approaches for possible future use. The activity could have gone in different directions. Neil could have made links to algebra by moving towards a generalised approach to looking at all possibilities. Instead, he deliberately chose to build the connection to probability.

PAUSE AND REFLECT

Consider the different decisions that Neil made, such as to follow up on Josie's story; collecting but not formalising the children's different recording processes; extending the scenario and his use of technology to do this; and his decision to emphasise probability and the beginnings of quantifying chance outcomes. Think about the arguments for and against these decisions.

1. What other courses of action might Neil have taken?
2. What factors might influence these decisions?
3. What mathematical ideas did the children need in order to undertake the activity that Neil organised?
4. What do you think was the key focus of the lesson?
5. How did Neil's use of technology impact on the teaching?

Combinatorial tasks, such as Bobbie Bear, can be solved successfully using concrete materials by children as young as 7 years, using systematic trials of all possible combinations (English 1991). Because of the mathematical potential of tasks of this type, they should be encouraged in the middle primary classroom.

It is also important that children in these middle years begin to understand randomness in its mathematical sense. Many children think that 'anything can happen' is random, and they need to learn that a random event, mathematically, can be quantified by considering all possible outcomes. Activities such as 'What's in the bag?', from NZmaths, <http://nzmaths.co.nz/resource/whats-bag> or Maths300, <http://www.maths300.com> – in which one child hides a number of different-coloured cubes in a paper bag, and the class guesses what combination of coloured cubes is in the bag by sampling one cube at a time, showing the class, and replacing it – are useful starting points for developing ideas about random samples.

Global

The activity can be made more or less complex according to the experience and development of the children involved. HOTmaths widgets Marble Jar Chance and Probability Marble Jar for upper primary children also support this idea, and by Year 6 link this to quantification of probability (see Figure 9.6). You can access these resources by logging into HOTmaths and entering their respective names in the search field.

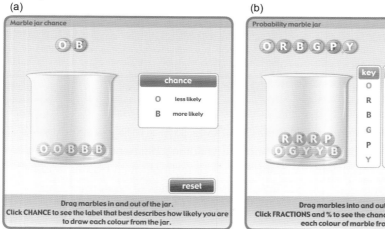

Figure 9.6 Screenshots of HOTmaths Marble Jar Chance (Years 3 and 4) and Probability Marble Jar (Year 6) widgets

Upper primary years

In the upper primary years, the emphasis in the Australian curriculum is on beginning to quantify uncertain events. Children use fractions, decimals and percentages to assign a likelihood to an event occurring. They also begin to conduct experiments using random generators such as spinners, coins and dice, and in so doing develop the more formal language of probability, such as sample space, frequency and outcomes. There is a useful piece of software called Probability Explorer, <http://www.probexplorer.com>, which can simulate experiments based on coins, dice or marbles. This software supports the investigative approach of New Zealand by allowing children the possibility of setting up investigations. The power of technology to run very large numbers of trials quickly is useful, and can be used to develop the idea that the larger the number of trials, the closer the experimental probability comes to the theoretical values.

Activities such as ordering chance words on a number line from 0–1 provide starting points for quantifying probability. Children in Year 5 or 6 can be encouraged to find phrases, sayings and words associated with probability in newspapers or on the internet. A class probability line can then be constructed, with each child placing a favourite word or phrase somewhere on the line. This activity could be done concretely, using string across the classroom and pegging the words on to it, or electronically, using a spreadsheet program. The Language of Chance Walkthrough in HOTmaths also helps to reinforce the learning from earlier years. This walkthrough can be accessed by logging into HOTmaths and selecting 'Cambridge Primary Maths' from the Course list and 'Aus Curric 5' from the Courses. Next, choose 'Chance' as the Topic, 'Describing chance' as the Lesson, and finally click on the 'Walkthroughs' tab.

Figure 9.7 Screenshot of HOTmaths Language of Chance Walkthrough

Once the words and phrases are ordered, children can begin to allocate a fraction to each one to quantify the chance of the event. As a follow-up, children could research the origins of phrases such as 'pigs might fly', 'once in a blue moon' and 'Hobson's choice', or write a story that uses chance language. The children's novel *Pigs Might Fly* (Rodda 1986) could provide a model, integrating mathematics and English in a meaningful way in the classroom. Children's literature is a source of rich language and opportunities to develop understanding (Watson 1993). It may be worth searching for e-books suitable for the relevant age group.

Simulation activities

In the upper primary years, children can begin to develop their own simulations. These may be concrete or technology-based, and support the investigative approaches to developing probabilistic ideas through experimentation.

Work on 'how many ways' should continue in the upper years of primary school, with an emphasis on efficient and effective recording to ensure that all possible outcomes are counted systematically. A worthwhile activity is to make a three-part (or more) flip-book, where different combinations are created by turning the top, middle or bottom sections of the page. Crazy Animals (available from <http://www.maths300.com>, is one example of a rich activity that has a number of potential learning outcomes. Children make a book with three animals divided into head, body and legs. They then explore all the possible outcomes that can be made by using one head, one body and one legs part. If the animals are, for example, a camel, a bear and a sheep, possible combinations could be listed as 'ceap' (camel head, bear middle and sheep legs), 'beeel' (bear head, sheep middle, camel legs) or 'shamr' (sheep head, camel middle, bear legs). How many different combinations are possible? Activities like this are problem-solving exercises that also link to literacy and arts.

Leading into more probabilistic thinking, children can roll a die to make an animal at random. Roll 1 is the head, roll 2 is the middle and roll 3 gives the legs, and the number on the die decides whether it is a camel part (1 or 2 rolled), a bear (3 or 4 rolled) or a sheep (5 or 6 rolled). From here there are several rich probability activities that could be followed, such as 'What is the chance of getting a three-part camel, a two-part camel, a one-part camel or a 0-part camel?' To answer this question children have to identify all the possible animals (outcomes). They can simulate the process and speed it up using virtual dice such as those available from <http://www.random.org/dice> or as part of the class IWB resources. A question that is more difficult to answer might be, 'How many rolls will it take to get a three-part camel?'

A similar activity is available in the HOTmaths Numble Folk widget, which can be accessed by logging into HOTmaths and entering its name in the search field. This widget allows children to control different attributes to make Numbles. After all the Numbles have been made, a simulation could be carried out using dice to identify the probabilities of getting a particular Numble if the generation were random.

Global

The unit of work 'Murphy's Law', <http://nzmaths.co.nz/resource/murphys-law>, provides an interesting approach to building experimental probability ideas. The ideas to be explored include buttered toast always landing buttered side down, your keys always being in the opposite pocket to your free hand and traffic lights always being red when you are in a hurry.

Children need considerable experience with different kinds of random generators. We mostly use dice, coins or spinners in the classroom, but spreadsheets will also produce random numbers, and could be used to simulate dice, for example. Spinners have the advantage that they can be partitioned to give unequal outcomes, which is not the case with dice. An adjustable spinner is available from the Illuminations website, <http://illuminations.nctm.org/ActivityDetail.aspx?ID=79>, which also produces a comparison between the experimental and theoretical outcomes. Asking children to predict the outcome from, say, 10 spins, and comparing it with the actual and theoretical outcomes, helps build useful intuitions about probability. Similar ideas are embedded into the learning object 'Random or not?' (available on Scootle in Australia and Digistor in New Zealand).

It is particularly important when dealing with probability that children are asked to explain their thinking. Often children will give the correct answer but have a deep misconception about the outcome. For example, if children are asked to predict the number of heads that will come up in four tosses of a coin, by the upper primary school years most children will give an answer of two because it is a 50/50 chance. If they are then asked to imagine doing that 100 times, and record the number of times in four throws that they get no heads, one head, two heads, three heads or four heads, a surprisingly large proportion of children will predict 20 for each outcome. They are confusing the equal chance of getting a head or a tail with the likelihood of getting each possible outcome. Using a tree diagram can be helpful for addressing this misunderstanding. Figure 9.8 shows the possible outcomes from tossing two coins together.

The four possible outcomes are HH, HT, TH, TT. Hence there is twice the chance of getting a head and a tail together than there is of getting either two heads or two tails. Such a diagram leads to the idea that the chance of getting two heads (or two tails) is 1 in 4, or $\frac{1}{4}$ or 25%, but the chance of getting a head

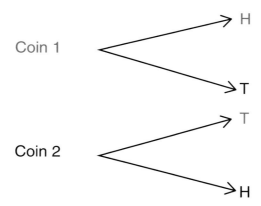

Figure 9.8 Tree diagram to show outcomes from tossing two coins

and a tail is 2 out of 4, or $\frac{1}{2}$ or 50%. If these outcomes are quantified, HH and TT have a 25% probability and the combined HT (HT, TH) outcome has a 50% probability. In this instance, the theoretical outcomes are not equally likely. If we then did an experiment in which two coins were tossed four times, we would not be surprised if the results were HH, HT, TT, HH, showing variation from the expected values.

The notion of simulation – that is, using a random generator to create a model of a situation – is a powerful idea that upper primary children should begin to experience. Many children find the abstraction from a real situation difficult to understand once it becomes more complicated than, for example, tossing a coin to predict whether a baby is a girl or a boy. Children may transfer the 50/50 chance given by a coin to other situations where the chance is not 50 per cent. To illustrate the use of simulations to solve problems, in Classroom snapshot 9.3, the problem is a famous one called the birth month problem: 'What is the chance that if any five of us meet by chance on the beach, at least two of us will have birthdays in the same month?'

CLASSROOM SNAPSHOT 9.3

The students in Di's Year 6 class have been discussing birthdays. She poses the birth month problem and asks the children to guess the answer. Each child secretly writes a guess on a piece of paper and then Di records these guesses on the IWB for later referral. The students in the class discuss how they might find out how accurate their guesses are. They decide to try it out for themselves using a random selection of names drawn from a container to create the groups. Di moves the class to a multi-purpose room where there is more space to move, and the children form groups of five as their names are drawn from the hat. They then compare their birth months and the number of groups in which there is a 'match' (two or more birthdays in the same month) is recorded. After doing this five times, it is clear that it will take quite a long time to generate enough data to be sure of the outcome.

The class returns to the classroom and Di introduces the idea of a simulation. Leading a discussion about what to simulate, she emphasises that there are 12 months and five people, so both need to be included. The children come up with different ideas, such as having a bag with 12 numbered counters, and drawing one counter five times; using 12 cards in the same way; using a spinner with 12 segments and spinning five times; and rolling a 12-sided die five times. They also discuss how best to record the matches and decide that a table of Match/No match recorded with tally marks would work.

The children are organised into groups of four and each group is asked to use its chosen method 10 times. After conducting this experiment, the children are asked whether they want to change their predictions. They decide that they still cannot be sure of the outcomes.

Di then goes to Roll Dice Online, <http://www.roll-dice-online.com>. She chooses to roll five 12-sided dice. She asks the children to explain why she is using five dice (to represent the five people) and why the dice are 12-sided (to represent the 12 months). These questions reinforce the links between the real-world situation and the simulation. She rolls the dice 10 times and the children record the number of matches in 10 rolls. The class repeats this until there are 100 rolls (10 lots of 10 rolls) completed. They then repeat the virtual experiment using an interactive spinner, <http://www.shodor.org/interactivate/activities/BasicSpinner>. They use trials of five spins, collecting the matches each time, and repeat this until there are 100 trials (500 spins) completed. The children then consider all the data and individually write a report about the activity and their conclusion about how likely it is that if five people meet by chance, at least two of them will have the same birth month.

There are several good reasons for Di to spend what appears to be wasted time getting the children first to model the problem by forming groups, and then to undertake a concrete simulation. A simulation is an abstraction from reality, and a virtual simulation – such as rolling five 12-sided dice or spinning a 12-segmented spinner – needs to be connected to the real problem in stages so that children can follow the line of reasoning. Using two different random generators – the dice and the spinner – provides an opportunity for discussion about the fact that these diverse methods will produce similar results, and that the more trials are conducted, the closer the outcomes will be to a theoretical value. Calculating the theoretical value is outside the scope of the average Year 6 student, but some advanced students may wish to look at this.

AC

Spinners are useful for children to experience unequal outcomes. HOTmaths has a useful printable HOTsheet with spinner blanks on it. You can access this resource by logging into HOTmaths, entering 'Designing Spinners' into the search field and then selecting the 'HOTsheets' tab. Spinners can be made by hooking a paper clip over a pencil that has its point in the centre of the circle. If the paper clip is flicked hard, it should spin freely around the pencil. Usable spinner pro formas can be made by enlarging the HOTmaths sheet on a photocopier.

ACTIVITY 9.2

The graph in Figure 9.9 shows the outcomes from spinning a four-colour spinner 20 times.
- Make a spinner that would most likely give you similar results.
- Test your spinner and record your results in any way that you wish.
- Explain why your results are the same as or different from the ones shown here.

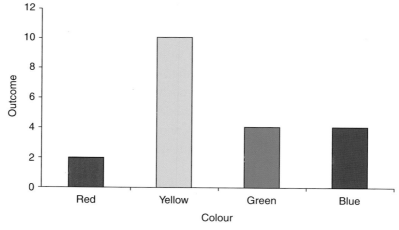

Figure 9.9 Graph of spinner outcomes

When reading a graph of the type in Activity 9.2, many children are likely to focus on the number of categories, rather than the proportion of each colour. They may make a spinner that has four equal quadrants, or one that has half coloured yellow, but the remaining proportions inappropriately distributed.

They will test their spinner, but their recording of results may be laborious and inefficient, such as listing the outcome from each spin, rather than using a summative approach, such as tally marks. Explanations for the difference between their results and those of the spinner shown on the graph are likely to focus on the physical characteristics of the spinner or the manner in which it is spun.

Later, children will develop a more sophisticated approach to reading data and interpreting the graph. They recognise the importance of the **relative frequencies**, and are likely to attempt to design a spinner by colouring half yellow, and the remaining half equally blue and green. Red may be placed inappropriately in either half so that the proportions are not exactly those shown in the graph. They will test their spinner and record the outcomes efficiently and systematically, and present their findings in appropriate ways such as a table or graph. Their explanations of the different outcomes, however, are likely to be based on chance only, and the relevance of sample space will not be appreciated.

Those children who have developed a good understanding of probability outcomes will complete the task efficiently and systematically. They will manipulate the data and understand the variation that is likely to occur in small samples of 20 spins. They may express this informally rather than in terms of theoretical probabilities, and are ready to learn theoretical probability. An appropriate spinner is shown in Figure 9.10.

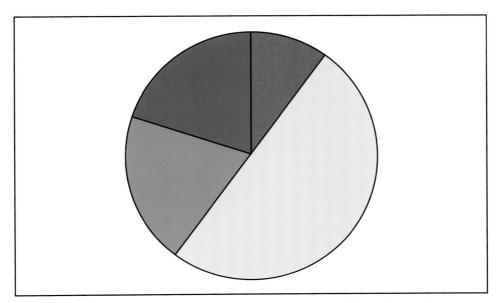

Figure 9.10 A spinner most likely to give the outcomes recorded on the graph in Figure 9.9

Activities such as the birth month problem and designing spinners from graphed or tabulated data make valuable links between probability and statistics, which become important later when inferential statistics are introduced.

In addition to experiments involving chance, children should have opportunities to recognise the uncertainty inherent in sampling. Discussions about the need for a sample, the difference between a sample and a population, and the notions of random and representative samples are needed before students design and carry out surveys. For example, if children want to survey the school, how should they select the sample? Should they have a random group selected from each year level – a random and representative sample – or simply rely on the chance of getting a representative group? These questions are less trivial than they may seem, because all these decisions are also affected by real-world constraints.

Representing probability

As well as connecting to the data aspects of statistics, probability connects to the Number and Algebra strand. The 'how many ways' activities can be extended to finding a generalisation, which is an important aspect of algebra. The representation of probability as a number can reinforce links between fractions, decimals and percentages. Mathematically, probability is represented as:

$$P \text{ of an outcome} = \frac{f}{n}$$

where P is the probability of a favourable outcome, f is the number of favourable outcomes, and n is the total number of outcomes. If we consider the probability of tossing heads, for example, on a fair coin this would be written as:

$$P_H = \frac{1}{2}$$

This representation is automatically in fraction form, and this can be changed to a decimal (0.5) and a percentage (50%).

Odds are widely used in everyday life. They are expressed as a ratio, and represent the relative probabilities of an event occurring and not occurring. For example, the odds of throwing a six on a single fair die are:

$$\text{Odds in favour of } 6 = \frac{\text{number of ways to get 6}}{\text{number of ways to fail}} = \frac{1}{5} \text{ or } 1:5$$

Compare this with the probability of getting a 6, expressed as:

$$\text{Probability of getting } 6 = \frac{\text{number of ways to get 6}}{\text{TOTAL number of outcomes}} = \frac{1}{6}$$

Because odds are regularly seen in newspapers, children may want to know about these. It is important to make it clear that odds and probability are obviously related, but do not express exactly the same quantity. Odds are also usually expressed as the odds *against* an event happening. So a sporting headline that says 'Reds at odds of 9:1' indicates that the reds have only a one in 10 chance of winning. If the headline had said 'Reds at 9:1 on', however, the word 'on' is used to indicate that the event is favourable – that is, the Reds have a nine in 10 chance of winning. Language is critical, and care is needed when reading newspaper articles or other texts that refer to probabilities or odds.

Representing probability as a fraction, decimal or percentage can provide good practice for children in moving between these representations. There are many useful interactive activities available, including HOTmaths widgets such as Probability Marble Jar, showing probability as a percentage, displayed earlier in the chapter in Figure 9.6. Although probability questions are often presented on a worksheet, encouraging children to express probabilities from the many concrete or virtual activities as fractions, decimals or percentages will make the representation more meaningful.

PAUSE AND REFLECT

From your reading and personal experience, why is probability so tricky for primary school children? We all live with uncertainty in our lives, but expressing this mathematically appears to present particular difficulties for children. How will you approach teaching probability in your classroom? How will you use technology to help you?

Chance activities can be linked to other areas of the curriculum. In many instances, these involve social situations that lead to subjective views of probability. Children are used to the inherent uncertainties of daily life, but may find it more difficult to accept that these can be quantified. An example is the bushfire danger sign that all Australian children see on a regular basis (Figure 9.11).

Behind this familiar object are some quite complex probabilistic ideas. The sign provides an indication of the risk – the likelihood of a bushfire

Figure 9.11 Bushfire warning sign

expressed as a fire danger from low moderate to extreme, and then to Code Red. In addition, it can be interpreted as a statement of conditional probability – if there is a bushfire, then it is likely to be dangerous, expressed as low moderate to extreme. The probability cannot be quantified in the way that counting outcomes on a die is possible but, based on experience and expert opinion, a generally agreed scale can be provided and interpreted. New Zealand uses similar ideas in relation to volcanoes, with alert levels from 0 to 5 – see the GeoNet: Volcanic Alert Levels website, <http://info.geonet.org.nz/display/volc/Volcanic+Alert+Levels>.

Australian children are used to the uncertain nature of bushfires, in which some trees are burnt and others survive. A cross-curriculum unit on natural disasters can be enhanced by use of tools such as 'Fire', <http://illuminations.nctm.org/ActivityDetail.aspx?ID=143>, a 'wildfire' (bushfire) simulator that allows children to set the probability of the fire spreading. The Bureau of Meteorology, <http://www.bom.gov.au/weather-services/bushfire/about-bushfire-weather.shtml>, has useful information about bushfires, and the National Rural Fire Authority site in New Zealand, <http://www.nrfa.org.nz>, provides a series of maps showing the likelihood of different types of fire. Such activities stress the cross-curriculum numeracy competency expected in the Australian Curriculum: Mathematics, and provide an opportunity to develop proficiencies – particularly reasoning and problem-solving.

Reflection

Probability is one aspect of the mathematics curriculum that creates difficulties for children. It is about uncertainty, and this is different from other areas of mathematics. Children have to come to terms with ideas like, 'We can calculate a theoretical probability but if we conduct an experiment the results are likely to differ from the expected values.' Making a guess about the likelihood of an event is not the same as predicting an outcome based on an expected value, and underlines the difference between subjective intuitions and objective predictions based on theory.

Despite the difficulties, this is an area of the curriculum that has potential for rich investigations and a high degree of engagement by children. As with all good teaching, it is important that children have opportunities to reflect on their learning, through discussion or by writing a report of an activity. Language is critically important and the more practice children get, the better.

Websites for exploration

Resources of different types are available by searching the probability topic link: <http://www.interactivemaths.net>.

Interactive games and real-world applications of probability (such as predicting avalanches) are available via searching a Canadian government education website: <http://www.learnalberta.ca>.

A wide variety of different resources, including interactive games, lessons and discussion-starters for class lessons are available at: <http://www.nrich.maths.org>.

A useful site, including some research articles and a flexible software tool that can be set up to create probability experiments, is: <http://www.probexplorer.com>.

Other useful sites include:

Count On: Games: <http://www.counton.org/games/virtualmathfest/snakesladders.html>

Funnysgames: <http://www.funnysgames.com/games/Virtual-ludo.html>

HOTmaths: <http://www.hotmaths.com.au>

Illuminations: <http://illuminations.nctm.org/ActivityDetail.aspx?ID=79>

Interactivate: Spinner: <http://www.shodor.org/interactivate/activities/BasicSpinner>

Maths 300: <http://www.maths300.com>

NZmaths: <http://nzmaths.co.nz/resource/whats-bag>

Probability Explorer: <http://www.probexplorer.com>

CHAPTER 10

Capitalising on assessment for, of and as learning

LEARNING OUTCOMES

By the end of this chapter, you will:

- understand the notion of assessment for, of and as learning, and how these forms of assessment work together in the mathematics classroom
- have a developmental framework to assist in designing an assessment item and assessing the quality of a student's response: the SOLO model
- be aware of the use of national testing data in a positive way to support growth in mathematical understanding
- be able to use various ICT tools to create valid assessment items.

KEY TERMS

- **Assessment:** To identify in measurable terms the knowledge, skills and beliefs of an individual or group
- **Assessment as learning:** The use of assessment information by teachers and students to guide curriculum planning that influences future student learning
- **Assessment for learning:** The central purposes of assessment for learning are to provide information on student achievement and progress and to set the direction for ongoing teaching and learning.
- **Assessment of learning:** The process of gathering information about student achievement and communicating this information

While the activities presented in the previous chapters all have the potential to be used for **assessment** purposes, this chapter specifically targets making the most of assessment in terms of identifying students' levels of understanding, better targeting future student activities and providing feedback to students and parents/caregivers. The most exciting fact about assessment in the primary classroom is that student teaching and learning tasks and assessment tasks can be united seamlessly. As teachers, we can make the most of a mathematics assessment task if it provides evidence of students' current levels of understanding, if this can be communicated positively to students and parents/caregivers, and if the students have some role to play in their own goal-setting of where to head next. For too long, assessment has been driven by commercially produced closed worksheets that do not provide the flexibility and open-endedness needed for students to enter the task at a variety of levels.

This chapter investigates the design of assessment for learning tasks, assessing and evaluating the tasks, and providing feedback to students, parents and caregivers. The chapter considers positive strategies for utilising the National Assessment Program – Literacy and Numeracy (NAPLAN) items, school results and individual student results as a diagnostic tool to move forward.

Assessment

Assessment generally means to identify in measurable terms the knowledge, skills and beliefs of an individual or group. While the definition appears simplistic, problems arise when one attempts to categorise assessment practices into distinct groups. We can fall into the trap of categorising assessment items as either summative or formative, and thus placing them in two separate groups. Summative means assessment at the end of a teaching period (such as a unit, term, semester or year), with the aim being to assess what the students know for reporting purposes. Formative, on the other hand, traditionally was perceived as less formal, often not warranting recording, and enabled feedback to the students and teacher to identify where they were at and where they needed to head. Educationally, we have developed our ideas of assessment greatly and are no longer so quick to categorise different forms of assessment as one or the other and to judge their value.

This multi-dimensional view of assessment has led to the terms **'assessment *of* learning'**, **'assessment *for* learning'** and **'assessment *as* learning'**, where the power of assessment lies in the overlap in the nature of these types. One of the most exciting times for assessment in the primary mathematics classroom was during the implementation of assessment policies that focused on assessment for learning. A significant example of this was the NSW Board of Studies, Teaching & Educational Standards (2012) *Assessment for Learning Practices* document. One of the key factors was the integration of assessment tasks into the teaching and learning sequence, whereby students may not be aware that they are being assessed. This is often described as assessment tasks and teaching and learning activities being seamlessly united. New Zealand's assessment strategy demonstrates a sound balance between assessment for, of and as learning. The NZmaths website, <http://www.nzmaths.co.nz/mathematics-assessment>, is designed to help mathematics teachers to select a tool that is appropriate to their needs at a particular point in time. The website asks teachers to consider a range of informal strategies, such as classroom observations and questioning, and formal assessment strategies. These formal and informal types are included in subsequent sections.

Assessment for learning

The Australian Curriculum: Mathematics adopts the principles of assessment for learning. Teachers are asked to integrate their assessment practices into their daily mathematics tasks in a manageable fashion. Assessment strategies should not pressure teachers or students. Goals should be articulated clearly to students, with the emphasis placed on gaining a deeper understanding as opposed to a higher mark or rank. Assessment should be inclusive and accessible to all learners, and provide valuable and timely feedback to students. This is supported by the Australian Association of Mathematics Teachers (AAMT 2008) in its *Position Paper on the Practice of Assessing Mathematics Learning*, which stresses that students' learning of mathematics should be assessed in ways that:

- are appropriate
- are fair and inclusive, and
- inform learning and action (AAMT 2008, p. 1).

Assessment for learning has also been high on the agenda of the Assessment Reform Group (2002) in the United Kingdom, which devised the following 10 principles of assessment for learning:

1. It is part of effective planning.
2. It focuses on how students learn.
3. It is central to classroom practice.
4. It is a key professional skill.
5. It has an emotional impact.

6   It affects learner motivation.
7   It promotes commitment to learning goals and assessment criteria.
8   It helps learners know how to improve.
9   It encourages self-assessment.
10 It recognises all achievements.

Alongside such documents and guidelines, materials have been produced across the primary and secondary spectrum that assist teachers in the design of such tasks. Examples of these include materials from the Assessment Resource Centre (ARC), <http://arc.boardofstudies.nsw.edu.au>, the Count Me in Too website, <http://www.curriculumsupport.education.nsw.gov.au/countmein/index.html> and the Australian Curriculum documents.

Classroom snapshot 10.1 provides an example of an assessment for learning task and the feedback provided by the teacher to the students. With the introduction of the Australian Curriculum: Mathematics, online materials have been released by ACARA that assist teachers in choosing suitable assessment for learning tasks within each of Foundation to Year 10 years of schooling. Links to these can be found on the Australian Curriculum v5.0 Mathematics Foundation to Year 10 Curriculum website at <http://www.australiancurriculum.edu.au/Mathematics/Curriculum/F-10>. The expected level of understanding is articulated for each year, with content descriptors, and there are work sample portfolios for each year.

The New Zealand Mathematics curriculum provides an extensive bank of resources, including a large collection of assessment for learning tasks titled the *Illustrations of National Standards of Mathematics*, <http://nzmaths.co.nz/national-standards-illustrations>. This site provides student tasks and student work samples. In addition, for levels one and two in the NZmaths framework, teachers are provided with a selection of 11 assessment modules known as Junior Assessment of Mathematics (JAM). For levels two to five, there is a link to the Assessment Resource Bank, <http://arb.nzcer.org.nz>. Of particular interest in NZmaths is the inclusion of an online bilingual assessment tool, known as e-asTTle, which was developed to assess students' achievement and progress in reading, mathematics, writing, and in *pānui, pāngarau* and *tuhituhi* (see <http://e-asttle.tki.org.nz/About-e-asTTle/assessment_for_learning>).

CLASSROOM SNAPSHOT 10.1

Alex's Year 1 class was working on the 'friends of 10' concept (see Figure 10.1). Using numeral necklaces, he previously had asked the children to find their 'friend of 10' – for example, if a student was wearing a '2' necklace, they needed to find

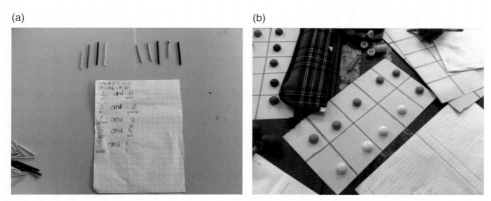

Figure 10.1 Friends of 10 activities

an 8 to match with. Alex decided he needed to assess the students' understanding of this concept before moving on. He usually wrote a series of numbers from 1 to 10 on the whiteboard and then asked the students to find the matching 'friend of 10'. This time, he decided to give the students a range of materials such as 10-frames and counters, coloured toothpicks and a 'Friends of 10' iPad application, <http://www.littlemonkeyapps.com/friendsoften>. He asked them to find all the 'friends of 10' and record them. Alex observed some very interesting differences in the way students tackled the task.

PAUSE AND REFLECT

Before reading on, what do you think Alex may have observed while watching the students solve the task?

CLASSROOM SNAPSHOT 10.2

As the children manipulated the variety of materials in groups, Alex made the following observations:

- Some students counted each square covered with a counter and then counted the squares without counters by touching each square.
- Some students used the counters for the first number, and then subitised to find the second number.

- Some students used the counters to find a pair such as 4 and 6, and later repeated the process using the counters for 6 and 4.
- Some students identified 3 and 7, for example, and then instantly recorded 7 and 3 without using counters.
- Some students did not use counters at all, and recorded each of the 'friends of 10' in a systematic order.

As Alex was observing the children while they worked through the task, he applied his pedagogical content knowledge to provide the students in two of the groups with a harder problem. He went on to ask these students to find all the 'friends of 16' and 'friends of 20' (see Figure 10.2).

Figure 10.2 Using counters to assist in finding friends of 16

This simple activity highlights the positive nature of the assessment for learning practices. The task was part of the everyday teaching and learning sequence, the students were unaware of being 'formally' assessed, the task could be varied to suit the different levels of understanding and the teacher could make decisions concerning the appropriate direction of the activity during the lesson. Students received positive and constructive feedback promptly, and the recorded work samples provided evidence of the students' thinking at this point.

The following activity has been explored in numerous classrooms for many years, and versions of it can be obtained from the Australian Curriculum website, the Assessment Resource Centre and the Maths300 website. These sites are examples only, as each state and other national documents have similar materials.

Classroom snapshot 10.3 targets a middle school content area involving the concept of finding the volume of a cylinder. The scenario in Classroom snapshot 10.3 is an authentic assessment for learning task, as the students had been exploring volume concepts and had focused upon the cylinder in the last week. While this concept is usually explored in the lower secondary years, it provides a useful teaching example for primary teachers to consider, as it puts us back in the role of student as we play around with our response to this task.

CLASSROOM SNAPSHOT 10.3

Megan wanted to assess each student's level of understanding at this stage before moving on to the next unit of work. In the past, she had produced a series of routine questions similar to those they had completed in class. The students were usually told that they would be doing an assessment task, and the teacher was aware that some students displayed anxiety when placed in a test situation.

After some searching, she found an assessment task on the Assessment Resource Centre website, <http://arc.boardofstudies.nsw.edu.au>, known as Mr Tall and Mr Short. Megan decided to use this activity as an assessment task, with the students being unaware of the importance of the activity for assessment purposes. The students entered the room as if for any other activity that week.

The activity began with a class vote about whether students thought a cylinder, formed by rolling an A4 piece of paper lengthways (Mr Tall), would have a greater volume than a cylinder formed by rolling the paper widthways (Mr Short), or whether the volumes of the cylinders would be the same. An anonymous vote was held, and most students thought they would be the same.

Megan had prepared cardboard versions of the cylinders with a circular face to attach to the base. The students observed as Megan filled Mr Tall with rice and then poured the rice into a measuring cylinder. Rice was then poured into Mr Short until it reached capacity, and that rice was poured into a second cylinder. Most students were amazed that Mr Short held more rice, and thought that Megan had tricked them somehow.

Megan made the most of the students' intrigue, stating, 'Your mission is to prove mathematically which of these containers will hold the most – Mr Tall or Mr Short. You may each have two pieces of paper, one for measuring and one for recording your response. I would like you to use your best mathematics communication skills and also use diagrams to assist your explanation. You may use a ruler and a calculator. For this activity, I would like you to work individually.'

ACTIVITY 10.1

Before reading any further, and using the materials outlined in the scenario so far, write your own solution to: 'Which will hold the most – Mr Tall or Mr Short? Or do they have the same volume?' Make sure you clearly show why you have reached that conclusion.

Quality of student responses

The next phase in the assessment process will be considered in the context of the cylinder task. This phase occurs upon completion of any assessment task or observed student response during any mathematics activity, and involves evaluating each student's response. It is generally straightforward when marking closed items that require a single response, but even this type of question requires the teacher to consider the complexity of the thinking required to answer the question and the process used to determine the solution. Some teachers opt to use pre-designed assessment rubrics; however, these are often general in nature and not specific to the content area being observed.

CLASSROOM SNAPSHOT 10.4

Megan considered a sample of student responses to the Mr Tall and Mr Short task, and placed them into the following groups.

- *Group 1:* These students attempted to draw the cylinders and labelled one Mr Short and one Mr Tall. A selection of unrelated measurements were recorded by the students, such as the height of each cylinder, for no specific purpose. There was no attempt to calculate the volume.
- *Group 2:* These students drew each of the cylinders. The students attempted to use the formula, $V = Ah$, but instead multiplied the base and height of the original A4 piece of paper and came to the conclusion that the cylinders did have the same volume.
- *Group 3:* These students focused upon substitution into the correct equation, $V = Ah$, but the radius was taken as half of the circumference (half the base length of the A4 piece of paper).

- *Group 4:* These students used the correct formula; however, the radius was not accurately found as these students attempted to position their ruler to find the centre of the circular base. These students did not use the circumference of the circle to find the radius by dividing the circumference of the base by 2π.
- *Group 5:* These students used the relationship between circumference and radius to find the radius accurately. The students used what is known as reversibility to find the radius (Figure 10.3), and correctly divided the circumference of the base of the cylinder by 2π. With the correct use of the formula $V = Ah$, the students identified and justified that Mr Short had a greater volume than Mr Tall.

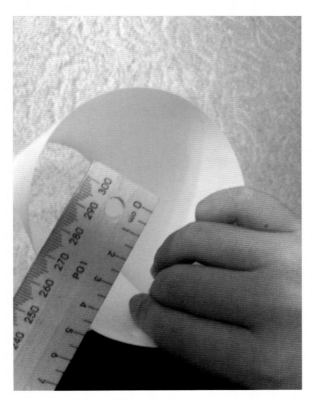

Figure 10.3 Estimating the radius of the circle

While this concept area is usually taught in the lower secondary setting, it provides a detailed example of the difference between quoting a right or wrong mark, or even a mark out of 10, and an analysis of the quality of the response indicating the level of understanding of a particular concept. One could place a numerical value on the responses of Groups 1 to 5 above, but the marks would mean nothing without the description of achievement alongside the responses.

All the previous chapters have used developmental frameworks or cognitive structures that have informed the teaching practice discussed. A variety of developmental models inform assessment. This section considers three of these: Newman's (1977) Error Analysis, Blank's (2002) questioning framework and the SOLO model (Biggs & Collis 1982). The next section provides an introduction to the developmental models to assist teachers in their interpretation of students' responses.

Across the education sector, teachers are discovering the work of Newman (1977), who suggests five prompts to assist in determining students' errors with attempting to solve word problems (White 2005). The prompts that can be used in primary and secondary contexts are shown in Table 10.1.

Table 10.1 Newman's prompts

Newman's prompts	Basic steps in finding a solution to a mathematics problem
1 Please read the question to me. If you don't know a word, leave it out.	1 Read the problem (R). (About 2 per cent of errors occur at this stage.)
2 Tell me what the question is asking you to do.	2 Comprehend what is read (C). (About 12 per cent of errors occur at this stage.)
3 Tell me how you are going to find the answer.	3 Transform the words of the problem into an appropriate mathematical strategy (T). (About 50 per cent of errors occur at this stage.)
4 Show me what to do to get the answer. 'Talk aloud' as you do it, so that I can understand how you are thinking.	4 Apply the process skills required for the chosen strategy (P).
5 Now, write down your answer to the question.	5 Encode a written answer in an acceptable form (E).

Source: Adapted from *NSW Department of Education and Training Curriculum* (2012).

After interviewing the student and working through the five prompts, the attempt can be coded as R, C, T, P or E, depending on where the student error took place. The sample word problem below demonstrates the use of the five prompts in the context of ducks on a pond.

SAMPLE WORD PROBLEM

There were nine ducks on a pond. They were joined by some other ducks, making a total of 15 ducks on the pond. How many ducks joined the group?

- *Step 1 – Decoding:* Can you read the problem to me? (picks up decoding issues)
- *Step 2 – Comprehension:* What is the problem asking? (retelling)
- *Student response:* I need to find out how many ducks came to the pond.
- *Step 3 – Mathematising:* What mathematics do you know that could help?
- *Student response:* adding up (child doesn't recognise that this is a subtraction problem – intervention needed).

OR

- *Student response:* If I take away the 9 from 15 it should tell me how many ducks joined. (Child can correctly turn the problem into mathematical strategy.)
- *Step 4 – Process skills:* Show me how you would do it.
- *Student response:* Child attempts to count back from 15 to 9 using fingers and answers '7'. (Child needs to move to a more efficient mental strategy.)

OR

- *Student response:* Child attempts to write a vertical algorithm but takes the 5 away from 9 and gets an answer of 14. (Work with number lines might help.)
- *Step 5 – Encoding:* What have you found out?

OR

- *Step 5 – Encoding:* Draw me a picture to show what you found (usually little intervention needed for straightforward word problems).

The work of Blank (2002) provides another model for questioning that describes four levels of question types:

- *Level 1:* Directly supplied information
- *Level 2:* Classification
- *Level 3:* Reorganisation
- *Level 4:* Abstraction and inference.

The patterns and algebra questions that follow provide an example for each level of questioning.

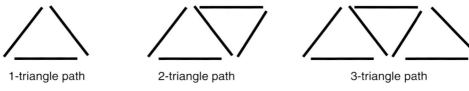

1-triangle path 2-triangle path 3-triangle path

Figure 10.4 Triangle path pattern

- Level 1: How many matches are used for each triangle?
- Level 2: What is the pattern?
- Level 3: How many more matches are needed each time?
- Level 4: How many matches are needed for 50 triangles?

ACTIVITY 10.2

Write an appropriate question for each level in Blank's (2002) question framework.

Level	Label	Example	Add one of your own
1	Directly supplied information	7 What number is this?	
2	Classification	27 − 29 = What kind of problem is this?	
3	Reorganisation	42 = 7 × 6 How could you write this as a division problem?	
4	Abstraction and inference	If you multiply an odd number by an even number, what kind of number do you get?	

The third model considers the quality of students' responses in terms of structure, and provides a guide to designing tasks. You will notice elements of the previous models within the description of the SOLO model that follows.

The SOLO model

The Structure of the Observed Learning Outcome (SOLO) model has been identified as extremely useful in assisting with assessment task design, evaluating the quality of the student response and determining the direction of the follow-up student activities. This model was developed by Biggs and Collis (1982) and extensively built on by researchers such as Pegg and Davey (1998), Serow (2007a) and Watson et al. (1995). It utilises the modes described by Piaget (see below), but acknowledges that previously developed modes are still available to learners as they mature. While the model has been researched to a very detailed level, only its introductory global key characteristics will be explored in this chapter.

The SOLO model grew from the desire of Biggs and Collis (1982) to explore and describe students' understandings in the light of criticisms of the work of Piaget. Rather than focus on the levels of thinking of students, the emphasis in the SOLO model is on the structure of students' responses. The framework consists of two main components: the modes of functioning and the levels of the model.

The five SOLO modes represent the level of abstraction of a response. The characterisation of the SOLO modes appears similar to the Piagetian developmental stages; however, there are some fundamental differences. The SOLO model characterises the mode of functioning utilised within a response, as opposed to characterising the person as at a single developmental mode; newly acquired modes do not subsume previously acquired modes; and multi-modal functioning is possible. The modes of functioning are described below. Each has been included, but you can expect frequent observation of the first three modes of functioning in the primary setting. Basically, this means that we acknowledge the support of reactions to the physical environment, and images and verbal language, when working with written symbols and solving problems using these symbols.

Modes of functioning

- *Sensori-motor*: The response involves a reaction to the physical environment. It is associated with motor activity, and can be described as tacit knowledge. Examples include a child learning to walk or an adult playing sport.
- *Ikonic*: The response involves the internalisation of images and linking them to language. There is a reliance on images and development of language, and thinking in this mode can be described as intuitive knowledge. Examples include a child developing words for images, and an adult's creation of science fiction images.
- *Concrete symbolic*: The response involves the application and use of a system of symbols, which can be related to real-world experiences. This abstraction enables concepts and operations that are applied to the environment to be manipulated through the medium of symbolic systems – for example, written language and number problems.

- *Formal:* The response involves the consideration of abstract concepts, as there is no longer a need for a real-world referent. The formal mode is characterised by a focus on an abstract system, based on principles in which concepts are embedded.
- *Post-formal:* The response involves the challenge or questioning of abstract concepts and theoretical perspectives of the formal mode.

The SOLO modes represent developmental growth, with the acquired SOLO modes remaining accessible and continuing to evolve while supporting other modes.

Levels of the model

The five levels of the SOLO model describing the complexity of the structure of a response are prestructural (P), unistructural (U), multistructural (M), relational (R) and extended abstract (EA). The levels can be identified by observing the structure of an individual's response to a given task. While unistructural, multistructural and relational responses appear in each of the modes of functioning, a prestructural response is typified as at a lower level of abstraction than is required for a task. An extended abstract response to a task requires higher synthesis and application than is reasonably expected. The five levels are described below:

- *Prestructural:* The response is below the target mode. In an attempt to provide a response, the learner is misled or distracted by irrelevant aspects of the task and responds in a lower mode. A typical response may be 'The square is like a box.'
- *Unistructural:* The response is characterised by a focus on a single aspect of the problem/task. Since only one relevant piece of information is utilised, the response may be inconsistent. A typical response may be 'A square has all sides equal.'
- *Multistructural:* The response is characterised by a focus on more than one independent aspect of the problem/task. No relationships are perceived between the components utilised. A lack of integration is evident and some inconsistency is apparent. A typical response may be 'A square has all sides equal, four right angles, all sides are parallel, and two pairs of opposite sides.'
- *Relational:* The response is characterised by a focus on the integration of the components of the problem/task. The relationships between the known aspects are evident, with consistency within this system. A typical response would be 'A square has all sides equal and one right angle. I don't need to say all of them, because if it has that, the others will be equal and they will also be parallel.'
- *Extended abstract:* The response is taken beyond the domain of the problem/task and into a new mode of reasoning.

Within each mode, there exist cycles of levels. For example, researchers have found two cycles within the concrete symbolic mode. These are acknowledged here, but are not discussed further in the context of this chapter.

If we revisit the five groups of responses to the Mr Tall and Mr Short assessment task in Classroom snapshot 10.3, we can interpret the responses using the SOLO model in the following way:

- *Group 3:* Concrete symbolic mode – unistructural level
- *Group 4:* Concrete symbolic mode – multistructural level
- *Group 5:* Concrete symbolic mode – relational level.

The SOLO model provides a framework through which we can design questions that elicit responses at a target level or are accessible to a range of levels.

Construction of assessment tasks

The categorisation of students' responses to tasks, rather than the categorisation of the individuals, requires careful construction of assessment items. Both closed and open assessment items are applicable to the SOLO model. However, the most appropriate method for eliciting optimum responses depends upon the type of investigation being carried out. In the same manner that some students think of the equals sign as meaning 'the answer is', there are many community members who believe that mathematics assessment items and test questions only have one answer. Some people find it difficult to believe that there can be many legitimate ways of solving a problem. This is particularly evident in relation to formal algorithms.

Open-ended items, also known as free-response items or constructed-response items, require the student to 'create a response rather than select it from a list' (Collis & Romberg 1991, p. 84). 'Open-ended and free response questions … require the student to generate the correct answer, not merely to recognise it. Such assessment items would … allow for more reliable inferences about the thought processes contributing to the answer' (Alexander & James 1987, p. 23). Added to this, open-ended assessment items are enhanced further by seeking justification and clarification from the student about the way they have tackled the task. With this notion in mind, the New Zealand assessment material includes a link to diagnostic interviews to provide a window for viewing students' levels of understanding and the strategies they are using. This is known as the Individual Knowledge Assessment of Number (IKAN) (see <http://www.nzmaths.co.nz/ikan-forms>).

An example of an extended open-ended activity involves students coming up with as many different designs for placing four identical cubes together as possible, with complete faces touching only and with the ability to stand freely. The Maths300 lesson The Architect's Puzzle, <http://www.maths300.com>, places the student in the context of being an architect with a design scope and costing schedule. This task also illustrates the development of literacy skills within a mathematics exploration.

Task construction is considered in the remaining sections of the chapter in the context of making the most of national testing feedback and items, syllabus support material and online resources.

National testing

In political circles, school classrooms, school staffrooms and students' homes, it is common to hear discussions concerning national testing. In Australia, the National Assessment Program – Literacy and Numeracy (NAPLAN) is an assessment that has been administered to Years 3, 5, 7 and 9 since 2008. While the NAPLAN assessment covers aspects of reading, writing, spelling and numeracy, this chapter is concerned with the numeracy section only. The same written assessment is undertaken across the nation each year. This in itself highlights issues of equity that have been debated extensively. The tests cover the main areas of the curriculum that are considered essential and common; these are described as number, algebra, function and pattern, measurement, chance and data, space and working mathematically. 'Working mathematically' is a term that describes components of working as a mathematician, such as communication, reasoning, reflection and technology application. The numeracy test contains multiple-choice items and short-response items. Generally, each item is not accessible to a range of abilities, and is recorded as correct or incorrect. The primary tests are non-calculator for all items, and the secondary test has calculator and non-calculator sections.

NAPLAN describes minimum standards for numeracy in terms of skills and understandings at each particular year of schooling. An example of a standard follows:

> Students recognise common two-dimensional shapes and three-dimensional objects, describing them using both everyday language and geometric names. They sort and group them using common characteristics, draw sketches and construct reasonable models using a range of materials, drawing tools and

other technology. They recognise angles both as parts of shapes and objects, and in turns. (<http://www.nap.edu.au/naplan/statements-of-learning.html>)

Figure 10.5 shows a sample of the NAPLAN student performance feedback to parents.

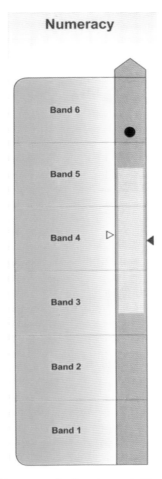

Figure 10.5 Sample NAPLAN communication to parents

The black dot indicates the individual student result, the filled triangle indicates the national average and the unfilled triangle indicates the school average. Alongside the standards, students are reported on scales used for comparison. These initially were developed as a means of measuring growth over time; however, such results have frequently been used in the political arena to compare schools and for funding purposes. These are presented through the MySchool website (<http://www.myschool.edu.au>). At this point, it needs to be made clear that AAMT (2008, p. 8) 'prohibits the publication of league tables of schools from their data' and that the information provided to schools should be statistically legitimate.

In contrast, for NAPLAN results to be used positively requires three levels of analysis: individual students, year trends and school trends. For example, the NAPLAN results may indicate differences between particular classes of students or general trends across a school.

CLASSROOM SNAPSHOT 10.5

In 2005, a primary school received its NAPLAN results. When compared with the rest of the state, it was performing just above the state average. The results could now be filed. The principal decided to take a closer look at the results before leaving the matter alone. When she did, she noticed that the test area that was the poorest in terms of performance across Year 5 was space and geometry. After discussions with staff, it was revealed that this was an area of shared concern, and teachers felt that they were not developing the geometrical concepts to a deeper level. This became a school focus in terms of professional development and school resourcing. The following two years of NAPLAN results indicated a growth in the area of geometrical concepts.

Figure 10.6–10.10 show some sample items from the NAPLAN website. You will notice that they are generally closed in nature, and can be adapted for classroom use.

$$43 - 27 = \boxed{}$$

Figure 10.6 Addition sample NAPLAN item

6 groups of 5 pens is the same number of pens as 3 groups of

| 10 | 6 | 5 | 3 |
| ○ | ○ | ○ | ○ |

Figure 10.7 Multiplication sample NAPLAN item

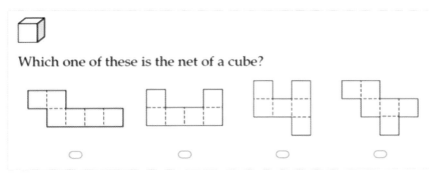

Figure 10.8 Nets sample NAPLAN item

A movie ticket for one adult costs $12.
A movie ticket for one child is three-quarters of the cost of an adult ticket.

What is the cost of tickets for **two** children?

$ ☐

Figure 10.9 Fraction sample NAPLAN item

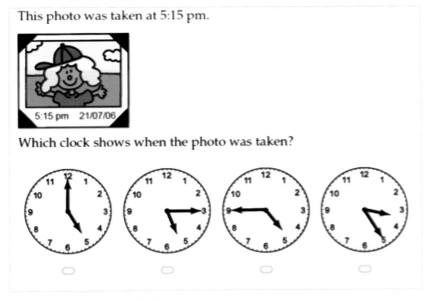

Figure 10.10 Time sample NAPLAN item

Some practical strategies for developing numeracy concepts without drilling students through NAPLAN question types include open-ended items for exploration, and providing opportunities for assessment for learning across all strands of the syllabus. This builds on the work of Sullivan and Lilburn (2004), who have developed an extensive collection of open-ended items that can be used in the mathematics classroom across all primary year levels. Examples of open-ended questions are:

- The area of the rectangle is 36 cm². What could the dimensions be?
- Draw as many different nets for a cube as you can.
- Provide the students with a graph. What could the graph represent?

A simple strategy for creating open items provided by Sullivan and Lilburn (2004) is to take a closed question, write the answer and construct a question that uses the answer in it. For example, instead of asking the students to find the area of a rectangle with certain dimensions, give them the answer and ask them to find the dimensions (as in the first example above). Another strategy involves taking a standard question and opening it up to have multiple answers (as in the third dot point example above). This strategy would also be useful in the context of opening up the multiple-choice assessment items included in the Progressive Achievement Tests (PAT) (see <http://www.nzcer.org.nz/tests/pats>), which are suitable for multiple year levels (Years 3 to 10) to assess student progress in the content areas of number knowledge, number strategies, algebra, geometry and measurement and statistics.

ACTIVITY 10.3

Take each of the closed sample NAPLAN items in Figures 10.6–10.10 and create an open-ended item to be used as an assessment item targeting the same concepts. You might like to match the concept covered with the appropriate outcome/s in your relevant syllabus.

The online Assessment Resource Centre, <http://arc.boardofstudies.nsw.edu.au>, provides a wealth of assessment for learning tasks that cover many of the primary topics. These are designed in a manner that is open to change to meet the interests of students in your class and the context of their community. The design of the items can also be adopted within different content areas. The discussion and activities below demonstrate the adaptation of the tasks.

ACTIVITY 10.4

Assessment task: Spotty Octopus

This activity, available at <http://arc.boardofstudies.nsw.edu.au/go/k/maths/activities/spotty-octopus>, is based on a scenario presented to the students. It is centred around an octopus with eight legs, 16 spots and the same number of spots on each leg. The children are asked to draw a picture to solve their problem and record their answer.

Play around with the context of this question:

1 Could you use something else instead of an octopus?
2 How could you have different students doing the same structured task but at different levels?
3 Is there a way you could have some students working on a response that involves groups of three, or a larger number in the total collection to be divided?

ACTIVITY 10.5

Assessment task: Halves

In this task, available at <http://arc.boardofstudies.nsw.edu.au/go/k/mathematics/activities/halves>, students are asked to make a ball with any available modelling materials that can be moulded by hand. Roll the material out, make a shape, then cut the shape in half. The important component of the task is that the students are asked to reflect on how they created a half and whether the shape they made first is actually cut in half.

1 What materials could you use if modelling materials such as play dough are not available?
2 As the children's understanding of fractions progresses, what fraction concepts could be targeted when adapting this activity?
3 How could you adapt this assessment task using an online geoboard app, such as the iPhone, iPad and web app developed by the Math Learning Center, <http://catalog.mathlearningcenter.org/apps/geoboard>?

ACTIVITY 10.6

Assessment task: Making number patterns

This activity, which can be found at <http://arc.boardofstudies.nsw.edu.au/go/stage-1/maths/activities/making-number-patterns>, is described as targeting lower primary patterns and algebra; however, the structure of the assessment task could be used in the context of number patterns on the decade, off the decade, counting up or down by any number, using fraction and decimal notations of various complexities, or even algebraic patterns in the secondary context.

It is important to bear in mind that a good assessment for learning task fits seamlessly into the teaching/learning sequence. The sequence is described on the website as follows:

- Students make a number pattern that increases or decreases.
- They explain their number pattern in words and record this explanation in writing.
- Students continue their number pattern.
- They then create another number pattern that has a particular number in it – for example, 10.

In the last step of the sequence, what elements could you ask the students to include in their number pattern at various stages of development?

ACTIVITY 10.7

Assessment task: What could the question be?

This assessment task, which can be found at <http://arc.boardofstudies.nsw.edu.au/go/stage-1/maths/activities/what-could-the-question-be>, is applicable to many concepts in mathematics. It is often described as providing the students with a scenario – such as, 'The answer is 16. What could the question be?' Students are asked to record as many questions or word problems as possible.

CLASSROOM SNAPSHOT 10.6

Miss Mack began a unit on chance by playing a class game of Snakes and Ladders, then asked the students to respond to a scenario in their journals. The scenario was, 'Do you think it is harder to throw a 6 or a 4 on a die?' Two of the students' responses are shown in Figures 10.11 and 10.12:

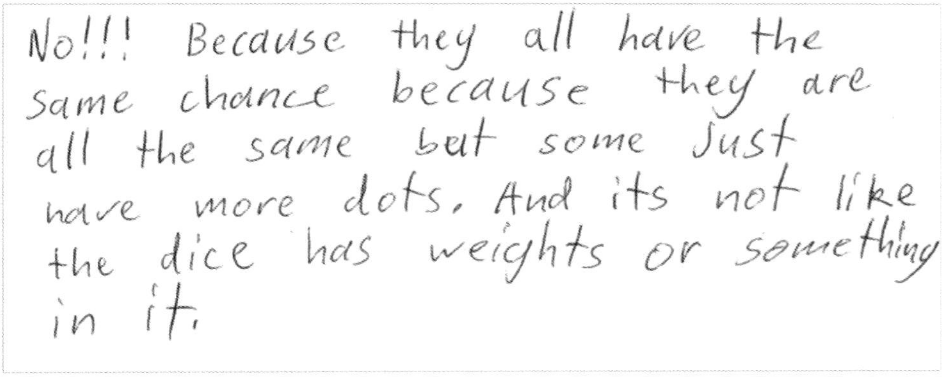

> No!!! Because they all have the same chance because they are all the same but some just have more dots. And its not like the dice has weights or something in it.

Figure 10.11 Chance student sample 1

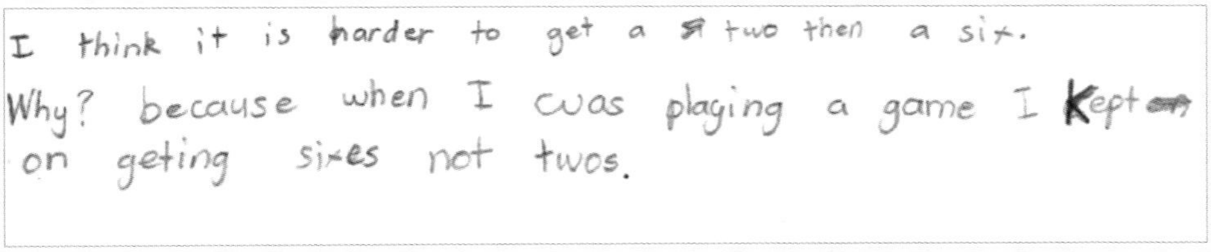

> I think it is harder to get a two then a six. Why? because when I was playing a game I kept on geting sixes not twos.

Figure 10.12 Chance student sample 2

Carefully consider the two responses to the task in Classroom snapshot 10.6. What do the responses tell you about the students' understanding in relation to this task?

How would you describe the quality of the students' responses? In the light of these two answers, what chance activities would you design to follow on with?

It is possible to source stimulus items from a wide range of technological tools and use these to create an interesting investigative assessment task. For example, in a student investigation of fish growth, students compared genetically enhanced fish with non-genetically enhanced fish. The students were required to display data appropriately using TinkerPlots software, and to make generalisations based on what they had identified. The students were also required to justify their answers.

Online assessment item banks

Another way of sourcing online assessment items is through an assessment item bank. The HOTmaths website provides a number of assessment item banks: assessment questions within each lesson, topic quizzes within each topic, and a Test Generator that allows items from the entire site to be accessed. These assessment tools provide opportunities for student self-assessment and for teacher-driven assessment, and give both students and teachers a means of tracking what has been attempted and achieved. Reports for tests created with the Test Generator include direct access to the actual marked papers for student and teacher review, as well as reports on student and class results.

The banks of assessment items are written at four levels of difficulty – Level 1, Level 2, Level 3 and Challenge level – for each concept. There are three standard item structures: input, multiple-choice and multi-multi-choice (where the student chooses all the correct options).

Descriptions of each level and example questions for the concept of a fraction are found below. These example questions are all from a HOTmaths Year 3 lesson. You can access this particular lesson by logging into HOTmaths and selecting 'Cambridge Primary Maths' as the Course list and then 'Aus Curric 3' as the Course. Then choose 'Fractions' as your Topic, 'Fractional parts' as your Lesson, and finally select the 'Questions' tab.

- *Level 1*: All answers found in the lesson notes and cover definitions and simple skills.

You can write one half as:

a. ○ $\frac{2}{3}$

b. ○ $\frac{1}{2}$

Figure 10.13 Level 1 question from HOTmaths (AC Year 3)

- *Level 2:* This level usually includes single-step questions covering the basic skills using familiar questions *only from this lesson*, including from walk-throughs and lesson notes. They are similar to questions found in the lesson notes and walkthroughs.

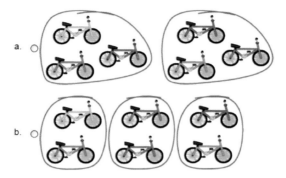

Figure 10.14 Level 2 question from HOTmaths (AC Year 3)

- *Level 3:* This level includes multi-step questions or unfamiliar contexts for the questions, including questions incorporating skills from *earlier lessons in this topic*. These questions may require the use of pencil and paper. Students will need to understand the maths in this lesson as well as aspects of earlier lessons in this topic.

This rectangle is divided into 5 equal parts.

What fraction is each part?

Figure 10.15 Level 3 question from HOTmaths (AC Year 3)

- *Challenge level:* This level includes problem-solving, integration of other topics or key learning areas, application and interpretation of working mathematically using concepts and extension to the next stage of a concept. These questions challenge you to think outside the square, bringing in skills from other topics (or working with others) to analyse the question and find a solution. These questions are similar to the volume of a cylinder task presented earlier in the chapter.

Joti was given $\frac{1}{4}$ of a packet of lollipops, which was 5 lollipops.

How many lollipops were in the packet?

☐ lollipops

Figure 10.16 Challenge level question from HOTmaths (AC Year 3)

The structure of building the complexity of a task through a series of connected items was explored by Collis and Romberg (1991), who produced a selection of super items in an attempt to design assessment items that addressed issues concerning open and closed questions. The tasks contain a stem with four questions specifically targeting the levels described in increasing order of conceptual difficulty. While each of the questions may be answered independently, it is expected that a correct response to a higher level question would be achieved after the successful completion of the earlier responses.

While the HOTmaths items don't align exactly with the super item system, the same principle is used. In the record section, students and teachers are given the correct number achieved out of the number attempted. On screen in the question section, students are shown the result of their last attempt at each question. Teachers can see the total correct out of the total number of students who have attempted the question in each class. Students achieve a star or a flame if they answer four or more questions correctly at that level. These totals are cumulative, and may be achieved over several sessions.

Various online programs exist that provide students with a bank of mathematics problems to motivate them to develop automaticity of responses. One example of this is the Scorcher section on the HOTmaths website. This can be accessed via the 'Scorcher' tab within any HOTmaths lesson. In terms of assessment for learning, this is an excellent resource to be used as a lesson starter to focus the students on the content area, and to promote discussion of the concepts to gain an understanding of where the students are at.

With the aid of a data projector, questions such as this work very well in a whole-class situation, where questions, processes and answers can be discussed. This brings these questions into a setting different from that of the individual test against the clock. A Scorcher example is provided in Figure 10.17.

Towards the end of the unit, it would be suitable to ask students to create triangles with a certain area. For example, provide the students with the scenario that a triangle has an area of 36 square units. Ask the students to draw

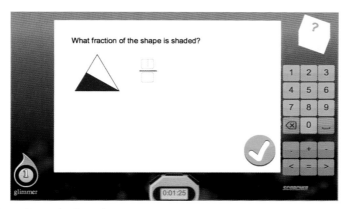

Figure 10.17 HOTmaths Scorcher question

as many different triangles with this area as they can, and to record the base and height of each triangle. This is an example of an assessment task that requires the students to utilise reversibility, and hence work relationally to find each of the triangles.

Assessment as learning

The notion of assessment as learning was flagged at the beginning of the chapter, and, like assessment of and for learning, assessment as learning should be an element of all assessment processes. Statutory educational bodies such as the Queensland Curriculum and Assessment Authority (QCAA), <http://www.qcaa.qld.edu.au/yr5-maths.html>, provide practical suggestions for scaffolding assessment as learning strategies in the mathematics classroom. They particularly target ways to develop confidence in students so they will actively become involved in the development of learning goals. These include:

- learning logs or learning journals
- 60 seconds to think about how your learning is going so far
- concept maps identifying areas requiring further exploration.

Reflection

In reading this chapter and completing the activity tasks, you will have explored the importance of seamlessly integrating assessment tasks into the teaching and learning sequence. You may have also noticed that it is often not possible to place assessment for learning and assessment of learning tasks into two separate baskets. It is how the teacher designs the task, presents the task and communicates the feedback that makes all the difference. NZmaths provides a sound model for providing curriculum documents and support material that adheres to assessment for learning as an underlying philosophy, thus demonstrating to many education systems that it is possible to have a valid assessment system without external national testing. We have explored NAPLAN items and communication of results, and considered cautiously how we allocate time for familiarisation with such tasks in the classroom without practising items to the extent that test anxiety is evident. In fact, we can use these items to our advantage by considering opening them up and aiming for a deeper understanding of the underlying concepts. A theoretical framework has been introduced from which we can analyse students' responses and design follow-up activities at an appropriate level. Various technological sources of assessment stimuli have also been explored.

Websites for exploration

Assessment Resource Bank (NZmaths): <http://arb.nzcer.org.nz>

Assessment Resource Centre: <http://arc.boardofstudies.nsw.edu.au>

Australian Curriculum (for work samples and portfolios): <http://www.australiancurriculum.edu.au/Mathematics/Curriculum/F-10>

Count Me in Too: <http://www.curriculumsupport.education.nsw.gov.au/primary/mathematics/countmeintoo/online>

e-asTTle <http://e-asttle.tki.org.nz/About-e-asTTle/assessment_for_learning>

Friends of 10 iPad application: <http://www.littlemonkeyapps.com/friendsoften>

HOTmaths: <http://www.hotmaths.com.au>

Maths300: <http://www.maths300.esa.edu.au>

National Assessment Program – Literacy and Numeracy (NAPLAN): <http://www.nap.edu.au/naplan/naplan.html>

New South Wales Department of Education and Training, Newman's prompts: <http://www.curriculumsupport.education.nsw.gov.au/primary/mathematics/numeracy/newman>

NZmaths Assessment Material: <http://www.nzmaths.co.nz/mathematics-assessment>

Progressive Achievement Tests (PAT): <http://www.nzcer.org.nz/tests/pats>

Queensland Curriculum and Assessment Authority: <http://www.qcaa.qld.edu.au/yr5-maths.html>

TinkerPlots: <http://www.tinkerplots.com>

CHAPTER 11

Capitalising on ICT in the mathematics classroom

LEARNING OUTCOMES

By the end of this chapter, you will:

- be able to consider the role that digital technologies play in today's classrooms
- have gained an understanding of how technology impacts upon mathematics teaching
- understand how ICT is described and integrated in Australian Curriculum: Mathematics and New Zealand documents
- be able to develop strategies for supporting students in their use of ICT
- have gained an insight into the way technology can be incorporated into classroom routines to enhance learning experiences for students.

KEY TERMS

- **Digital technologies:** Any technology controlled using digital logic, including computer hardware and software, digital media and media devices, digital toys and accessories, and contemporary and emerging communication technologies
- **General capabilities:** A key dimension of the Australian Curriculum: Mathematics; addressed explicitly in the content of the learning areas
- **HOTmaths:** An interactive online mathematics learning and teaching resource for students, teachers and parents
- **Maths300:** An online resource containing over 170 inquiry-based lesson ideas for teachers of students from Foundation to Year 12

- **Technological pedagogical content knowledge (TPACK):** A framework developed by Koehler and Mishra (2008a), which describes how teachers' understandings of technology and pedagogical content knowledge interact with one another to produce effective teaching with technology
- **Technology metaphors:** A way of describing students' use of technology and technological tools as proposed by Galbraith et al. (2000)
- **TinkerPlots:** Dynamic data software that aims to develop students' understanding of data, number, probability and graphs

This chapter looks at how the use of technology in the mathematics classroom can be planned for and capitalised on to support students' learning. The chapter begins with a consideration of how **digital technologies** can and should be used to deepen students' understanding of mathematical concepts. Links are made with relevant Australian and New Zealand curriculum documents, and the ways in which students make use of technology are examined. The latter part of the chapter uses classroom snapshots, **HOTmaths** and case studies to show how mathematical skills, knowledge and understanding can be developed through using ICT in mathematics lessons.

Digital technologies offer new ways of teaching and learning mathematics that can help to deepen students' understanding and influence the nature of the tasks selected, classroom interactions and the subject of mathematics itself (Goos 2012). Technology can change the nature of school mathematics through engaging students in active mathematical practices such as experimenting, investigating and problem-solving. Pierce and Stacey (2010) used a pedagogical map to represent ways in which technology transformed teachers' mathematical practices. They found that technology provided teachers with opportunities to think about:

- *tasks set for students* – for example, using technology to improve speed, accuracy and access to a variety of mathematical representations
- *classroom interactions* – for example, using technology to change classroom dynamics and the roles of teachers and students
- *the subject taught* – for example, using technology to provoke mathematical thinking or change the sequencing and treatment of mathematical topics.

The Australian Curriculum: Technologies is divided into two subjects – Design and Technologies, and Digital Technologies – and two strands – Knowledge and Understanding, and Processes and Production Skills. Links are also made with numeracy as a **general capability**:

The Technologies curriculum provides opportunities for students to interpret and use mathematical knowledge and skills in a range of real-life situations.

Students use number to calculate, measure and estimate; interpret and draw conclusions from statistics; measure and record throughout the process of generating ideas; develop, refine and test concepts; and cost and sequence when making products and managing projects. In using software, materials, tools and equipment, students work with the concepts of number, geometry, scale, proportion, measurement and volume. They use three-dimensional models, create accurate technical drawings, work with digital models and use algorithmic thinking in decision-making processes when designing and creating best-fit solutions (<http://www.australiancurriculum.edu.au/technologies>).

In New Zealand, technology is not included as a key competency in the New Zealand Curriculum, but is included as a Learning Area (see <http://nzcurriculum.tki.org.nz/The-New-Zealand-Curriculum/Learning-areas/Technology/Learning-area-structure>). It comprises three strands:

- *Technological practice:* Within this strand, students examine the practice of others and undertake their own. They investigate issues and existing outcomes and use understandings gained to inform their own practice.
- *Technological knowledge:* Within this strand, students develop knowledge related to technological enterprises and environments, along with an understanding of how and why systems operate in the way they do.
- *Nature of technology:* Within this strand, students learn to critique the impact of technology on societies and come to appreciate the socially embedded nature of technology.

PAUSE AND REFLECT

An expectation of the Australian Year 8 mathematics curriculum is that students can solve a range of problems involving rates and ratios, with and without digital technologies (ACMNA188). Look for other instances where the use of digital technologies is mentioned. How do you think the learning would be different when and if digital technologies were included?

Access the achievement standards for Level 4 Mathematics and Statistics from <http://nzcurriculum.tki.org.nz/The-New-Zealand-Curriculum/Learning-areas/Mathematics-and-statistics/Achievement-objectives> and Level 4 Technology from <http://nzcurriculum.tki.org.nz/The-New-Zealand-Curriculum/Learning-areas/Technology/Achievement-objectives>. Compare the two. Select an achievement standard for Mathematics and Statistics and identify an appropriate way to incorporate technological knowledge into the standard.

Supporting students' learning in technology

Galbraith and colleagues (2000) developed four **technology metaphors** for the way in which ICT can mediate learning. They provide a useful framework for thinking about the degrees of sophistication with which students and teachers work with technology. A description of the metaphors is provided below.

- *Technology as master*: The student is subservient to the technology – a relationship induced by technological or mathematical dependence; if mathematical understanding is absent, the student is reduced to blind consumption of whatever output is generated, irrespective of its accuracy or worth.
- *Technology as servant*: Technology is used as a reliable time-saving replacement for mental or pen-and-paper computations (e.g. calculator).
- *Technology as partner:* Technology is used creatively to increase the power that students have over their learning; students often appear to interact directly with the technology (e.g. use of graphical calculator), treating it almost as a human partner that responds to their commands.
- *Technology as an extension of self*: Students incorporate technological expertise as an integral part of their mathematical repertoire; technology is used to support mathematical argumentation as naturally as intellectual resources; this is the highest level of functioning.

Goos (2012) provides an example of how the framework can be used to interpret teachers' and students' use of technology in the classroom. In her vignette, she describes how students investigated their level of physical activity over a week using a pedometer. Each day, they entered the data (the number of paces they walked or ran) into a shared Excel spreadsheet and then analysed the data by using the facilities within Excel. They then used charts from Excel to display their data and make comparisons. Observations showed that the students used the spreadsheet as a *servant*, to record, analyse and represent the data. It was also used as a *partner*, in that it mediated discussion between students in relation to differences they observed as they critically compared their own results with those of others and attempted to explain the differences.

> **PAUSE AND REFLECT**
>
> Consider the four metaphors above in connection with your own relationship with technology. How would you characterise your technological skills and expertise?

Considering teacher knowledge

Together with consulting the relevant curriculum frameworks, and bearing in mind how we want students to use ICT, the ability of teachers to capitalise on ICT opportunities in the classroom is also influenced by their own content knowledge and **technological pedagogical content knowledge (TPACK)**. TPACK is described in Chapters 1 and 15, and can be used to look at how teachers' understanding of technology and pedagogical content knowledge interact with one another to produce effective teaching with technology. It includes seven components: technical knowledge (TK), content knowledge (CK), pedagogical knowledge (PK), pedagogical content knowledge (PCK), technological content knowledge (TCK), technological pedagogical knowledge (TPK) and technological pedagogical content knowledge (TPACK). In essence, TPACK represents:

- an understanding of the representation of concepts using technologies
- pedagogical techniques that use technologies in constructive ways to teach content
- knowledge of what makes concepts difficult or easy to learn, and how technology can redress some of the problems faced by students
- knowledge of students' prior knowledge and theories of epistemology
- knowledge of how technologies can be used to build on existing knowledge and to develop new epistemologies or strengthen old ones (Koehler & Mishra 2008b, pp. 17–18).

The first part of this chapter looked at the Australian and New Zealand curricula for mathematics and technology, and discussed some of the factors to be considered when capitalising on ICT within the mathematics classroom. The next part of the chapter uses classroom snapshots, HOTmaths and case studies to show how mathematical skills, knowledge and understanding can be developed through using ICT.

Developing an understanding of place value

Developing an understanding of **place value** and the base-10 number system is a focus in early childhood and primary classrooms, with students in Year 1, for example, expected to recognise, model, read, write and order numbers to at least 100 and locate these numbers on a number line (ACMNA013), or know the forward and backward counting sequences of whole numbers to 100 (see <http://nzcurriculum.tki.org.nz/The-New-Zealand-Curriculum/Learning-areas/Mathematics-and-statistics/Achievement-objectives>). Classroom snapshot 11.1 describes how one teacher, Mrs Knight, uses a variety of classroom routines, planned activities and ICT to develop a conceptual understanding of place value in her Year 1 class.

CLASSROOM SNAPSHOT 11.1

The students in Mrs Knight's class look forward each day to recording how many days they have spent in school for the year. The days are recorded in a place value chart at the front of the classroom, with icy-pole or bundling sticks used to represent each day. When 10 days are reached, these are bundled into 10; when 100 days are reached, the bundles of 10 are bundled into 100. Each day the total is recorded symbolically and a discussion based around the numbers is held.

Another daily routine involves the use of the hundreds chart. Each day, an interactive hundreds board is displayed on the IWB. Mrs Knight uses a variety of daily activities to focus attention on the base-10 number system. For example, on Tuesdays students play a cover-up game where they have to identify the hidden number on the chart by using the numbers surrounding it as clues (see Figure 11.1).

1	2	3	4	5	6	7	8	9	10
11	12	13	14	15	16	17	18	19	20
21	22	23	24	25	26	27	28	29	30
31	32	33	34	35	36	37	38	39	40
41	42	43	☐	45	46	47	48	49	50
51	52	53	54	55	56	57	58	59	60
61	62	63	64	65	66	67	68	69	70
71	72	73	74	75	76	77	78	79	80
81	82	83	84	85	86	87	88	89	90
91	92	93	94	95	96	97	98	99	100

Figure 11.1 Hundreds chart with covered number

Students are expected to explain how they identified the hidden number using place value terminology (for example, 'I know that number 44 is hidden because the number before is 43, the number after is 1 more than 44, 45, and the number 10 below it is 34 and the number 10 above it is 54.' Through using the cut and paste function in Microsoft Word, Mrs Knight can prepare a number of different versions, producing different numbers each day. Alternatively, she can use EasiTeach to set up randomly selected numbers.

HOTmaths has a widget called Hundred Grid that can be used in the same way. You can access this widget by logging into HOTmaths and entering its name in the search field.

Figure 11.2 Screenshot of HOTmaths Hundred Grid widget

Number lineup

Mrs Knight also makes regular use of games to teach place value concepts and to encourage students to work with partners or in small groups. The following game, sourced from <http://mathwire.com/numbersense/placevalue.html>, encourages students to think about how the relative value of a digit changes, depending on its placement.

Materials

- Deck of digit cards for each set of partners (2–4 each of 0–9, depending on the level of students and the size of the numbers they will create). *Note:* Spinners with 0–9 may be used instead of cards, if desired.
- Place value mat for each player.
- Recording sheet, if desired.

Directions

- Partner A turns over the first card and decides where to place that card on their place value mat. Once the card is placed, it may not be moved.
- Partner B turns over a card and decides where to place that card on their place value mat. Again, the card may not be moved once it is placed.

- Play continues, with each partner turning over a card and deciding where to place it on their place value mat in the hope of building the largest number.
- When all slots are filled on the place value mats, partners compare numbers to see who created the larger number. That partner wins a point for the round.
- Partners record both numbers on their recording sheet and circle the larger number.
- Students clear their mats, shuffle the cards and play additional rounds, as time allows.

Variations

Students try to form numbers to meet specified criteria (which will vary from these suggestions, based on the number of digits used):

- Students try to form the smallest number.
- Students try to form a number that is closest to 500 (or 2000 or …).
- Students try to form a number that is less than 1000.
- Students form numbers and earn a different number of points, depending on the range within which the number falls (for example, one point for numbers from 0–500, two points for numbers from 501–1000, etc.)

The interactive game Number Top-it, <http://media.emgames.com/emgames/demosite/demolevel1.html>, reinforces this concept and players can challenge the computer to form the smallest five-digit number (see Figure 11.3). Interestingly, the winner is the player with the smallest number, rather than the one with the largest number.

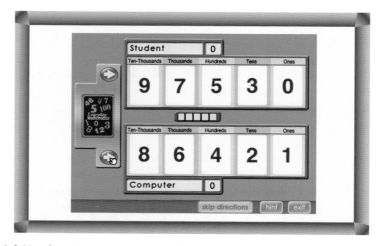

Figure 11.3 Number Top-it game screenshot

Nasty Games

Nasty Games is another favourite of Mrs Knight's class. In this game, students use dice to form three-digit numbers, with the twist being that they can place each rolled digit in any player's space (see Figure 11.4 for game sheet).

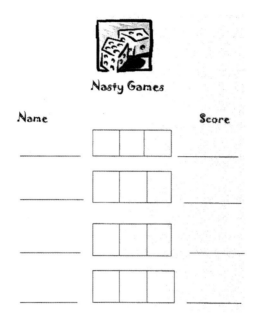

Figure 11.4 Nasty Games game sheet

It is an ideal game to play in small groups, and Mrs Knight often begins these sessions by projecting the game board on the IWB and selecting five students to play against each other while the rest of the class watches and offers suggestions. The whole-class focus allows discussion to occur around 'strategies' and the relative value of the digits. Students can then play the game in small groups and keep a tally of their scores.

PAUSE AND REFLECT

Each of the examples provided in Classroom snapshot 11.1 was designed to provide students with opportunities to engage in exploring the key ideas associated with the place value system. Think about how ICT was used in each instance. Was it used to introduce the concept? Was it used as reinforcement or consolidation? How effective do you think the activities would be without the use of ICT?

Other suggestions for capitalising on ICT in the classroom

Electronic examples of technological activities for each of the grade bands pre-K–2, 3–5, 6–8 and 9–12 that support the US National Council of Teachers of Mathematics' (NCTM) principles and standards are available on the NCTM website at <http://www.nctm.org/standards/content.aspx?id=16909>. Five or six electronic examples for each grade band illustrate for parents and teachers ways in which technology can enhance mathematics learning. Some of the examples include applets that students can use directly, including tangram explorations suitable for the foundation years to Year 2.

E-example 4.4 deals with developing geometry understanding and spatial skills through puzzle-like problems with tangrams. Students manipulate tangram pieces on the screen to fill an outline of a picture, using spatial visualisation and problem-solving skills to flip or rotate shapes (see Figure 11.5).

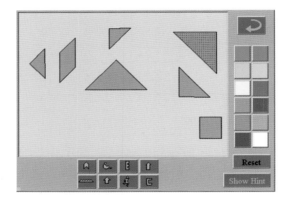

Figure 11.5 Screenshot showing NCTM tangram applet

As the site explains, completing the same or similar puzzles with both physical and computer manipulatives may help students to generalise their experiences. The computer environment is also likely to encourage them to think about how they need to manipulate the tangram pieces rather than approach the task mainly by trial and error. Working with a partner at the computer to complete puzzles also encourages students to become more precise in their use of vocabulary about space. Teachers can enrich students' vocabulary in class discussions by commenting on students' actions, such as, 'I see you are rotating the parallelogram' or 'What difference would flipping make?'

Maths300, <http://www.maths300.com>, contains over 175 lesson ideas suitable for early childhood classes through to Year 12. Approximately 75 of these have supporting software that reinforces and extends the original lesson. For example, the Fraction Estimation lesson uses a clothes line and pegs to encourage students to estimate parts of a nominated whole (length of the clothes line). The teacher distributes a number of pegs to the students and asks them to place their peg on the line where they think, for instance, one-third would go. After a few pegs have been placed, the rest of the class is asked to vote on which peg is closest to one-third.

The task can obviously be adapted to accommodate different fractions and discussions can take place on how to 'measure' who is closest. Estimation of fractions improves the more times the activity is played. Individual practice can then occur through accessing the Fraction Estimation software. Figure 11.6 illustrates a fraction strip that can be manipulated by the Fraction Estimation software to show the nominated fraction. To provide added interest, data show how close estimates are, and students are encouraged to refine their estimates and improve their percentages. Full access to the Maths300 site requires an annual subscription, but a free sample tour is available.

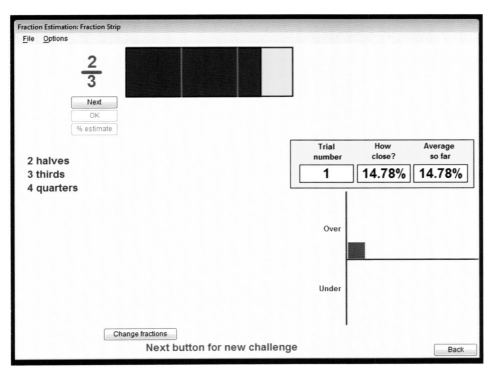

Figure 11.6 Fraction Estimation software from Maths300

> **PAUSE AND REFLECT**
>
> Chapter 15 contains a framework for evaluating digital teaching resources (see Figure 15.3). Refer to this framework to consider the two websites named above:
>
> 1 To what extent are they:
>   - relevant to your country's curriculum
>   - focused on substantive mathematics
>   - able to provide differentiation
>   - engaging or motivating
>   - easy to implement?
> 2 Are teacher support materials included?
> 3 Do you think it is important to evaluate web-based resources in this way? How critical are you of the resources you access online?

HOTmaths as a one-stop shop

The following sequence of activities considers using the HOTmaths website as a one-stop shop. Take note of the nature of the student tasks and the level of support available to teachers to support the design of quality learning sequences. This sequence targets fraction concepts and in the Australian Curriculum: Mathematics addresses the outcome, 'Model and represent unit fractions including $\frac{1}{2}, \frac{1}{4}, \frac{1}{3}, \frac{1}{5}$ and their multiples to a complete whole' (ACMNA058). In the New Zealand Curriculum, this would be relevant to the Year 2 level, 'Know simple fractions in everyday use'.

The sequence of activities can be found by logging into HOTmaths and selecting 'Cambridge Primary Maths as the Course list' and then 'Aus Curric 3' as the Course. Then choose 'Fractions' as your Topic, and finally 'Fractions parts' as your Lesson. When you open the lesson, you will notice that you have notes on the left-hand side divided into 'Student notes' and 'Teacher notes'. When designing a sequence of activities, it is useful to note the tabs on the right of the screen, which provide links to Resources, Walkthroughs, Scorcher and Questions for the particular concept. Following the student notes as a starting point provides an avenue for reviewing, and sometimes developing, mathematical content knowledge (MCK) for the chosen topic. A sample from the student notes for Fractional Parts is provided in Figure 11.7.

This herd of elephants is divided into thirds.

This herd is not divided into thirds because the groups are not equal.

Print *Fractions with groups* 📄 Hotsheet and find fractions of groups of objects.

Click ☑ Solutions to check answers.

Figure 11.7 HOTmaths lesson excerpt showing herds of elephants from Fractional Parts lesson notes

The Teacher Notes section begins with the focus of the sequence. In this case, it is 'Explore fractions as equal parts of regular shapes or as equal parts of collections of groups. Use a variety of models such as the area and collection models.'

This is followed by a list of resources you may need and a language section listing the key terms with links to the dictionary for particular concepts (blue font), shown in Figure 11.8.

Resources you may need

- Print out:
  Shaping fractions 📄 Hotsheet
  Fractions with groups 📄 Hotsheet
- Counters, paper squares, paddle pop (popsicle) sticks, pattern blocks

Language

fractions, **parts, whole,** partitioning, area model, **collection model, fraction symbol notation, halves, thirds, quarters, fifths, sixths, sevenths, eighths, ninths, tenths,** numerator, denominator

Figure 11.8 List of resources and language screenshot

The sequence of activities describes strategies for integrating online tools with opportunities for exploration and class discussions. The section headings provide a guide for the focus of each activity, the first being Partitioning: Area Model, as shown in Figure 11.9.

Partitioning: Area model
EQUAL AND UNEQUAL PARTS

Open the *Shape cutter* ✲ Widget. Model using the ✂ tool to partition a selected shape, the C tool to rotate shapes and the colour palette to colour in parts of shapes and so on.

- Select a shape and cut into equal parts, eg quarters.
- Focus students' attention on the equal size of parts.
- Then select a shape and divide into 4 unequal parts. Ask 'Would each part be one quarter if I cut the shape like this?'
- Review fractions as equal parts.

Figure 11.9 Partitioning: Area Model screenshot

When the widget icon is clicked, the 'shape cutter' shown in Figure 11.10 appears. This cutter could be utilised in a whole-class, small-group or individual activity, and used to explore the area model, rotating model, and compare fractions and symbol notation (Figure 11.11).

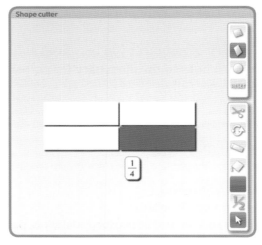

Figure 11.10 Screenshot of HOTmaths Shape Cutter widget

The Teacher Notes suggest student tasks that use concrete materials as an essential component of the sequence. The fraction example shown in Figure 11.12 uses paper-folding techniques to explore fraction concepts. An alternative concrete materials activity is also provided (Figure 11.3).

ROTATING MODEL

Open the *Shape cutter* ✱ Widget.

- Model a range of fractions.
- Use a vertical cut to divide a shape into two equal parts.
- Colour $\frac{1}{2}$ and then rotate the shape to various positions.
- Discuss how the fraction remains constant.
- Use a combination of vertical, horizontal and diagonal cuts to model various fractions and colour the fractional parts.
- Discuss the name of the fractions displayed at each step.
- To label a fractional part, click the button marked $\frac{1}{2}$, then click the part and enter the numerator and denominator, eg $\frac{1}{8}$.

COMPARE FRACTIONS

Open the *Shape cutter* ✱ Widget.

- Select the rectangle shape and divide it into two equal parts.
- Colour one part and ask students what fraction has been created.
- Clear the division, partition the shape into three equal parts and then ask the students what fraction has been created (thirds).
- Focus students' attention on the size of the fraction: an **increase** in the number of parts results in the **decreasing** size of the fraction.
- Ask students which fraction is bigger, $\frac{1}{4}$ or $\frac{1}{2}$.
- Use the widget to model a solution to this task.
- Allow students to create their own scenarios comparing the size of fractions that they can share with the class and investigate on the widget.

Figure 11.11 Rotating Model and Compare Fractions screenshots

USE CONCRETE MATERIALS

Use paper squares and allow students to experiment with folding the paper into equal parts. Students show different ways to fold their shape into:

- halves
- thirds
- quarters
- sixths
- eighths

Modification: Create fraction flags using paper squares and paddle pop (popsicle) sticks stuck to the rear of the flag. Students can describe the fraction created.

$\frac{1}{2}$ of this flag is red and $\frac{1}{2}$ is yellow.

Figure 11.12 Use Concrete Materials screenshot

Alternative activity: Make fractions with pattern blocks.

Each triangle is $\frac{1}{4}$ of the square.

Print *Shaping fractions* 📄 *Hotsheet* to consolidate the ability to represent and recognise fractions as equal-sized pieces.

Click ☑ Solutions to check answers.

Figure 11.13 Alternative Activity screenshot

The first task of the consolidation activity in this 'Use Concrete Materials' section is accessed by clicking on the HOTsheet icon. This HOTsheet is shown in Figure 11.14. The HOTsheet solutions can be accessed by clicking on the solutions icon in the lesson notes and can be found in the resource list on the right-hand side of the page.

HOTsheet

SHAPING FRACTIONS

TASK 1   Draw the fraction

Draw the fraction three ways using 2D shapes. The first row has been done for you.

Fraction	Drawing 1	Drawing 2	Drawing 3
$\frac{1}{2}$	(circle half-shaded)	(square with triangle shaded)	(rectangle with top half shaded)
$\frac{1}{4}$			
$\frac{1}{8}$			
$\frac{2}{3}$			

Figure 11.14 HOTmaths Shaping Fractions HOTsheet

The lesson ends by providing activities exploring symbol notation and the comparison of fractions using groups (Figure 11.5).

> **Symbol notation**
>
> Open *Shape cutter* Widget.
>
> - Focus on the symbolic representation of a fraction, eg $\frac{1}{2}$, $\frac{3}{4}$.
> - Create shapes and write the symbols. Introduce the terms 'numerator' and 'denominator' and discuss their meanings.
>
> **Compare fractions of groups**
>
> Use counters to make groups.
>
> - Ask students to make a group of 5 counters where $\frac{1}{5}$ of the counters are blue. Ask the students 'What fraction of your group is not blue?'
>   *Extension*: Discuss the fact that you can have a group of 10 counters and make 2 blue. Explain that 1 in every 5 is blue.
> - Make other groups of counters and ask students to draw a picture of their group and write sentences describing the fractions, eg $\frac{7}{10}$ of my counters are green, $\frac{2}{10}$ are blue and $\frac{1}{10}$ are red.

Figure 11.15 Symbol notation screenshot

Each sequence of activities is described in detail and can be used as the basic structure of a unit of work. Alongside the lesson activities you may choose to use the Walkthroughs, Scorchers and Questions.

Capitalising on ICT through an integrated unit

The following case study provides an overview of one approach to developing mathematics skills, knowledge and understanding through an integrated unit. It includes a teaching sequence that develops understanding of statistics and probability, in which the use of technology in various ways is integral to the teaching.

James has a Year 5/6 class of 23 students in a country school. He decides that he wants to address the content descriptors of data representation and interpretation using a large data set from an outside source to broaden the students' experience. He goes to Census at School New Zealand, <http://new.censusatschool.org.nz>, to find a data set that provides information about students in four countries: Australia, New Zealand, the United Kingdom and Canada. He downloads this as an Excel file. Looking at the data set, and the other resources available on the websites, James realises that an integrated unit will also address the geography learning area and aspects of literacy.

Although the Census at School data set is in a spreadsheet format, James decides that his class is not sufficiently fluent in the use of Excel to be able to use it efficiently. The school has purchased **TinkerPlots**, <http://www.tinkerplots.com>, and he realises that the data set can be imported into this specialised educational software, which is more intuitive for his class to use, and with which they are already familiar. James creates a new TinkerPlots file from the Census at School data. This file provides 300 cases including data about the importance of reducing pollution, saving water and ways in which children travel to school.

Lesson 1

Objectives

Lesson objectives are:

- familiarisation with the Census at School data and a reminder of how to use TinkerPlots
- the use of inquiry questions that the children will explore
- the effective use of digital technologies.

James shows the children the Census at School data set in Excel and in Tinkerplots using the IWB. Using a TinkerPlots data card (see Figure 11.16), he discusses the types of information provided by the file, focusing on the nature of each attribute and the units, and demonstrates that the two formats, Excel and TinkerPlots, contain the same information. The children are particularly interested in the travel time to school, and he promises that they will look at this in more detail later.

He then turns to a page on which he has prepared some guiding questions for the children to explore the data set for themselves (see below). Working in pairs, the children open the TinkerPlots data file (Figure 11.16) from the school network and work through the guiding questions. James moves around the room helping children who are having problems with the software and encouraging students to share their findings, and their ways of going about answering the questions, with the neighbouring group.

Figure 11.16 TinkerPlots data card from Census at School data set

Guiding questions

1. How many countries are there in the data set and what are their names?
2. Which student is the tallest in the data set and how tall is this student?
3. How many male and how many female students are there? What percentage of the whole group is female?
4. How many male students and how many female students are from Canada?
5. Find two new interesting facts about the students in this data set and write these into your TinkerPlots file using the text tool.

As the children work at answering these questions, James moves around the room helping individuals with the technology where needed, and making suggestions about efficient ways to find the answers to the questions. At the end of the lesson, the children report back the interesting facts they have found, and save their files to their own folders. James promises the children that they may continue to work on their investigations during free-choice time – unstructured time when the children have some control over what they work on.

Lesson 2

Objectives

Lesson objectives are:

- communicate findings, using appropriate displays
- compare distributions visually.

(from <http://nzcurriculum.tki.org.nz/The-New-Zealand-Curriculum/Learning-areas/Mathematics-and-statistics/Achievement-objectives>).

In this lesson, James increases the difficulty of the tasks that he gives the children. He wants them to recognise differences between groups and to create data displays to compare two or more groups. He also wants them to be able to use filters in TinkerPlots, and sets up the questions so that he can show the children how to do this. He has deliberately chosen these kinds of activities to build on the children's previous work, and their emerging interest in the differences among the students from the four countries. In geography, the children have been researching information about students in the other countries, using a variety of internet resources. They have begun to create an electronic portfolio of their own findings about the ways in which students in these other countries live. This work has increased their interest in the data set.

James uses a question and answer session to remind the children of their previous work and the stories they have researched in the meantime. The children work in small groups, and open their own TinkerPlots files to work on during the lesson. James puts a series of guiding questions on the IWB that will assist the children to construct data displays. The first guiding question repeats one from the last session, but adds to the complexity by asking children to show this in different ways. As he moves around the room, he gives the children tips for improving their data displays, such as using the Order tool to group cases with the same attributes together.

Guiding questions

1. Find out how many male and how many female students are in each country. Show this information in two different ways.
2. Choose one of the 'Importance' variables. Find out two interesting facts about the students who answered this question. You might look at their age, where they come from, male and female differences and so on. Display your findings so that someone else can understand them.

3   Compare the information that you have found with the information that another group has found. How is it the same? How is it different?

The children each create a short report of their findings and save this in their electronic portfolio.

Lessons 3 and 4

In these lessons, James decides to work on the measurement variables of height, foot length and so on.

Objectives

Lesson objectives are:

- to understand the difference between measurement (continuous) variables and discrete (categorical) variables
- to explore the relationships between two variables.

James does not expect his young class to develop a statistician's view of the types of variable or of the relationships between the variables. He knows that the children have intuitively started to try to summarise and compare variables and wants to help them do this in ways that are more systematic and statistically sound but that do not confuse them. He also wants to see how well they can use the Data Detective PPDAC cycle (see Chapter 8) with a data set that they have not collected themselves.

Guiding questions

1   What is the smallest arm span?
2   What is the difference between the shortest and longest foot lengths?
3   Do children with the biggest feet have the longest arm span?
4   Write some 'I wonder …' questions about the data that you have, and find ways of answering one of these questions.
5   Write a report about similarities and differences among children in the four countries.

The children record their answers to these questions in their electronic portfolios.

Lessons 5 and 6

Before these lessons, James reviews the children's portfolios to check that they are all familiar with TinkerPlots and have completed all the work to date. He wants to introduce the idea of average as a way of summarising the data. Although formal ideas of central tendency and spread of data are not expected until later in schooling, James recognises that having some background will be helpful to the children, and they have encountered the terms 'average', 'mean' and 'median' when looking at media reports.

Objectives

Lesson objectives are:

- to understand average as a way of summarising data
- to explore ideas about distribution of data
- to find the mean and median of a data set using TinkerPlots.

James starts the lesson with a concrete example, using the IWB. He has created a number line with the numbers 0 to 5 equally spaced along it. He asks each child to come up and make a dot above the number that shows how many brothers and sisters they have. In this way, he creates a dot plot like the ones in TinkerPlots. Most children have one sibling and no children have five siblings. He asks the children to suggest a way of summarising the data. Ideas emerge, such as taking the most common number or using the middle number. James picks up the idea of the middle number. One child says it should be 3 because it is the middle of 0 to 5. Other children object because there are more children with one sibling and only one with four siblings.

James gradually leads the discussion to the idea of finding the data point that lies in the middle of all the data. By counting, the children establish that the middle point is 2. James explains that this is the median, and brings up the Maths Dictionary website, <http://www.amathsdictionaryforkids.com>, to help him discuss the meaning. He points out that ordering the data was done through the dot plot, and that the class counted to the middle. He then asks about other ways of summarising the data: 'Is there another kind of "middle"?'

One person says the word 'average'. James picks this up and describes this as a balance point. He demonstrates the idea using a number balance that he has borrowed from Year 1, moving the weights until they are at the middle and the balance is even. He then asks different children to come to the IWB and move the dots from the ends towards the middle, moving one place at a time. He ends up with a representation that looks similar to the one in Figure 11.17.

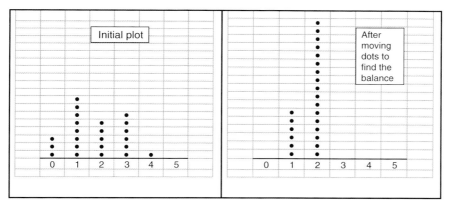

Figure 11.17 Representation of the number of siblings in James' class

He explains that the balance point is not a whole number, but lies between 1 and 2, and through questioning gets the children to recognise that it would be closer to 2. He chooses not to develop the computation aspect of the mean at this time, but does tell the class that this average is called a mean.

The children then move to their TinkerPlots files. They create a stacked dot plot of the age of the children in the data set. James asks them to describe the distribution of the data and to predict what the average might be. He collects their predictions on the IWB. Again using the step-by-step approach, James shows them how to find the middle 50 per cent value of the age using a hat plot and says that the median and the mean values will lie somewhere in the 'crown' of the hat He then shows the children where to find the median and mode in the software. He stresses that what the program is doing is very like what they have just seen from their class, but it is more accurate and also more hidden. The children then explore the ideas for a short time on their own. When James is happy that the groups can all create a hat plot and put in the median and mean, he poses some further guiding questions.

Guiding questions

1 Find the mean and median of the heights of all students.
2 Which country had the tallest students?

Lessons 7 and 8

James wants the last two lessons in this sequence to provide an opportunity for the children to consolidate what they have learnt in mathematics. He sets up an investigation of the Census at School data, which they will present on a poster created in PowerPoint using a template that they have used before. The investigation is designed to complement the work they have also done in history.

Objectives

Lesson objectives are to:

- provide an opportunity for children to apply their new knowledge
- give children the opportunity to extend their understanding through an open task
- assess children's understanding of data representations.

James asks the children to continue to work in pairs, and asks each pair to complete an investigation and prepare a poster for publication. He stresses that he will be looking for data representations through graphs, tables and use of statistics such as means and medians, and interpretations of what they see referring to the information they have created. He reminds them of the work they have done and puts a summary page on the whiteboard of all they have explored. He also says that they can ask him to remind them about anything they may have forgotten how to do. He gives each child a printed sheet containing several guiding questions (shown below).

INVESTIGATION OF THE CENSUS AT SCHOOL DATA

These questions are guiding questions only. You are free to explore the data in any way that you like but you must make sure that you use graphs, tables, a mean, a median, and make some comparisons between our country and at least one other country. Present your work on a poster using the PowerPoint template in the class folder. You have two mathematics sessions to complete this task.

- What is a typical student like?
- Investigate the students from one country (use a filter).
- Compare the students in our country with students in a different country.
- What can you say about the things that matter to students in our country? Are these the same things that matter to students in the United Kingdom?

> **PAUSE AND REFLECT**
>
> Before reading further, think about how James' approach could be interpreted through using the TPACK framework. Identify aspects of his approach that you found to be particularly effective in terms of engaging students. How do you think the students were using the technology in relation to the four metaphors discussed earlier?

Pedagogical commentary

This section provides a commentary on James' teaching sequence and relates his approach to the TPACK framework (see Chapter 1).

By choosing a large data set for his class to work on, James has to include technology in his planning. It is evident that the school is reasonably well set up, because James uses an IWB and networked folders for every child quite effectively. He has good *technology knowledge (TK)* within his school's context, and is able to make effective use of the technological resources available, including software and file storage. He also demonstrates good *technical content knowledge (TCK)* in choosing to use TinkerPlots rather than Excel for his class to work with. Excel is a powerful data-management and analysis tool, but it is not designed as a teaching instrument. Older students should learn to use programs such as Excel, but younger children often benefit from specialist software so that the focus can be on the learning intentions rather than on learning to use the software.

In the early lessons, James also demonstrates his *technical pedagogical knowledge (TPK)* because, rather than trying to get all the children to carry out identical tasks in the same way, he recognises the advantage of the software in allowing children to explore the data set. From experience, he knows that his students will 'play' with the software and discover a range of facts that they find interesting. He provides guiding questions to get them started, but allows them choice and control over where this goes. The questions he gives the class increase in complexity as they become more familiar with the information. Initially, he asks them to identify a single case, or a single attribute; in Lesson 2, he deliberately focuses on more than one group or attribute, such as male/female and ships. The questions are deliberately more open, and he also asks for several representations, knowing that the students are familiar with data presentation in the form of column graphs, pie charts and tables.

James also shows good *pedagogical knowledge (PK)*. He develops sound lesson plans, and has thought through his unit of work. He has recognised the efficiency of using the same stimulus material – the Census at School data set – for history and geography, and also sees that he can develop literacy and possibly other areas such as visual art. He demonstrates understanding of his class and what the children are likely to be capable of, but also wants to stretch their capacity by using a large data set. He generally manages the classroom well, having children working in pairs at the computers, and timing his lessons so that they can complete a substantial piece of work but can also continue exploring in the free-choice time that he builds into most days. He responds to the children's curiosity in the money values, understanding the importance of interest in engaging and motivating students. The importance of having a comprehensive summary, or opportunity to reflect on every lesson, is also evident. James closes every lesson with either a sharing discussion or the children writing about what they have learned. This lesson ending is important for children to consolidate their understanding, and also sets up the next lesson.

James shows good mathematical *content knowledge (CK)* when he sees that the nature of the data set provides an opportunity to discuss the effects of an outlier on the different types of average. Although this topic is not expected in the curriculum until Year 7, James knows that it will require considerable consolidation before it is understood, and aims to provide a platform for future development. He sees that conversion of money can be helpful with some mental computation practice, and has an underpinning relationship to which he might be able to refer back later, when he considers some of the algebra aspects of the curriculum. The idea of identifying the median through counting is sound, and can be reproduced exactly in TinkerPlots, although it works best with a small data set.

In teaching the idea of the mean as a balance point, however, although James shows good mathematical understanding, these lessons (Lessons 5 and 6) are probably only partly successful. The idea of demonstrating the balance idea using concrete materials shows *pedagogical content knowledge (PCK)*, but in the context in which it is used, James might be criticised for trying to do too much. He does not choose the best data to use (number of siblings) to show the idea of moving to the middle to identify the balance, because the result is not a whole number, and this in itself can be confusing to children who cannot understand the idea of 'part' of a child. It might have been better to take a couple of lessons to establish this conceptual understanding of the mean prior to working with the larger data set.

The aim of using a concrete representation before moving onto the computer is important, and James evidently understands this. However, the lesson has too much in it, and important concepts may be lost in the rush to move on to the Census at School data. The careful development of increasingly complex questions about the Census at School data set also demonstrates high levels of PCK, recognising that children need to move from describing single groups and attributes to comparing groups and attributes, using percentages and other forms of proportional reasoning to make comparisons.

James demonstrates *technological pedagogical content knowledge (TPACK)* in several ways. He knows that he cannot use the larger data set without technology, and makes a considered decision about the type of software to use. He makes effective use of appropriate internet resources for teaching, and of the technology available in the school, through the use of e-portfolios and shared files. The technology is embedded in James' teaching approach, but he goes beyond using it as an electronic textbook. He aims to involve his class actively, by getting individual children to interact with the IWB. He allows his class some control over their exploration using the power of the technology, but sets parameters around this so that they don't go off task because of the novelty of the software.

James' teaching is a blend of explicit skills teaching – such as when he uses a step-by-step approach with the whole class to teach the class – and guided exploration. He also recognises that learning happens through dialogue, because the children work in pairs and he encourages the sharing of ideas. He works hard during the lessons, moving round the room supporting, questioning and helping children with their work, but not doing it for them. He also implicitly realises the limits of technology, and uses a blend of concrete materials, worksheets in hard copy and technology to develop children's understanding.

Reflection

This chapter has provided some suggestions about how to capitalise on ICT in the mathematics classroom. Through reading the chapter and completing the activities, you will have explored the impact of TPACK upon the role of the teacher and considered some different teaching approaches. HOTmaths has been provided as a resource that can be used to facilitate students' understanding of fractions. The case study about James and his lessons has provided an example of how mathematics and ICT can be used in an integrated unit of work, especially when the teacher has solid TPACK.

Websites for exploration

Maths300: <http://www.maths300.esa.edu.au>

Mathwire: Place Value: <http://mathwire.com/numbersense/placevalue.html>

NCTM: Principles and Standards for School Mathematics: <http://www.nctm.org/standards/content.aspx?id=16909>

New Zealand Curriculum Online: Technology: <http://nzcurriculum.tki.org.nz/The-New-Zealand-Curriculum/Learning-areas/Technology/Learning-area-structure>

TinkerPlots: <http://www.tinkerplots.com>

CHAPTER 12

Diversity in the primary mathematics classroom

LEARNING OUTCOMES

By the end of this chapter, you will:

- understand the complexity of primary mathematics classrooms
- recognise a range of potential barriers to learning mathematics
- be able to plan for diversity in the mathematics classroom
- recognise the potential of technology to meet all learners' mathematical needs.

KEY TERM

- **Inclusion:** Policy of including all students in mainstream classrooms wherever possible

Australian and New Zealand classrooms are becoming increasingly diverse. In addition to a wide range of mathematical abilities, primary teachers of mathematics have to meet the needs of children from numerous cultural backgrounds, many of whom have had different mathematical experiences. Children who have physical, intellectual, social or emotional difficulties may be included in the mainstream classroom, but all have the capacity to learn mathematics. There may also be children who are classified as gifted and talented in one or more domains. Technology can be helpful in meeting some of these varied mathematical needs.

Although this chapter addresses issues relating to **inclusion** in primary mathematics classrooms, it does not pretend to provide a special education focus. The needs of children with specific disabilities can be highly technical, and it is well beyond the scope of this chapter to try to deal with all the detailed requirements and concerns that may be encountered. Rather, the chapter aims to help the primary teacher deal with the reality of mathematics teaching in modern classrooms, where there may be children with very diverse learning needs, and to consider the role that technology may play in this.

A fundamental tenet of both the Australian and New Zealand curricula is that all learners should have access to the full mathematics curriculum, regardless of background (Commonwealth of Australia 2009). The rationale included in the Australian Curriculum: Mathematics begins with the statement, 'Learning mathematics creates opportunities for and enriches the lives of all Australians.' Later the rationale includes the statement:

> The curriculum anticipates that schools will ensure all students benefit from access to the power of mathematical reasoning and learn to apply their mathematical understanding creatively and efficiently. The mathematics curriculum provides students with carefully paced, in-depth study of critical skills and concepts. It encourages teachers to help students become self-motivated, confident learners through inquiry and active participation in challenging and engaging experiences (<http://www.australiancurriculum.edu.au>).

The same principle is stated clearly in the New Zealand curriculum:

> The curriculum is non-sexist, non-racist, and non-discriminatory; it ensures that students' identities, languages, abilities, and talents are recognised and affirmed and that their learning needs are addressed (New Zealand Curriculum Online, <http://nzcurriculum.tki.org.nz>).

Both countries have recognised the need for the curriculum to be adapted in some instances, and offer advice and help to use the curriculum effectively to provide for the diverse needs of students – including those with a disability, gifted and talented students, and students for whom English is an additional language or dialect (EAL/D). The Australian Curriculum: Mathematics, in line with the overall curriculum structure, covers mathematics content, general capabilities (including literacy, numeracy, and personal and social capability) and cross-curriculum priorities, including Aboriginal and Torres Strait Islander cultures, Asia and engagement with Asia, and sustainability. The capabilities and cross-curriculum priorities can become the basis for age-appropriate adaptations to the mathematics content. Some approaches to achieving this goal are the focus of this chapter. In New Zealand, there is a particular focus on students from Māori or Pasifika backgrounds and their cultural needs, and the curriculum documents are available in Māori language (see <http://tmoa.tki.org.nz>).

ACTIVITY 12.1

Go to the curriculum website appropriate to your country – Australia: <http://www.australiancurriculum.edu.au>; New Zealand: <http://nzcurriculum.tki.org.nz> – and access the resources available regarding students with disabilities. Although there is advice about curriculum adaptation, this is fairly general and may not specifically target mathematics. Using these resources and any others you might find, consider what adaptations might be needed to undertake activities in Year 4 that address learning outcomes about measuring lengths, areas or volumes using formal and informal units. Think about children with physical disabilities such as vision or hearing impairment, or an inability to manipulate physical objects; children who do not speak English as a first language; children who have intellectual disabilities or developmental delay; children with social and emotional disabilities; and children who might be gifted and talented.

1 Which of these children require specific adaptations to the curriculum itself?
2 Which children require a careful consideration of the nature of the activities that you might want to use?
3 How can technology help with any of the adaptations that you might need to make?

The challenges posed by the range of different children in a single classroom are far from trivial. The philosophy of inclusion can be uncomfortable and difficult to implement in practice. It is the professional responsibility of teachers to ensure that all children they teach have access to the powerful learning available through mathematics.

Many children with special needs have teacher aide time allocated for classroom support. It is important that the teacher is in control of the learning program and that, as far as possible, this is in line with what the rest of the class is doing. Teachers must explain the learning objectives and aims of the lesson to the allocated aides and, wherever possible, the aide should ensure that the child participates fully with the class.

In addition to classroom support, there may be a range of other professional services working with individual children. These services may include speech pathologists, occupational therapists, guidance officers or school counsellors, social workers and specialists in English language support. All of these people are working towards the same goal of ensuring that educational outcomes for the child in question are the best that they can be. The teacher is responsible for directing the mathematical learning of the child, and should discuss specific adaptations and needs with the relevant person.

For example, framing a mathematics question in terms of football teams and using 'Bombers' as a team name may be very appropriate if the school is located in an area where the Bombers are a local team. For children from refugee backgrounds, however, the name may have emotional connotations that could interfere with their understanding and response to the question. Similarly, talking about negative numbers in terms of temperature in a remote Indigenous community in the Northern Territory is likely to have less meaning than using a rope swing over the creek as an example, where the knots below the water level show negative numbers. Undertaking the activity of collecting data about balancing on one leg (see Chapter 8) is not appropriate for a class that includes a child in a wheelchair. Using colour as a discriminator ('Group all the red triangles together') is not suitable for children with vision impairment. Being sensitive to context is an important aspect of effective mathematics teaching, particularly when considering classroom diversity.

Children with special needs may also become tired more quickly than other class members. High expectations should always be the norm, but these should be tempered by each child's individual needs. With appropriate supports in place, all children can learn mathematics, and this is the expectation of the Australian Curriculum: Mathematics.

Teachers' beliefs

It is important to acknowledge that teachers' beliefs about mathematics and its teaching and learning are factors to consider. For example, DeSimone and Parmar (2006) found that Year 6, 7 and 8 teachers who had students with learning disabilities in their classrooms had limited understanding of the students' mathematics learning needs. They also judged collaboration with other teachers to be the most helpful way of dealing with diverse needs and felt inadequately prepared to deal with diversity. Beswick (2007–08) reports similar findings in an Australian context, and also shows that many beliefs – such as 'Some people have a maths mind and some don't' – were remarkably persistent, even after appropriate professional learning.

All students benefit from well-taught mathematics that has a focus on strategy use rather than recall, and includes relevant, contextualised mathematics in which high expectations by the teacher are the norm and children are actively engaged in their learning (Boaler 2008). Language is important, but it is possible for children with severe language impairment to learn to calculate accurately, even when they have difficulty with counting tasks (Donlan et al. 2007). Nevertheless, the complex language of mathematics cannot be ignored. For example, the distance round a 2D shape is referred to as the perimeter, but when applied to a circle it is the circumference. Neither of the two mathematical terms is easy for children to understand, and for those with language delays or intellectual disabilities, the words may be completely incomprehensible. Paul Swan has written a useful, easy-to-read article about mathematical language, which is available from the Online Teachers' Resource Network at <http://members.iinet.net.au/~markobri/Paul_Swan/Paul_Swans_homepage.html>. A Maths Dictionary for Kids, <http://www.amathsdictionaryforkids.com>, is an interactive resource that is accessible to children and explains mathematical terms with diagrams and examples. HOTmaths also has a dictionary resource that teachers could use. This resource can be accessed by logging into HOTmaths and clicking on the Dictionary icon on the Dashboard.

PAUSE AND REFLECT

Why is it important to ensure that all children learn mathematics in meaningful ways so that they develop mathematically to their greatest potential? Is this simply being 'politically correct', or is it an important principle that all teachers of mathematics should follow? Share your views with another person.

All teachers should think through their beliefs about children's learning. It is important to have a strong philosophy that underpins your approach to teaching. There are no rights or wrongs – teachers are as diverse as the children they teach, and differing views should be respected. Teachers are, however, expected to work within the policies in their particular system and school. At times, these may be at odds with your personal beliefs, but you will be expected to implement the policies regardless. This chapter aims to help you adapt the curriculum and its delivery as appropriate for a range of special needs within a mainstream classroom.

Children with physical disabilities

Many children with physical disabilities can participate fully in the curriculum intellectually if suitable activities are used. Children confined to a wheelchair, for example, often need little adaptation, but some children with cerebral palsy cannot manipulate physical objects easily. Large rulers or measuring tools, calculators with large buttons and magnetic boards with numbers and symbols are all appropriate tools to allow children with physical disabilities to access mathematics learning. For example, mental computation questions can be answered by pressing calculator keys. When a teacher asks 'What is 7×6?', the response can be made by pressing the 4 and 2 keys, thus using the calculator as a communication tool rather than for computation. The calculator can be used in the same way for playing shops when learning about money, and this can also allow children for whom English is limited (EAL/D) to participate.

It is worth asking the relevant support person in your school or system what adaptive technology is available, and investing time in learning to use this effectively. It is beyond the scope of this chapter to detail specialised adaptive technology, but readily available mainstream resources include speech-recognition software, tablet computers, virtual manipulatives, apps for tablets and audio-technology that allows a child to listen to the teacher. Children who have difficulty writing or drawing, for example, can record their mathematical thinking using a variety of audio resources including smartphones, IWB facilities or computer software. Such adaptations are particularly important for children with vision impairment.

Problem-solving activities that are suitable for children with disabilities may be equally effective for non-disabled children. For example, the geometry challenge Design Paper Money for the Visually Impaired (Robicheaux 1993) is an excellent task for a Year 6 class. The aim is to develop some way of differentiating paper money by snipping off corners so that people with vision impairment can easily identify the different notes. The corner cuts have to be congruent, right-angled isosceles triangles, and this alone can lead to a discussion about meaning. On the surface, this appears to be a simple task, but if it is remembered that it is not possible to distinguish between the front and back of the note, then the task becomes more complex.

Not only does a task of this kind develop visualisation skills, and understanding of symmetry and congruence; it should also lead to a discussion about the difficulties experienced by people with vision impairment, building a more inclusive classroom. This task clearly addresses problem-solving and reasoning proficiencies in the Australian Curriculum: Mathematics, and can enhance understanding of mathematical transformations. A similar task is to design and build a scale model of a ramp for wheelchair access to the school hall, for example, with requirements for height, width, slope and so on. Tasks of this type are challenging mathematical problems that can have immediate relevance to children who share their classrooms with fellow students with physical disabilities, and they meet the goals of the mathematics curriculum.

CLASSROOM SNAPSHOT 12.1

Sara has a Year 1 class that is learning about ordering numbers from 1 to 9. She wants to use an activity that she knows is successful. It involves children role-playing Mrs Number's children, who get mixed up. The class has to reorder them by giving directions to the children who have placards with numbers 1 to 9 around their necks. HOTmaths has a printable resource for this activity. You can access this resource by logging into HOTmaths, entering 'Digit cards' in the search field and then selecting the 'HOTsheets' tab.

In the class this year is Sophie, who is intellectually very capable, but is confined to a wheelchair. She cannot be one of Mrs Number's children easily, so Sara wants to find an alternative approach that will allow Sophie to participate fully. Before the lesson, she prepares some pages for her IWB showing Mrs Number and her children all mixed up. She starts the lesson as she has always done by handing out number cards to different children and lining them in order from 1 to 9. Then they mix themselves up and the rest of the children have to give them clues. Sophie says, '3 must be next to 4', and the other children agree and give similar clues.

Then Sara moves to the IWB. First she asks the children with the cards to line up the numbers on the IWB exactly as they have just done. Then she hands the cards to other children so that every child will get a turn at being one of Mrs Number's children. Sophie has the card showing 7. The class plays the game in exactly the same way, with the children who have the numbered cards moving the numbers on the IWB. The IWB is fixed low on the wall so that Sophie can access it easily. This time, as the children are playing the game, Sara asks the children to give a reason for their clue.

Sara ends the lesson by asking the children to draw a picture of Mrs Number's children, all lined up in order. For those children who clearly understand, including Sophie, Sara sets the additional challenge of drawing the children in the reverse order.

Sara is making use of technology to ensure that Sophie is able to participate in the same way that the other children do. She can follow up this activity with others such as Caterpillar Ordering, <http://www.topmarks.co.uk>. An activity such as Birds on a Wire to 10 (see Figure 12.1) provides another resource in which the numbers are represented in different ways, and would make a useful follow-up to Mrs Number. This activity can be found by logging into HOTmaths and clicking on the FUNdamentals icon on the Dashboard. Next, click on 'Numbers' and then 'Numbers to 10'.

Figure 12.1 Screenshot of HOTmaths Birds on a Wire to 10 FUNdamentals

Hearing impairment

Children with hearing impairment are more common than may be apparent. Many children suffer from intermittent hearing loss associated with 'glue ear', and children in busy classrooms may also use the support of lip reading to some extent. In mathematics, when there is a focus on mental computation, these children may miss important words and give incorrect responses because they have not heard clearly what is being said. When giving instructions, providing explanations or posing problems, it is sensible to stand so that light from the window falls onto your face rather than standing with your back to the window so that your face is in shadow.

For mental computation practice, there are any number of technology resources available that could be used as an alternative to oral questioning for children with hearing impairment. The important aspect is to ensure that all children can develop strategies that lead to powerful computation. To support this goal, some dialogue with the child or within groups of children is important. Voice-recognition software might be useful for one-to-one conversations with the child concerned. If a child uses Auslan and has an interpreter in the classroom, discuss the lesson with the interpreter beforehand, and slow the pace of the lesson a little to allow the hearing-impaired child to watch the interpreter as well as any work that is being completed on the whiteboard. These suggestions are practical ways to support all children.

Formal signs for numbers may be in conflict with the ways in which hands and fingers are used to signify numbers informally. Signing even a fairly simple algorithmic problem, such as 'How many more marbles does Sally have than Harry?', becomes very complex. Deaf children do not do as well in mathematics as children without any hearing impairment (Gregory 1998). One possible reason could be the difficulty with language – especially words such as 'if' and 'because', which are common in mathematics. In addition, the specialised language – such as isosceles, hypotenuse, numerator, denominator and so on – is complex, and may be difficult for hearing-impaired children to read or process. There are also words that have everyday meanings but are used in a specialist way in mathematics. Whole, for example, is used to denote a type of number (whole numbers) and the complete part of a fraction ('If this is one-quarter, what is the whole?'). Developing vocabulary around concepts rather than objects is more difficult for hearing-impaired and deaf children. Resources such as mathematics dictionaries can be helpful; however, deaf children may have better non-verbal reasoning skills than hearing people (Braden 1994). This finding has implications for mathematics teaching because it suggests that introducing ideas using non-verbal signs could be helpful. Technology can support this idea by providing virtual manipulatives, such as interactive number lines, shapes or patterns.

Vision impairment

Children with vision impairment may have difficulty with specific mathematical tasks. For example, concepts such as geometrical transformation are predominantly recognised visually and described using visual language (for example, flip, slide, turn). Tactile materials will be important to use in developing these ideas, using a straight stick as a reference.

Learning about right angles may require tactile frames that can be used with a set square to check whether an angle is a right angle. Circles can be represented by CDs or plates, books are rectangular and so on. Making reference to everyday objects and allowing the child to feel them is important, as is often done in the early years of schooling; these kinds of adaptation may need to be continued longer for children who are vision impaired (Figure 12.2). It is critical, however, that these are seen by the child and the rest of the class as 'adaptations', not babyish aids. Tablets can be very helpful because they can magnify print to make it easier to read. This provides access to the normal range of online resources, although be aware that many mathematics sites make use of Flash™ technology that may not be accessible on tablets.

A wide variety of tactile materials are available, including measuring tapes and rulers, dice, clocks, talking scales and so on. However, as children progress through school, and the mathematics content becomes more abstract and symbol based, children with severe vision impairment may be disadvantaged. Mathematical text is tricky on ordinary computers, but the handwriting-recognition software for use with tablets and computers has some potential for visually impaired children. Some children read Braille, and for these children access to the mathematics curriculum may require thinking ahead to ensure that Braille versions of texts or other material are available. It is important to seek specialist advice, and to talk to the child and parents/caregivers to identify the best approaches for the individual.

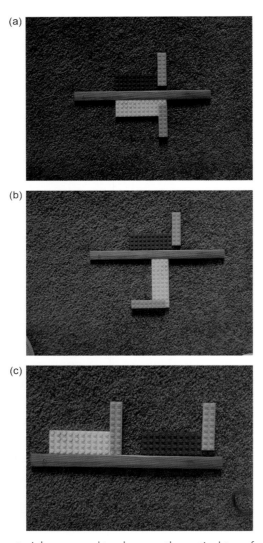

Figure 12.2 Tactile materials arranged to show mathematical transformations

CLASSROOM SNAPSHOT 12.2

Tom's Year 6 class is undertaking a mathematical investigation into circles, with the aim of developing understanding of the relationship between diameter and circumference.

Dan is a student in Tom's class who is visually impaired. He uses a tablet with a keyboard, and is skilled with voice-recognition, handwriting-recognition and audio-recording software for recording mathematical ideas.

Tom organises the class into groups of three people, and gives each group a box containing a variety of circular objects and several measuring devices, including rulers, tapes and string. The groups have the task of measuring across the diameter and around the edge of each object, and every child is required both to measure and to create a recording system for each measurement.

Dan's group is provided with a tactile ruler in addition to the materials that the other groups are given. With support from the other group members, Dan is able to measure the concrete objects using string and the tactile ruler. He creates a table on his tablet using the touch-sensitive screen to scribe notes, and calculates the ratio of the diameter to circumference using a tactile calculator. Finally, he records a summary of his findings using the audio-recording facility on the tablet.

The next day, the class repeats the activity in the playground, this time creating large circles with string and chalk. The purpose is to discuss measurement error and how this can have a big impact on small measurements but a lesser impact on larger ones. Dan participates fully by using a trundle wheel and counting the clicks that this makes, with the members of his group helping him to stay on the circle accurately.

The adaptations made for Dan are those necessary to ensure that he has access to the learning activity. The curriculum itself is not compromised or changed in any way, and the classroom expectations are the same for Dan as for any other child. In addition to mathematical outcomes, there are social outcomes for Dan and the other class members in terms of cooperation and understanding that people can be different but achieve in the same way.

Children with intellectual disabilities

Mathematics classrooms have always included a wide range of student cognitive abilities. Increasingly, however, children with specific difficulties such as Down syndrome are learning mathematics alongside their age peers in mainstream classrooms. All teachers should be prepared to manage a wide range of mathematical abilities within the classroom.

The research on teaching mathematics to students with learning or cognitive disabilities suggests that visual and concrete approaches can be effective (Brady, Clarke & Gervasoni 2008). These children often need direct instruction, and there is some evidence that such children do not learn from mistakes, but rather that making the same mistake can reinforce the error (Moyer & Moyer 1985). Traditionally, students with learning disabilities have received a curriculum focused on traditional algorithms based on a drill and practice model. Evidence suggests, however, that children with learning disabilities do benefit from being actively engaged in hands-on tasks that involve discussion and thinking about mathematics.

Some children with learning disabilities – especially those with language delay – respond to visual approaches to mathematics learning, and these can be supported by technology. For example, the use of HOTmaths widgets such as Dot Pictures and Leaping Frog is appropriate for all students, and the challenge level can be altered as required (see Figure 12.3). You can access these widgets by logging into HOTmaths and entering their respective names in the search field.

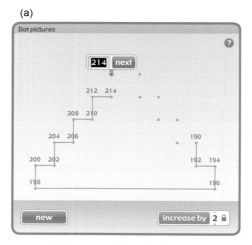
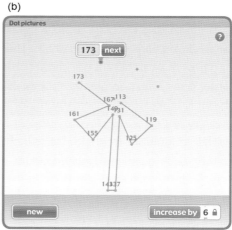

Figure 12.3 Screenshot of HOTmaths Dot Pictures widget, showing the variation possible within the one type of activity

There are other widgets in the FUNdamentals section of HOTmaths, located on the Dashboard, that can be used in the same way. For children with learning difficulties, the starting numbers can be made smaller and the jumps less difficult. All children are thus participating in identical activities, but the intended outcomes are chosen to be appropriate for children with special needs. There is a watch-point, however: children with learning difficulties may not respond to 'wrong' answers in the same way that other children do. They are likely to continue to repeat the same error. When working on technology tasks, it is

CLASSROOM SNAPSHOT 12.3

Mary is in Year 5 and has Down syndrome. She is working with her aide, Katrina, with the HOTmaths Split-and-Add Robot widget.

Before the lesson, Mary's teacher has provided Katrina with a sequence of problems to start the session and then some starting numbers so that Mary can choose the second number to add. Katrina explains the task to Mary and demonstrates what the robot does using 14 + 13. She then enters the number 13 and asks Mary, 'How will the robot split that number up?' Mary correctly answers, 'Into a 10 and a 3.' Katrina then enters the second number as 24 and asks Mary the same question. This time Mary says, 'Into a 10 and a 4.'

Katrina then starts a dialogue with Mary about the meaning of the number 24. She asks first, 'What is the number?', which Mary answers correctly. Then she asks about the meaning of the digits with questions such as, 'Where is the 20?', 'What does 20 mean?', 'What does the 4 show?' When she is satisfied that Mary knows what 24 represents, she returns to the robot and asks Mary how the robot splits numbers up, returning to the 13 and then, finally, to the 24. This time Mary correctly identifies that the robot will give a 20 and a 4. Mary then presses the release button and, as the robot splits the number up, Katrina and Mary together say the numbers that are produced (see Figure 12.4).

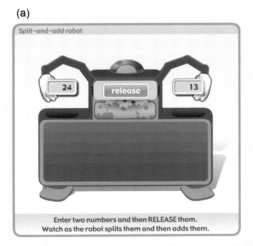
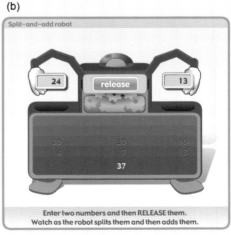

Figure 12.4 Screenshot of HOTmaths Split-and-Add Robot widget

When Mary has worked through the questions provided by the teacher, Katrina gives her a worksheet that the teacher has prepared showing the robot before the numbers are entered (use a screenshot and provide a set of similar two-digit add two-digit problems to reinforce the strategy). Mary starts the worksheet in class and takes it home for homework to discuss with her parents what she has been doing.

There are several points about this lesson that add to its effectiveness for Mary's mathematical development. First, although she is in Year 5, the task chosen is suitable for Year 4 as well as Year 5. Mary's teacher has clearly planned a sequence of numbers that she knows Mary can cope with, and has constructed the early activities so that the sum does not go over the next 10, so that the structure of the number is the focus at this initial stage. She has discussed with Katrina a suitable series of questions to ask, and the importance of not letting Mary simply play with the widget. The long dialogue may seem tedious, but it is through the discussion and questions that Mary is learning. The repetition of the questions and saying the numbers aloud as they are produced also add to the learning experience for Mary. The worksheet provides reinforcement both visually in using the robot picture and mathematically by having a set of similar problems. The curriculum has been modified slightly because the expected outcomes are closer to Year 2. Mary is using strategies to combine and partition whole numbers leading ultimately to efficient approaches to addition. The activity clearly fits the Australian Curriculum numeracy capability element of estimating and calculating with whole numbers (see <http://www.australiancurriculum.edu.au>). Although Mary works closely with the aide, Katrina, Mary's teacher is in control of the learning experience.

Tasks that are multi-layered are valuable tools for developing mathematical understanding in children with intellectual disabilities. The circles activity described in Classroom snapshot 12.2 is accessible to children with Down syndrome, but the purpose of the activity for such children would be to develop measuring skills at about a Year 4 level of using measuring instruments and recording the information systematically. Using the circles activity with children who have an intellectual disability is an example of the curriculum itself being adapted by changing the expected outcomes. The activity remains the same because it is an age-appropriate activity for the class. Many activities described in this book can be used in this way.

Children with social, emotional and behavioural disabilities

The number of children within this category of disability appears to be growing. For example, an estimated one in 100 children in Australia and New Zealand is diagnosed as having an autism spectrum disorder (ASD), and boys are four times more likely to be diagnosed than girls. ASD impacts on the ways in which children interact with the world around them, and hence may have a big impact on their schooling (Aspect 2013).

There is a growing body of research suggesting that children who have difficulty with social interactions and communication may interact effectively with technology. In the United Kingdom, for example, software is being developed that allows children with social difficulties to interact with virtual characters to develop behaviours that they can translate into the classroom (Guldberg et al. 2010). Despite these initiatives, at present there is little specialised mathematics software available, and teachers have to use currently available resources or create their own (Leach 2010).

Routines are very important for many of these children. They need certainty, not unpredictability, so lessons should have a familiar structure. Often this is a 'whole-part-whole' structure, where the whole class is introduced to the focus of the lesson, then students work individually or in small groups on the lesson content, with a final reflection session involving the whole class. Within this structure, there is scope for many different activities. Lessons do not have to be confined to explication of the idea, drill-and-practice and a final summary. The lessons described in the classroom snapshots in this chapter could all be taught within this framework – for children with ASD, it is the predictability of the overall routine that is critical.

Because many children with social difficulties interact productively with technology, online resources can be helpful. When using resources from the internet, however, some specific considerations are needed – particularly for children with ASD. Many of these children are sensitive to loud sounds. Some online resources are accompanied by noise effects. When using these, it would be sensible to turn off the sound. For example, a virtual pinboard – such as that at <http://www.crickweb.co.uk> – can be used to set a variety of challenges at different stages of schooling, but it is accompanied by a popping sound that can cause anxiety in over-sensitive children. Illuminations, <http://illuminations.nctm.org> and FUNdamentals in HOTmaths have a variety of interactive resources that have limited sounds, and these can be turned off. Personalised activities can be created by using IWB resources (Leach 2010) and incorporating familiar situations, such as pictures or photographs from the child's environment. Mathematics activities that are accompanied by sounds and graphics when the answer is incorrect can be distracting because the reward for getting an incorrect answer is greater than that for getting it correct. For example, a drill-and-practice activity for number bonds asked children to save their teacher from getting a custard pie thrown at them and the accompanying graphic showed a cartoon teacher covered in custard when the answer was incorrect – hardly an incentive to try to get a correct answer!

Technology activities that require group interactions may also be helpful for children with social difficulties. The nrich website, <http://nrich.maths.org>, has a number of cooperative activities in which each member of the group has a clue that cannot be shown to anyone else. Group members may describe their clue and ask clarification questions, but no one person ever sees all the clues at once. The group works together to solve the problem. Examples from nrich include Arranging Cubes (activity 6973) and What Shape? (activity 6986). What's My Number?, <http://www.teachingideas.co.uk>, may also be useful, as is the HOTmaths A Mystical Map HOTsheet.

Global

Such activities can help children with social difficulties because, although they have to interact, it is within a structured framework and not too much is asked of them. The activities are educationally sound for all children, but because some limited interaction is required they can provide other social outcomes for children with special needs, addressing the social management element of the personal and social capability in the Australian Curriculum: Mathematics.

Behavioural problems, such as Attention Deficit Hyperactivity Disorder (ADHD), can be tricky to manage in the primary mathematics classroom. Again, routines are important, as is controlling the amount of work presented at one time. Providing one investigative task rather than an overwhelming number of mathematics problems to solve can aid in keeping these children engaged. The same principles apply to children with ADHD as to all children. They respond to a rich mathematics curriculum where they are actively involved and have some control over their activity. Because some drill-and-practice is needed, and children with ADHD find it difficult to sustain attention for long periods, computer activities can help to provide engaging situations. Using a variety of online activities that provide suitable practice activities may be more effective than giving a child with ADHD worksheets or rapid-response mental computation practice. There is some indication that technology use helps children with attention deficits to focus for longer (Lucangeli & Cabrele 2006), but the research is limited.

Multicultural classrooms

There are many children in Australian and New Zealand primary classrooms who do not speak English as a first language. These children include many Aboriginal and Torres Strait Islander Australians and Māori people, as well as immigrants and refugees from other countries. The principles of treating these children with respect, and acknowledging their personal mathematical backgrounds, are important. Successful teachers of mathematics not only understand mathematics and its pedagogy; they also demonstrate a culturally relevant pedagogy (Morris & Matthews 2011).

In the early years of schooling, when mathematical language is critical, it is important for children with limited English that mathematics contexts relevant to their situation are emphasised. If they know how to count in their first language, this can be celebrated and shared with the class. There are websites where counting in other languages can be explored – for example, <http://www.marijn.org>.

Older children can carry out research projects into counting systems from other parts of the world, which are rich and diverse, and can provide a basis for developing understanding about why mathematics has developed in different ways across the world, and how these systems have impacted on a society. For example, see the HOTmaths lesson on Roman numerals. Log into HOTmaths and select 'Australian Curriculum' as the Course list and then 'AC Year 6' as the Course. Select 'Number systems (Enrichment)' as the Topic and then 'Introducing Roman numerals' as the Lesson.
AC

Geometry is another area rich in culture, from patchwork quilts from North America to Islamic tiling patterns and Pasifika weaving patterns.

Figure 12.5 A woven screen and basket in a Māori *marae*

When teaching mathematics, the cultural backgrounds from which the mathematics has evolved can be emphasised. It is worth asking children from an EAL/D background what their words are for particular mathematics ideas. Some cultures, for example, have no words for left and right; others label 2D shapes according to the number of edges so that a question such as 'How many sides has a pentagon?' translates as 'How many sides has a five-sided shape?' Sharing these ideas with the class provides an interesting starting point for a discussion about the mathematical properties. Activities that do not have a language basis may also be useful. Figure 12.6 shows an activity from the NLVM called Ladybug Leaf. The aim is to hide the ladybug under a leaf using the directions provided by the symbols. Such an activity could be undertaken by children in pairs, without the need for high levels of language, because they could point to the movement needed.

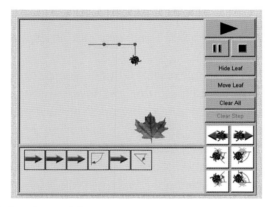

Figure 12.6 NLVM direction activity

Examples that deliberately encompass diverse cultural backgrounds can be used as a basis for investigations – for example, in statistics. Gapminder, <http://www.gapminder.org>, is an interactive resource suitable for older children that allows them to investigate a variety of statistics across the world.

Questioning children from other cultures

In mathematics classrooms, teachers aim to ask questions and to encourage children to ask questions about the mathematics. In some cultures, however, it is not appropriate to ask, or respond to, direct questions. Children have to be taught explicitly that it is okay to ask and respond to questions in the classroom. Establishing culturally responsive classroom norms may take time and considerable reinforcement, but is essential if all children are to participate in mathematical learning (Thompson & Hunter 2015). One approach is to consider the mathematics classroom as a 'family' in which members can have friendly disagreements and discussions about different approaches to solving mathematical problems (Bills & Hunter 2015). Good resources for developing mathematics in culturally responsive ways can be found at the Make It Count website, <http://makeitcount.aamt.edu.au> and at the NZmaths website, <http://nzmaths.co.nz>. From a mathematical perspective, all children – regardless of social and cultural backgrounds – learn mathematics best when the problems are relevant to their context and understanding, there is some degree of freedom in how they respond to the mathematical problems and they are encouraged to talk about mathematics and ask questions about mathematics.

Children who are gifted and talented

Children who are gifted in mathematics may need stimulation and extension beyond the mathematics suitable for their year group. Often teachers will carry out some pre-assessment before starting an activity, and then continue to teach regardless of the fact that some children can already do all the work. Such a practice leads to boredom and sometimes behaviour problems. There are two common approaches to teaching children who are gifted in mathematics: accelerating the curriculum or providing more depth and breadth. Multi-layered tasks provide opportunities for both approaches.

CLASSROOM SNAPSHOT 12.4

Pat has posed the question to her Year 1 class, 'I was walking down the corridor when I heard a child say "19" and the teacher said, "That's correct." What question might have been asked?'

After some discussion, the children are working individually to create questions for which the answer is 19. They are recording their answers on their tablets. Simon, who is gifted mathematically, initially provides random problems such as 57 ÷ 3 and 1199 – 1180. Gradually, he begins to work systematically, working from 19 + 0, 18 + 1, and so on down to 0 + 19. At this point, he continues –1 + 20, –2 + 21 and so on. When Pat asks him about what he has written, he says, 'It's a pattern that goes on forever. It's the add-ons backwards.' Pat asks him to save his work and encourages him to share his findings with the rest of the class.

The activity that Pat chose was very open-ended, and children could use any numbers with which they were comfortable. The activity allowed Simon to explore his understanding and Pat validated this through her response of sharing it with the class.

Many activities, such as those on the nrich or Illuminations websites, provide good starting points for developing a challenging program for children who are gifted in mathematics. Once again, the approach of creating a rich and varied mathematics classroom with multi-layered tasks offers opportunities for all children to participate and learn.

PAUSE AND REFLECT

From what you have read in this chapter, any additional research that you have done and what you have seen in classrooms, how well do you think the diversity of students' needs are met in the primary mathematics classroom? What approaches have you seen or tried that appeared to have been successful? How can technology use enrich the mathematics classroom so that all children can participate in meaningful mathematics learning?

Reflection

Australian and New Zealand primary mathematics classrooms are probably some of the most diverse anywhere in the world. The underlying equity focus is important for social and economic reasons, and should not be dismissed as either unrealistic or unnecessary. Mathematics is critical to future success, and many children are turned off mathematics very early in their schooling. The approaches to teaching mathematics to cater for diversity that are suggested in this chapter are grounded in research that indicates that all children benefit from a rich and meaningful mathematics curriculum. Technology can play an important part in providing tailored resources to allow all children to participate meaningfully in mathematics.

Further reading/websites for exploration

Further reading

Boaler, J. 2010, *The elephant in the classroom. Helping children learn & love maths*, London: Souvenir Press.

Websites

ABC: Splash: <http://splash.abc.net.au>. An Australian Broadcasting Corporation education site containing a wide variety of resources

Aspect: <http://www.autismspectrum.org.au>: Information about ASD and advice for working with children with ASD

Make It Count: <http://makeitcount.aamt.edu.au>

Maths Dictionary for Kids: <http://www.amathsdictionaryforkids.com>

New Zealand Curriculum Online: Kia Ora: <http://nzcurriculum.tki.org.nz>

Nrich: <http://nrich.maths.org>

NZmaths: <http://nzmaths.co.nz>

Paul Swan: <http://members.iinet.net.au/~markobri/Paul_Swan/Paul_Swans_homepage.html>

Pinterest: Educational – Math: <http://www.pinterest.com/tonsoffun/educational-math>. An online pinboard with a variety of early mathematics ideas

Pinterest: Tactile for the Visually Impaired: <http://www.pinterest.com/ab3867/tactile-for-the-visually-impaired>. An online pinboard with a variety of ideas for tactile materials to use with children who are visually impaired

Scope: <http://www.scope.org.uk/Support/professionals/learning-together/subject-areas/maths/key-stage/1/physical>: A UK site with information and advice about children with physical disabilities in mathematics classrooms

TKI: <http://tmoa.tki.org.nz>

CHAPTER 13

Surviving as an 'out of field' teacher of mathematics

LEARNING OUTCOMES

By the end of this chapter, you will:

- be aware of issues pertinent to 'out of field' teachers of mathematics in Australia, New Zealand and further afield
- know about strategies offered to assist 'out of field' teachers of mathematics
- realise the positive side to teaching mathematics as a primary-trained teacher
- be aware of examples of teachers' critical pedagogical content knowledge that has a high impact on students' growth and development across the main strands of the curriculum.

KEY TERMS

- **Mathematical content knowledge:** Understanding the necessary mathematics beyond the immediate demands of the curriculum
- **'Out of field' teacher of mathematics:** A teacher of mathematics who has not completed the minimum requirements to be registered to teach secondary mathematics. These teachers may be qualified to teach in the primary years or in secondary areas other than mathematics.
- **Pedagogical content knowledge:** The complex blend of content and pedagogical knowledge that provides the basis for classroom decisions made on a day-to-day basis

Shortage of secondary mathematics teachers and associated issues

It is not unusual to find a regional, rural or remote school without a qualified secondary mathematics teacher. In fact, such a dire shortage exists in the supply of qualified secondary mathematics teachers in our schools that, in Australia, nearly 50 per cent of teachers employed to teach junior secondary mathematics do not possess the appropriate qualifications and 32 per cent of teachers teaching senior secondary mathematics to not meet the tertiary mathematics content requirements to teach in the area (Vale 2010, p. 17). New Zealand schools share this problem, as they also have a serious shortage of teachers in the areas of mathematics and science. To make matters worse, the percentage of students studying advanced courses in mathematics at senior secondary levels is declining. While governments are introducing policies to reduce the higher education fees of mathematics programs, the benefits will not be felt for a considerable time, and a large number of secondary schools will have to accept that they will need to support 'out of field' mathematics teachers. This chapter tackles some essential pedagogical content knowledge (PCK) to assist 'out of field' teachers of mathematics in the junior secondary context.

CLASSROOM SNAPSHOT 13.1

Ella completed her Bachelor of Education (Primary) course and had prepared herself for the possibility of not finding employment in Sydney. In the final semester, she began applying for positions in rural areas of New South Wales. In mid-January the following year, Ella received notification that she had been granted an interview at a central school (K–10, ages 5 to 16 years) as a primary teacher. She did some investigating and found that the school was organised across two campuses, with K–6 students on one site and a 7–10 campus approximately 2 kilometres away on the other side of town. The primary campus had one class per year group and appeared to be adequately resourced, with a mix of experienced and early-career teachers.

During the interview, Ella communicated her philosophy of teaching and her professional teaching experiences during her studies. In relation to her course, members of the interview panel were keen to know more about her primary mathematics curriculum major and her elective major in environmental science. The following week, the principal of the school phoned Ella and began by stating that the interview panel had decided not to give her the primary position. The principal went on to add that the school would be delighted if she would take the position of mathematics/science teacher in the secondary campus. While Ella was surprised, her thoughts were on the fact that she had been offered a teaching position for her 'first year out'. She hadn't stopped to think about the learning curve she was about to face, teaching outside her primary teaching area.

This classroom snapshot to be continued …

The difficulty faced by teachers in Ella's position is often described in terms of a combination of limited **mathematical content knowledge** that is closely linked to pedagogical content knowledge. While primary teachers often come into the secondary sector with a kit of sound practical teaching strategies, their existing qualifications have not given them the opportunity to develop a deep conceptual understanding where the connections among concepts are the focus (Askew 2008; Vale 2010). It is interesting to note that programs that have been specifically designed to support **'out of field' teachers of mathematics** through further professional learning 'drew on their general **pedagogical content knowledge** as practising teachers and was enhanced by classroom and school-based enquiry' (Vale 2010, p. 19).

PAUSE AND REFLECT

When faced with the proposition of teaching in an unfamiliar teaching area, teachers often resort to teaching in the same manner as they were taught. How would you describe your own mathematics experiences in the secondary setting? What can you remember about the strategies used by your own secondary mathematics teachers?

The disappointing reality is that many members of the general public believe that secondary mathematics teaching means working through a textbook from beginning to end, with the most important classroom instruction being similar to 'Class, please turn to page 102 and complete questions 1 to 6 showing all working.' This provides a sad contrast to Sullivan's (2011, pp. 60–1) strategies to be considered for systematic planning for teacher learning in mathematics. The first of the four strategies targets 'creating possibilities for engaging students in mathematics', and the issues identified by Sullivan are summarised as follows:

- examining 'big ideas' that underpin curriculum strands
- examining the proficiencies – namely, understanding, fluency, problem-solving and reasoning – and creating opportunities to experience and explore these
- emphasising numeracy and practical student tasks in teaching and assessment
- promoting decision-making, connected experiences and usefulness
- selecting engaging student tasks and building these into a lesson
- exploring the mathematical content within mathematics tasks and developing strategies for learning knowledge as required
- examining strategies for heterogeneous classes that will enable all students to engage and develop, including specific ways to support students having difficulty and students requiring extending.

If the issues summarised above were addressed, the mathematics classroom would be a far cry from the textbook-reliant classrooms that still exist. However, this requires a paradigm shift to move from instructional top-down mathematics teaching to student-centred teaching, where the students are empowered to develop their own mathematical ideas. This procedural approach is particularly apparent in junior secondary mathematics classrooms, as many schools are allocating their experienced teachers of mathematics to senior classes, 'leaving teachers from other subject areas such as science to cover the gaps in junior and middle school mathematics teaching' (Australian Council of Deans of Science 2006, p. 58).

For 'out of field' teachers, there is frequently a considerable amount of mathematical content knowledge, pedagogical content knowledge and knowledge of students to explore while teaching full time. It is essential for these teachers to find a suitable 'mentor' in the school, who will take the time to have professional discussions concerning these issues and other school-based issues that arise. Preferably, the mentor should be a mathematics teacher; however, this may not be possible. Professional relationships can extend beyond your own key learning area, and in some schools this has led to programs targeting numeracy and literacy within all key learning areas.

In remote areas, professional associations such as the Association of Australian Mathematics Teachers (AAMT), <http://www.aamt.edu.au>, and affiliated state-based associations, such as the Mathematics Association of New South Wales (MANSW), <http://www.mansw.nsw.edu.au>, and the New Zealand Association of Mathematics Teachers (NZAMT), <http://www.nzamt.org.nz>, are invaluable. Professional reading concerning curriculum issues, research, teaching strategies (including ICT), professional development opportunities and conference information can be accessed online.

CLASSROOM SNAPSHOT 13.2

With only two weeks to prepare for her new classes, Ella contacted the school to find out which classes she would be teaching. She was surprised to hear that she had been allocated Years 7–10 mathematics, Years 7–10 science, one Year 7 music class, an interest elective (two hours per week) in home science and a Year 9 home room group. The reality of the situation had now hit, and on closer inspection she realised that her Year 8 and 9 classes were multi-grade. Ella's first port of call was the syllabus documents. In relation to mathematics, there appeared to be many topics. Ella was confused about where she should start and for how long she needed to teach each unit. Where would it end?

When Ella went to her new school to begin getting her room organised, she spoke to the principal and was encouraged to discover that he was a former maths/science teacher. He suggested that she contact him whenever she had any questions related to what she was teaching and how she was teaching, no matter how trivial she thought the questions seemed. Ella survived the first term by having an on-site mentor, a school environment that encouraged working as a team and the professional support provided by her mathematics association.

CLASSROOM SNAPSHOT 13.3

The members of a small group of teachers in a regional community were asked to join a professional learning study where groups of mathematics teachers were requested to develop their own professional learning project based upon the group's identified needs. The teachers were familiar with the term 'dynamic geometry software (DGS)'. They had opened DGS software and had a little play; however, they had no idea how to incorporate DGS into the teaching/learning sequence. As a group, they decided to make 'the meaningful use of DGS in the mathematics classroom' the aim of their professional learning proposal. The proposal involved three phases:

1. a professional development session with a DGS expert to facilitate engagement in student-centred DGS activities as a group and discuss the experience

2. as a team, teachers incorporating three DGS activities into their next two- to three-week unit
3. sharing of students' work samples, the teaching experience, evaluation and planning suitable DGS activities for the following unit.

With this shared focus among the team of secondary teachers, they had moved from 'retreatism' (Serow & Callingham 2011) to using DGS as a key component of student-centred tasks. Over the period of two school terms (20 weeks), the teachers were growing in their confidence to:

- develop lessons that explored GeoGebra basics, such as writing your name, designing a picture using the reflection tool and exploring the resultant quadrilateral when joining the mid-points of the sides of any quadrilateral using the segment tool
- construct figures using their properties to form 'robust' figures (when dragging points on the shape, the shape remains in that class of figures)
- turn a written open-ended task into a student-centred DGS task
- create templates that could be opened by students to lead them through an exploratory task. This is illustrated in Figure 13.1, showing all text that would appear when the students click on the action buttons. The italicised text shows what the students wrote after the constructions, 'dragging' and measuring had been completed.

Senior secondary teachers introduced the first derivative by exploring the change in gradients of tangents to the curve using DGS. For the first time in their experienced teaching career, these teachers witnessed an understanding of the gradient function.

Algebraic relationships were taken a step further, and represented on DGS by the students when suitable. This didn't require moving to the computer room; instead, the teachers began using the data projector as a catalyst for student discussion.

On reflection, this group of four teachers – of whom two were 'out of field' – developed in their content knowledge and pedagogical knowledge (including using technology as a teaching tool) as the intervention targeted a teacher-recognised area of need. It was embedded in their classroom practice over a six-month period.

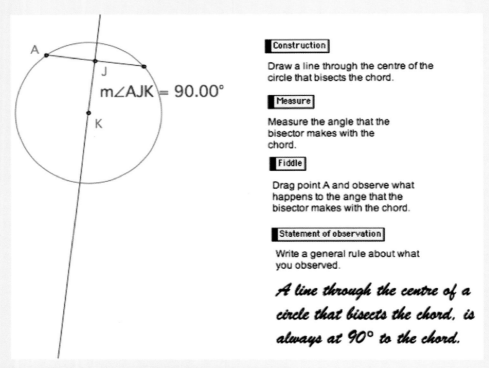

Figure 13.1 Circle geometry student work sample
Source: Serow & Inglis (2010, p. 13).

Secondary lesson structures

For the 'out of field' teacher of mathematics, it is heartening to know that lesson-planning skills developed in other settings can be transferred to the secondary mathematics situation. Keeping this in mind reduces the risk of falling into the trap of teaching secondary mathematics in the instructional procedural manner we sometimes hold in the memory banks of our own experiences at school. Although a multitude of combinations of lesson components exist, a few key techniques will help you get started. Considering typical secondary mathematics lesson components, we have formed the following six categories to assist lesson planning:

- short questions at the start of the lesson or the presentation of a scenario
- a clear introduction to focus the lesson
- student tasks

- sharing/discussion
- homework setting/correcting
- the conclusion to the lesson.

Short questions at the start of the lesson or the presentation of a scenario

This lesson component has a few functions, with the short questions usually taking five to 10 minutes to complete. The short questions can assist in:

- transitioning the students from their activities prior to beginning the mathematics lesson
- gathering information concerning students' current level of understanding of concepts that will be developed (assessment for learning)
- focusing the students on the topic to be explored
- leading into the introduction of the lesson
- automaticity of mathematical facts to relieve the short-term memory.

Some teachers have found the use of HOTmaths Scorcher, widgets and quizzes to be a useful source for short questions (see the upper right tabs of any HOTmaths lesson). Scenarios can lead into a contextual investigation grounded in real-life situations.

A clear introduction to focus the lesson

While it may take only one minute, this is a very important component of the lesson. In an attempt to lead the students to the development of the mathematical idea, it is important to communicate the intention of the lesson clearly, and hopefully in an intriguing manner. This could be as simple as, 'Today we are going to explore one of the techniques that builders use in their constructions. Let's begin by looking at a few pictures of builders at work.' This could be the introductory component to a lesson exploring Pythagoras's theorem.

Student tasks

The student tasks can take on various forms and have a range of purposes. The tasks have been divided into three main categories. Three secondary HOTsheet tasks have been included to illustrate the different categories. You can access these HOTsheets by logging into HOTmaths, entering their respective names in the search field and then selecting the 'HOTsheets' tab.

Tasks that develop the mathematical idea

INVESTIGATING PARABOLAS

Use the widget from this lesson to complete these tasks.

TASK 1   Explore positive integer coefficients of x^2

The diagram shows the graph of $y = x^2$.

Use the widget to create the graph for each of these equations and then sketch the curves onto the graph, labelling each curve with its equation:

- $y = 2x^2$
- $y = 5x^2$

Describe what happens to the curve as the positive coefficient of x^2 increases.

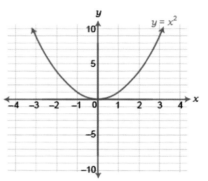

_____

_____

_____

TASK 2   Explore negative coefficients of x^2

The diagram shows the graph of $y = x^2$.

Use the widget to create the graph for each of these equations then sketch the curves onto the graph, labelling each curve with its equation:

- $y = -x^2$
- $y = -2x^2$
- $y = -5x^2$

Describe the difference between the graphs of $y = x^2$ and $y = -x^2$ or $y = 2x^2$ and $y = -2x^2$

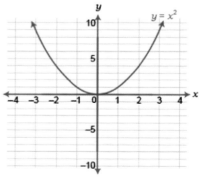

_____

_____

_____

Figure 13.2 HOTmaths Investigating Parabolas HOTsheet

Chapter 13: Surviving as an 'out of field' teacher of mathematics

Tasks 3 and 4 of this HOTsheet go on to explore coefficients of x^2 between -1 and 1, and the constant term. It is evident in the design of this student task that the students have ownership of the mathematical idea as opposed to explaining the relationships and procedures and asking the students to repeat the process.

Tasks that further develop the mathematical idea

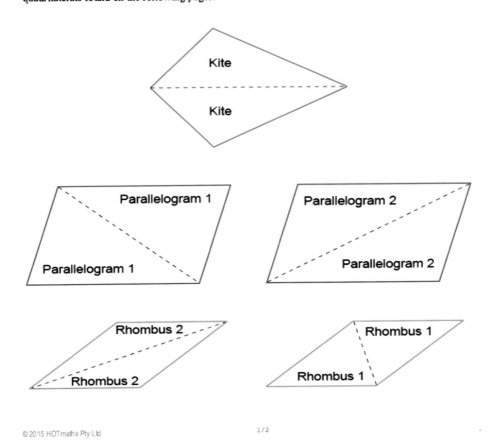

Figure 13.3 HOTmaths Triangles Inside Quadrilaterals HOTsheet

Tasks that utilise the mathematical idea in a problem-solving context

THE ANZAC BISCUITS BUSINESS

You decide to make Anzac biscuits to sell for a fundraising drive. This recipe makes 10 large biscuits.

ANZAC BISCUITS
- 1 cup plain flour (200 g)
- 1 cup rolled oats (regular oatmeal) uncooked (150 g)
- 1 cup desiccated coconut (150 g)
- 1 cup brown sugar (200 g)
- 1/2 cup butter (100 g)
- 2 tbsp golden syrup or honey (40 g)
- 1 tsp bicarbonate of soda (bicarb) (5 g)
- 2 tbsp boiling water

TASK 1   Decide on the numbers

Determine the quantity of ingredients needed. Number of biscuits you want to cook = _____

TASK 2   Research the costs

Use supermarket brochures, the internet or visit a supermarket to find the cost for each ingredient.

Ingredient	Amount needed	Cost
Flour		
Oats		
Coconut		
Brown sugar		

Ingredient	Amount needed	Cost
Butter		
Golden syrup		
Bicarb		

TASK 3   Calculate the cost

Calculate the total cost of biscuits.

TASK 4   Calculate the profit

You will need to make a profit of at least 40% to make the venture worth your while.
How much do you need to sell each biscuit for to make at least 40% profit?

TASK 5   Break even

Calculate the number of biscuits you need to sell to *break even* (to make back the costs).

TASK 6   Profit

If you sell all your biscuits, how much profit will you make?

© 2015 HOTmaths Pty Ltd

Figure 13.4 HOTmaths The ANZAC Biscuits Business HOTsheet

Sharing/discussion

This component of the lesson is just as important as the student tasks, and it is often not included. Due to the period structure of most secondary school timetables, running out of time can mean that the discussion simply doesn't happen. One way around this is to make the discussion of the key ideas of the previous lesson flow after the short questions in the next lesson. Each student in a class will benefit from the sharing of ideas in the mathematics classroom. Even if more able students have mastered the concepts previously, they are given the opportunity to express their findings and converse at their level of understanding. You may witness other students moving from informal to formal descriptions of mathematical concepts during pair, small-group or whole-class discussions. Moving students from an implicit to an explicit focus is facilitated by mathematical conversations where students are prepared to take mathematical risks.

Homework setting/correcting

It can be difficult to determine the stage of the lesson at which it is best to introduce this component. There is no hard and fast rule, but some teachers find it easier to state homework expectations before the conclusion of the lesson. This avoids problems related to students packing up their equipment as the teacher attempts to get them to record homework details. While many schools have homework policies, there is no need to set mathematics homework each night if it is not suitable for the next lesson or it is not a useful revision task. Mathematics homework for 'homework's sake' creates negativity and boredom. When the homework is achievable, is related to classroom experiences, engages students in varied contexts and/or is a form of communication between school and home, homework tends to have a positive impact.

Conclusion to the lesson

The conclusion is the 'punch' in the lesson. It should be a time for the students to clarify their thoughts briefly on the 'big mathematical ideas' of the lesson. It can take many forms, including a journal entry, a discussion, a brainstorm, a concept map or a diagram, to name only a few. The HOTmaths website includes an illustrative dictionary of mathematics concepts that can provide an interesting stimulus for lesson conclusions (see Figure 13.5). Mathematics dictionaries are useful to clarify language and can be conversation starters.

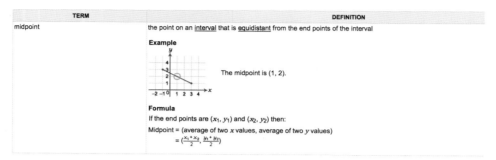

Figure 13.5 HOTmaths illustrated dictionary example

Issues to think about in the secondary context within each strand

When a teacher first begins targeting a new key learning area, it takes quite some time to get a handle on the important issues. This section highlights a range of concepts that require meaningful consideration in the classroom.

Chance and data

Watson (2006) has carried out extensive explorations of students' statistical and probability understanding. Based on this work, the following concepts require some explicit attention at the secondary level:

- sampling in terms of size, representation and randomness
- graphing and data representation to tell the story beyond the mechanical labelling
- averages in terms of what they actually represent
- summarising data
- chance and likelihood concepts, including language used to describe likelihood, exploring single and compound events and early conditional elements

- the beginnings of statistical inference in the form of justifying conclusions by referring to data
- variation concepts, which are recognised as underpinning many statistical relationships in terms of describing distributions.

Students at the secondary level yearn for opportunities to explore mathematical concepts related to real-life examples. The exploration of data and statistics concepts is open to a wealth of examples. For example, even a trip to a decommissioned prison could enable the collection of stimulus material captured as an image on a mobile device. A secondary mathematics teacher recently collected images that included presentations of data relating to the ageing population in prison, titled 'Growing old behind bars: Why it matters', and the hidden costs of incarceration, titled 'Full impact of incarceration'. Both topics provided opportunities for in-depth mathematical investigations while fostering interest in current debates at a local and international level.

All teachers of mathematics should aim to contextualise mathematics tasks to draw out the key mathematical ideas from real-world problems as much as possible. Simple strategies that make use of extra-curricular school activities provide a good source of stimulus material. These could be drawn from the mathematics behind running a stall at the school fete, measurements of a running track, football scores and league tables, or local-interest stories such as problem traffic areas near the school.

Geometry

At this stage of schooling, the students are required to focus upon the relationships among figures and properties; however, many secondary students are still at the level of understanding where they focus on individual properties. Some students will focus on visual cues, and activities will need to target properties before relationships among them become the focus.

Students at this stage benefit from tasks in which they identify the relationships themselves. As in the circle geometry example in Figure 13.1, if exploring relationships among parallel lines and the angles formed by a transversal, it would not be suitable to explain the relationships, provide worked examples and ask students to complete 20 similar questions. The students will gain more by being asked to draw five different pairs of parallel lines (different lengths, different distances apart and in different orientations) and cut each pair with a transversal. The students can then investigate all the angles formed and describe any relationships they find. After sharing what they have found, formal names for the relationships – such as alternate angles, corresponding angles and co-interior angles – can be introduced. At the same time, students will be revising angles at a point and angles on a straight line. This task could be done on paper or using a form of dynamic geometry software on personal devices.

Number

Just as we have witnessed the shift to a greater emphasis on mental computation in the primary context, this is also pertinent to the secondary context. When visiting secondary classrooms, it can be disappointing to find students relying completely on their calculators for simple computations. As mentioned earlier, students' working memory is better placed on the 'big idea' at hand. It is essential for secondary students to continue developing their mental computation skills across each of the strands.

ACTIVITY 13.1

Consider strategies for using HOTmaths Scorcher and Topic Quizzes to enhance mental computation in your class. What could a regular schedule look like in your classroom?

Patterns and algebra

Again, focusing on the relationships is key to the development of algebraic concepts. It is interesting to note that the strand that uses symbols to represent values relies strongly on concrete materials and explorations to develop these notions. Taking the time in the early years of secondary to work with real-life algebraic relationships where the students explore the number patterns, describe the relationships, represent them and manipulate them will have more impact than completing a multitude of repetitive similar questions. Solving equations with the equal arm balance and then moving to the visual is worth the time it takes to organise.

Measurement

Measurement in the secondary context gradually becomes a web of connections requiring knowledge of concepts across all strands. It is useful to draw students' attention to the connections and give them time to reflect on all the strategies and knowledge that they have brought to the task. Presenting questions in an interesting context enhances engagement in measurement lessons. Taking the time to physically explore relationships is important. A HOTmaths example of this is provided in Figure 13.6. You can access this HOTsheet by logging into HOTmaths, entering its name in the search field and then selecting the 'HOTsheets' tab.

AC

ACTIVITY 13.2

When do you think it would be appropriate to use a HOTmaths widget that demonstrates the relationship between volumes of prisms and volumes of pyramids? During which lesson component/s would you incorporate it?

In all content strands, tasks that ask the students to use reversibility techniques require them to focus on the relationships among the elements rather than follow procedures. These tasks often begin with what we may see as the answer to a traditional problem, where the students are required to reverse their thinking in order to solve the problem. An example of this comes from the Maths300 website, <http://www.maths300.com.au>, and targets the concept of the area of a triangle. The students are provided with the area of a triangle and are required to find the possible dimensions of the triangle that will result in that area. The task enables movement of the corners of the triangle, followed by an animation to show the rectangle to which the area of the triangle is equal.

Sample secondary sequence

The teaching sequence described in Serow (2007a) was designed with two main elements in mind: using the van Hiele (1986) teaching phases (introduced in Chapter 4) as the teaching/learning framework, and the embedding of dynamic geometry software and other suitable technology. The two-week teaching sequence (eight sessions of 40 minutes' duration) is appropriate to a secondary mathematics class targeting classifying, constructing and determining quadrilateral properties. An outline of the teaching sequence and sample student responses to tasks is provided next.

COMPARING VOLUMES: PRISMS AND PYRAMIDS

On the following pages are the nets of an open square prism and two open square pyramids.

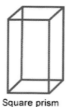

Square prism   Right square pyramid   Oblique square pyramid

The prism is *right*, that is, its sides are perpendicular to its base.

One pyramid is also right, so its apex is above the centre of the base. The other is an *oblique pyramid*.

All three solids have the same bases and the same perpendicular heights.

Paste these nets onto thick cardboard then cut them out. Fold along the dotted lines and use sticky tape to join the edges to form the solids.

TASK 1   Compare oblique and right pyramids

Use sand or rice to test whether the volume of both pyramids is the same. Describe your findings.

_____

_____

TASK 2   Compare right prism and pyramid

Fill the pyramid with sand or rice. Pour this carefully into the prism and keep refilling the pyramid to determine how many pyramids full of rice are required to fill the prism.

What fraction of the space inside the prism would be taken up by the pyramid?

Volume of pyramid = $\dfrac{\Box}{\Box}$ of volume of prism

Figure 13.6 HOTmaths Comparing Volumes of Prisms and Pyramids HOTsheet

Session 1: Information phase – mechanics and recall

Students work through simple constructions using a form of dynamic geometry software (DGS) and brainstorm known quadrilaterals. Constructions involve the following tasks.

- Write your name using GeoGebra.
- Create a person and reflect the person. Measure a selection of corresponding sides and angles. What do you notice when you drag one of your people?
- Create a house design using the six quadrilaterals: kite, trapezium, square, rectangle, rhombus and parallelogram.

At this stage, in most cases, the students will construct their figures using the line tool. This will be extended in later phases. When the students are asked to drag (drag test) the quadrilaterals they have formed in this way, they will notice that the constructions are not robust. This becomes the motivation for the following phase.

Session 2: Explicitation phase – robust templates and recording

Students create robust templates for each of the six quadrilaterals on separate DGS pages. If the drag test allows the figure to remain as intended, the construction will involve known properties of each figure. Discussions will begin to occur on relationships among figures – for example, comments such as 'this is really strange – when I drag the parallelogram it is sometimes a rectangle, square or rhombus'. This activity will involve constructions such as parallel lines, perpendicular lines and transformations. It is essential for the students in this phase to describe their construction within a textbox on the DGS page and to record the properties for each quadrilateral in a teacher-designed table.

Session 3: Directed orientation phase – irregular quadrilateral and mid-point construction

Students are instructed to:

- create any irregular quadrilateral using the line tool
- construct the mid-points

- join the mid-points to construct another quadrilateral
- answer the question 'What do you notice?'
- investigate the properties of this shape to justify what they have found, and record their justification in a text box.

Session 4: Free orientation phase – further exploration of properties and figures

Students design a spreadsheet where the six quadrilaterals are contained in the first column and the first row contains all possible properties of quadrilaterals. Particular care needs to be taken to include diagonal properties such as 'diagonals meet at right angles'. The students record the properties of each figure by ticking the appropriate cell. There is an element of surprise in the classroom when the students notice that the square has the maximum number of ticks.

Session 5: Free orientation phase – diagonal starters game design

This activity is designed to reinforce diagonals as a property and not merely a feature of the quadrilaterals. Students are given the challenge to create the diagonal formation needed for each of the quadrilaterals. The aim is for the students to construct templates for younger students to complete the figure and explore the properties.

Session 6: Free orientation phase – concept maps and flowcharts

Students create concept maps and flowcharts using suitable software. The chosen software will vary according to school accessibility. Figures 13.7 and 13.8 show student samples of a concept map and flowchart using Inspiration software, <http://www.inspiration.com>.

Session 7: Free orientation phase – property relationships consolidation

Using DGS, students explore and record the relationships among properties. For example, students may record the relationship among opposite sides

Chapter 13: Surviving as an 'out of field' teacher of mathematics

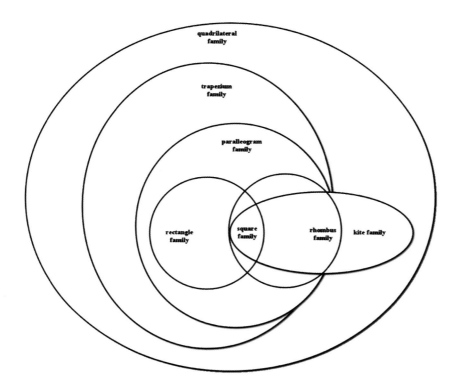

Figure 13.7 Sample of student's concept map

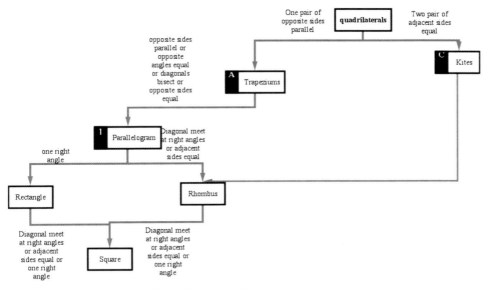

Figure 13.8 Sample of student's flowchart diagram

parallel, opposite sides equal and opposite angles equal. Figure 13.9 shows an example of a student's record of property relationships.

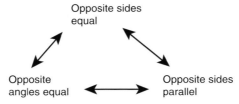

Figure 13.9 Sample of student's property relationships summary

Session 8: Integration phase – information booklet design

Students organise the constructions they have made, justifications, tables, spreadsheets, concept maps and flowcharts to produce an information booklet to explain what they know about the relationships among quadrilaterals and quadrilateral figures. Students are instructed to include an overall summary of their findings.

Session 9: Integration phase – sharing and routine questions

This phase involves class sharing of booklet designs and routine questions involving known properties and relationships. A main feature of the teaching sequence presented is the integration of dynamic geometry software using the van Hiele (1986) teaching phases as a framework for maintaining students' ownership of their mathematical ideas. This is facilitated via student-centred tasks that acknowledge students' individual experiences and the progression from informal to formal language use. The teaching sequence combines a range of effective teaching practices that make use of the potential of the technological tools currently available to secondary students.

Reflection

While it is essential for governments and universities to consider the long-term effect of the demand for teachers with a strong mathematics background, the education community can provide avenues for assisting 'out of field' teachers of mathematics in the secondary setting. Suitable mentoring programs, in-service education that directly impacts on classroom practice and accessing online information and tools will assist in creating a community of practice where mathematical and pedagogical content knowledge is developed and shared. Essentially, societal attitudes towards mathematics teaching are impacting the number of people considering a career as a mathematics teacher. For this attitude to change, we need to begin by considering the experience that the school-age generation is having in the mathematics classroom. With a student-centred approach based on investigation, exploration and relevance to the world around them, motivation to continue in a mathematics career should keep growing.

Websites for exploration

GeoGebra: <http://www.geogebra.org>

Geometer's Sketchpad Resource Centre: <http://www.dynamicgeometry.com>

HOTmaths: <http://www.hotmaths.com.au>

Inspiration Software: <http://www.inspiration.com>

Maths300: <http://www.maths300.esa.edu.au>

CHAPTER 14

Teaching mathematics beyond the urban areas

LEARNING OUTCOMES

By the end of this chapter, you will:

- explore challenges and rewards when working in remote locations in Australia, rural New Zealand, and small Pacific Island nations
- discover teaching strategies that assist in maintaining a positive learning environment in remote and small Pacific nation classrooms
- understand the importance of the relationships among and between parents, students, teachers and other community members
- discover how to make the most of available resources, including online sources of teacher reference material.

KEY TERM

- **Remote area:** A location far away from an urban centre or large regional town.

One of the most rewarding and professionally invigorating experiences a teacher can have is working beyond the city, in regional, rural and remote communities. Whether these communities are situated in rural and remote Australia, the rural areas of New Zealand or a developing Pacific Island nation, the teaching of mathematics requires a thoughtful responsiveness to culture and context.

If you have ever taught in a **remote area**, one of the first things you may have noticed is the importance of community and parental involvement. Relationship-building is the key to success at all levels. While mathematical content knowledge (MCK) and pedagogical content knowledge (PCK) remain important ingredients, elements of PCK are further extended to enable concept development when working in a climate of limited resources, reduced availability of technological tools and access, and fewer role models in terms of mathematics career paths. Often, lower rates of school attendance impact on individual student growth in mathematics, and it is essential to implement strategies to improve attendance rates and close the gap in mathematical opportunities.

This chapter considers issues common in the mathematics classroom in rural and remote areas of countries similar to Australia and New Zealand, and the Pacific Island country context. Classroom snapshots capture real classroom situations within the context of primary and early secondary mathematics teaching. Many of the issues can also be found in suburban and regional centres; however, their prevalence and the need to address them are often exacerbated in remote areas. This chapter explores strategies for making the most of available resources and the invaluable professional experience of working in these areas.

Rural and remote areas

While approximately two-thirds of Australia's population lives in major cities, outside the major cities Australia has the lowest population density compared with the rest of the world, according to the Australian Institute of Family Studies (AIFS 2011). Despite this, it can be a shock for some early career teachers to find that they have received a teaching position in areas often described as 'the bush', 'the country' or any place out of 'the city' that requires venturing into an unknown environment. For statistical purposes, Australia's areas are divided into major cities (69 per cent), inner regional (20 per cent), outer regional (9 per cent), remote (1.5 per cent) and very remote regions (8 per cent). While there is little difference in terms of family composition, similar to Pacific Island households, very remote areas have a higher proportion of their households described as multi-family. Everyday family life involves the extended family, and this should also include the school community and communication.

The AIFS (2011) has identified that, 'Access to services and educational aspirations are generally more limited in outer regional areas'. In addition, it has found that geographic area does affect the type of after-school activities in which students engage, with children from remote areas spending more time outdoors. Children from urban areas spend more time on a range of extra-curricular activities. There are also differences in children's developmental outcomes across geographic areas, with children in major cities doing better on physical development and learning outcomes measures than children in other areas. Issues such as sleep deprivation, appropriate school clothing, transportation and lunch provision are daily concerns faced by some students in many areas of the world.

While the geographical size of New Zealand is relatively small in comparison with other nations, there are some areas that are considered remote and that are also sparsely populated. A review of inclusive education in New Zealand (Powell 2012) suggests that challenges remain in relation to equity in access to education due to remoteness. As educators, we are asked to respond to different settings and different needs of learners by adapting the content of our teaching, as well as the strategies we employ to enable all children to have access to quality education.

While equity in access issues have been reported in varied arenas, it is not our task as teachers in these areas to placidly accept the gap in educational opportunity; rather, our role is to respond in a culturally responsive manner, developing and using strategies that reduce the gap. The solutions are not one-size-fits-all. We do know that sitting back and accepting the difficulties as unworkable is not the answer, though. For example, it would not be acceptable for a school to acknowledge that a community does not have a resident female scientist when noticing the scientific interest shown by a female student. Instead, the school should grasp the opportunity and search for an online mentoring/partnership program where female science professionals are linked with students who have an interest or potential in this area.

Being flexible

Many schools beyond the inner regional centres do not have access to key learning area specialists, and you will most likely find that music lessons, information skills, physical education and languages are planned and delivered to your class by you alone. As a consequence, the amount of planning required is increased, and release from face-to-face teaching is sometimes minimised.

There is also a need to be more flexible with playground supervision responsibilities in smaller schools. One strategy for making the most of teacher time outside is an edible school garden project that can extend to a wide range of key learning areas. In mathematics, for example, a range of number, measurement, statistics, chance, patterns and algebra, and geometry concepts can be explored with the school garden as a catalyst for student activities. Student tasks of this nature lend themselves to education for sustainability (EfS) links in the mathematics classroom (Serow 2015). These could include:

- developing algebraic relationships through tile patterns on square garden beds
- pen and paper-scaled garden designs
- dynamic geometry software constructions to create garden designs
- experimental chance activities through seed germination counting from the original number of seeds
- experimental chance activities through the analysis of produce from a transect of plants
- array activities in the context of plant rows and columns
- design and collection of data concerning students' fruit and vegetable preferences
- following a recipe and measuring substances to create a meal using fresh produce from the garden
- comparing fast food preparation time with meal options that focus on fresh foods.

An integrated program presents many challenges and rewards in out-of-city settings. They particularly lend themselves to making links to personal stories, the land and the community, and to varied ways of tackling problems. In some remote settings, English will be a second language, and it is not uncommon for some children to have up to four languages at their disposal by the age of 5. Bilingual and multilingual classrooms have the potential to be rich and engaging in terms of culture, and all languages should be celebrated and incorporated into classroom activities.

In remote schools, teacher turnover is high, and it is difficult to recruit experienced teachers (Sullivan 2011). Due to these two key factors, schools are often faced with professional knowledge moving out of a school each time a teacher vacates a position. Sustainable practices are required to reduce the impact on the school community, which often will have put considerable effort into professional development and positive change. Strategies for reducing the impact of teacher turnover will be explored in Chapter 15, which targets mathematics key learning area coordination. Classroom snapshot 14.1 provides a scenario of a school that recently tackled this issue.

CLASSROOM SNAPSHOT 14.1

A primary school in an outer regional area of Australia ventured on a 12-year journey of change that is ongoing. This journey involved being part of a larger project titled Make It Count: Numeracy, Mathematics, and Indigenous Learners, which was led by the Australian Association of Mathematics Teachers with federal funding (2009–12). During 1998, the school community had noted that Aboriginal students' numeracy outcomes were below those of non-Aboriginal students and an Aboriginal Action Plan was initiated that challenged teachers to consider their approaches and develop a sense of cultural awareness. The staff began acknowledging what the principal described as 'everyday culture' and 'started looking at individuals – no longer labelling'. This was only the beginning.

The school took a closer look at teacher and student engagement, and worked with Chris Matthews and Tyson Yunkaporta, Aboriginal academics who are educational leaders in this field. As community engagement was high on the school priorities, they began with community links. To illustrate, one of these took the form of building communication of the current mathematics focus between school and home. For example, when the students were targeting a particular mathematics topic at school, a small, related, practical task would be sent home for completion that involved family members and discussion. The teachers shared the Eight Ways approach (Yunkaporta 2009) with the students, and always began lessons with a 'Yarn Up' (story sharing) that often included a pre-assessment. They had begun to teach 'through culture and not about culture'.

Yunkaporta's 'eight ways came from Indigenous research, which is research done by and for Aboriginal people within Aboriginal communities, drawing on knowledge and protocol from communities, Elders, land, language, ancestors and spirit' (Yunkaporta 2009, p. 1).

The Eight Ways model (Yunkaporta 2009, pp. 4–7) includes the following elements:

- story sharing
- learning maps
- non-verbal learning
- symbols and images
- land links
- non-linear processes
- deconstructing/reconstructing
- community links.

The *Make It Count Correspondence Newsletter* became a regular publication; this was requested by Aboriginal parents. There was a sense of a 'learning together culture' evolving, where a common purpose had been struck. While there are over 600 students on two campuses, through Eight Ways they have a common language that has enabled them to continue to grow.

This project has brought about careful planning of each mathematics unit (K–6) using the Eight Ways pedagogical structure. The school has removed individual student texts, and is adopting a student-centred approach that uses varied materials (including ICT). As each unit was designed by classroom teachers for their particular groups of students, the teachers have noticed that the students are engaged and generally enjoying mathematics. Each unit is stored on an online database to enable transfer of information, further development and individualisation.

Figure 14.1 Using concrete materials and everyday items

Teachers made the following observations about their own changes in practice:

- *Teacher 1:* The thing that I've enjoyed about it is that the way we've been teaching it works for all kids … I like the yarn-up, for example – we give them a bit of an understanding of what it's about in everyday terms, if you're talking about angles, you talk, you know, where angles are, or volume, where they have volume in their everyday life, where it … you know, where it happens, where it occurs … How it, what it means to them, so I suppose then they're constructing that knowledge, building on that knowledge that they already have … therefore it's deeper based, I suppose.
- *Teacher 2:* When we were developing the program we did a lot of collaborative planning … that really helped me to understand how to connect the outcomes and indicators with the activities, and using the Eight Ways of learning to create different ways of teaching maths, I guess … we do a lot of hands-on assessment, which I find is really good with Kindergarten. And … I guess it's just really helped me to understand – I found assessment a really hard thing to sort of understand while I was at university, because we just didn't do much of it, so I didn't really know how it [would] fit in … with the 3D shapes for the assessment, it might be like sorting them into the different groups, or making them out of play dough, things like that, so it's really hands-on …
- *Teacher 3:* For me, a lot more hands-on activities. Maths really moved away from worksheets and … concentrated on different – differentiation and also hands-on activities and concrete materials and things like that. So that's what it has meant. It's changed, it's really changed a lot of the ways we're teaching maths … Also involved a lot of technology, and incorporating things like that – which is using the smartboards more, making up sort of … notebooks on a concept, so you still have it there at your fingertips … and taking them outside, and doing really good yarn-ups and things like that – which were always things we were doing in maths before, but really consolidating that … and having it there ready at your fingertips as soon as you're ready to go.

In relation to communication and professional sharing, the following comment was made:

- *Teacher 4:* You know, really [high-]quality assessments, and we're emailing those to each other. Which is … that's probably the first time in my career that we've really shared stuff like this too. Like, teachers always did it, but they'd do it on their own and … it just wasn't a culture of sharing. It wasn't that people wouldn't share, it's just that you were doing different things at different times and it didn't happen so much. So that's been really positive, I think.

Recognising the 'performance gap between Indigenous young people and their non-Indigenous peers', the Australian Association of Mathematics Teachers' *Make It Count Blueprint for Supporting Best Teaching of Mathematics for Indigenous Learners* (AAMT 2013) summarises some approaches to improving mathematics outcomes for Indigenous students that have been generated by research. These include:

- establishing a classroom environment that is predictable for students
- working with integrated curriculum in which there is a clear focus of mathematics development
- student learning groups
- cross-age tutoring in which older students are cast in the familiar role of helping educate younger children
- 'dialogic' teaching in which educators have a dialogue with students, and build narratives about and with mathematics.

AAMT supports the notion that there is no single approach that leads to successful mathematical development for all students. Partnerships are stated as being key to future progression. They are described as occuring on two levels, 'school, parent, family, and community' and 'professional learning communities as educators' (AAMT 2013).

Very young children take many stories from school to home and vice versa. We can make the most of this communication medium by linking our classroom explorations with simple home tasks. Figure 14.2 shows children's collections of four objects brought to school from home while targeting the numeral 4 that week. As a class, they added four exercise books and four pencils to the collections.

Figure 14.2 Communication between school and home

Considering classroom structures

When most of a teacher's personal experiences have been in single-grade classes, it can be a challenge to begin teaching in a multi-grade or multi-age class structure. Multi-grade classes have students from different grades within the same classroom. They may range from two grades together up to seven grades together, depending on the size of the school and how the school classes are structured. The determining factor for which classes will be grouped together in small schools is often the number of students in each grade. Multi-grade classes are very common in the remote areas of Australia. Cornish (2006, p. 10) outlines a number of factors that impact on teaching in a multi-grade class:

- number of grades in the class
- number of different classes (where one teacher is responsible for more than one class)
- size of the class
- teacher's experience
- teacher's preparation to teach a multi-grade class
- teacher's attitude to teaching a multi-grade class
- resources available
- curriculum documents
- mandatory requirements in terms of assessment and curriculum
- amount of flexibility available to the teacher
- support from educational professionals
- community support
- absenteeism of students
- absenteeism of teachers.

While absenteeism of students in remote areas and in many Pacific Islands is a frustrating and a continuing issue, absenteeism of teachers affects all members of the community. In many remote areas, there is no such thing as a supply list of available casual teachers. In fact, if a teacher is away, it usually means allocating the students to other classrooms until the teacher returns to work. If this happens at a crucial time within the sequence of the activities, it can impact upon the continuity of the unit of work. When this occurs, the class sizes are generally so large that planned resources do not 'stretch' to cater for the additional students in the class and the planned activity is abandoned.

ACTIVITY 14.1

HOTsheet

TOOTHPICK PERIMETERS

Both these shapes have a perimeter of 8 toothpicks.

Draw different 2D shapes with right-angled corners that can be made from 12 toothpicks.

Use the grid below to draw your shapes. One shape has been drawn for you.

Be careful that you don't repeat a shape by drawing it in a different orientation.

For example these two shapes are the same:

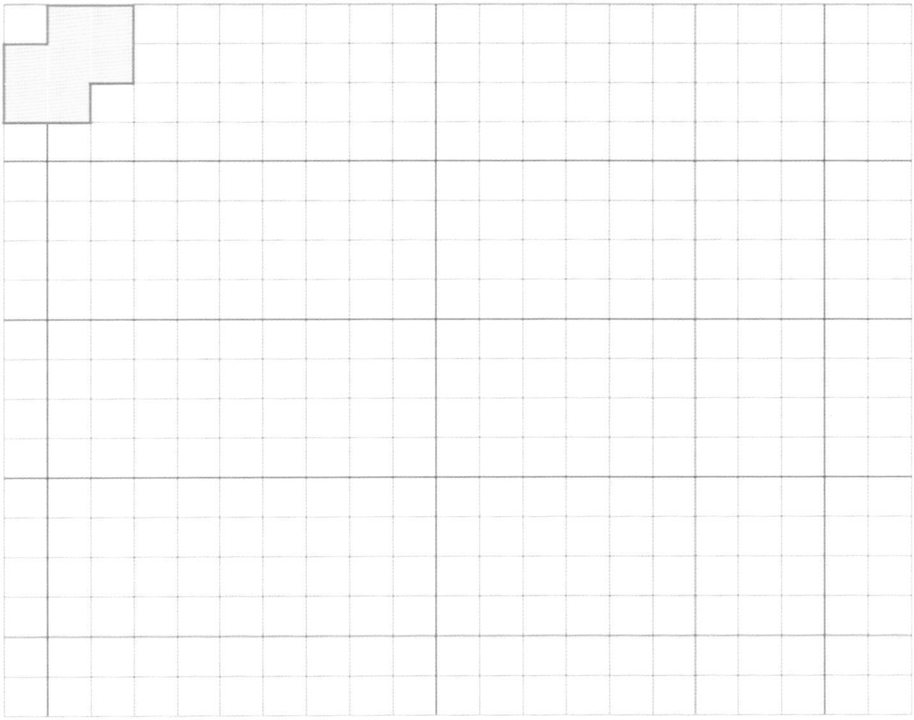

© 2015 HOTmaths Pty Ltd

Figure 14.3 HOTmaths Toothpick Perimeters HOTsheet

AC

Think about the scenario of teaching a Year 5/6 class of students in a remote school where casual teachers are not available. The Year 3/4 teacher is sick and her class of students will be joining your class. You have been in this position before, so you have started compiling a collection of open-ended mathematics tasks that can be the centre of an innovative large-class lesson sequence. Figure 4.3 is an example of a HOTmaths HOTsheet activity that is included in your file. Find five other open-ended tasks that could be added to the collection of HOTsheets. As detailed in Chapter 1, links are incorporated into each lesson within each topic. The HOTsheets are also found in the lesson resources located on the left-hand side of each lesson page. HOTsheet solutions are also provided within the lesson resource section.

In contrast to multi-grade classes, Cornish (2006) describes multi-age classes as a 'mixed-age by choice, by philosophical preference, and reflecting a well-thought-out approach to teaching and learning based on the idea that diversity leads to educational benefits'. This notion is not reflected in mono-grade systems, where 'the most effective way to organise a schooling system is to reduce diversity by grouping the students into classes based on their age and assumed similarity of development'. While it is not necessary to go down the track of multi-age classrooms, teachers of mathematics can organise multi-grade and mono-grade classes from the principle of targeting the individual needs of each student and carefully considering the structure of tasks and how developmentally appropriate the tasks are for each student in the class.

There are various ways of structuring the class to cater for students' individual needs. Bear in mind that teachers can use a combination of strategies based on the children in the class, the concepts being targeted and the context of the school community. Common ways of structuring mathematics lessons for multi-grade classes are described below (with each having its positive and negative elements):

- splitting the timetable to have different subjects for different grades – for example, mathematics activities for the Year 1 students while the Year 2 students are doing English activities
- mathematics at the same time for each grade
- a combination of whole-class teaching followed by separate mathematics tasks for each grade
- running topics on a cycle and teaching to the whole class, considering developmental differences among the students
- placing students into smaller learning groups – ability or non-ability groupings – across or within grades
- peer-tutoring strategies.

Tensions exist in each of these structures. When splitting the timetable, the teacher has multiple planning and resourcing responsibilities. Also, due to the increase in organisation, teachers can find that they spend more time organising and administrating than facilitating mathematics discussion and enabling the students' mathematical ideas to be expressed explicitly by the students. One of the most important skills the teacher of mathematics has is the ability to 'draw out' the key mathematical ideas from the students. The gradual shift from informal to formal language requires investigation, reflection and discussion. This is difficult to maintain across different groups doing different subjects at the same time.

Bear in mind that students' mathematical developmental journeys can be upset by curriculum mapping that is not sequential – for example, if topics are mapped over a two-year sequence, every second year the concepts have the potential to be out of sequence for many of the students. One way forward is to take a more flexible, mixed-methods approach. In multi-grade classes, it may be suitable to have some strands where sub-strands are the same for each grade due to the accessibility of open-ended tasks to be completed at different levels. Some sub-strands also lend themselves to task modification, where the central idea of the task is the same but the tasks can be attempted at varied levels of understanding in the classroom.

ACTIVITY 14.2

Have a close look at the concepts mapped out for each year group in the mathematics syllabus applicable to your setting. Choose any content strand, and look for follow-on key ideas that, if they were swapped, would impact upon the developmental sequence. Two examples are provided below. The first example is from the Australian Curriculum: Mathematics, in the fractions and decimals context:

- Recognise that the place value system can be extended to tenths and hundredths. Make connections between fractions and decimal notation (ACMNA079).
- Recognise that the place value system can be extended beyond hundredths (ACMNA104).
- Compare, order and represent decimals (ACMNA105).

The second example is from the NZmaths framework of key ideas for the development of probability concepts:

- At Level 1, students are beginning to explore chance situations. Students discuss possible outcomes, and develop early probability language to describe outcomes.

- At Level 2, students recognize from practical situations and games that some events are more likely to occur than other events. They begin to use more sophisticated language to describe likelihood, such as 'more likely' or 'less likely'.
- At Level 3, students explore one-stage chance situations through experimental activities. Students are able to list possible outcomes, record frequencies of outcomes, graph the outcomes of a chance experiment and derive the probability of an outcome using a theoretical model. The students develop an awareness of 'fair' and 'unfair' events.

National testing

As flagged in Chapter 10, national testing in Australia (NAPLAN) has been linked to teaching to tests, narrowed pedagogy, de-skilling of teachers, anxiety among participants, distribution of test results and the interpretation of results. Some interpretations have led to social comparisons that are detrimental to many school communities. The wider community should consider snapshots in context, along with the appropriateness of having one measure to determine numeracy levels across the extremes that exist in Australia and other nations (Jorgensen 2010). Australia is geographically vast, and the population is widely dispersed, so one needs to question the suitability of a one-size-fits-all policy and national testing. Skills tested in NAPLAN should be incorporated into teaching and learning experiences beyond preparation for the test to ensure that they are transferable to different contexts that lead to a deeper understanding.

Professional development in remote areas requires considerable attention, particularly in the area of mathematics education in the secondary context, where most teachers are teaching 'out of field'. Issues surrounding this reality were presented in Chapter 13. Beswick and Jones (2011) report on the need to provide a professional learning model 'designed to meet the specific needs of teachers and schools in the cluster and to establish relationships that would form the basis of ongoing contacts'. The importance of professional development is also emphasised by Jenkins, Reitano and Taylor (2011). Nicol, Archibald and Baker (2010) identify the critical aspects of culturally responsive mathematics education as:

- grounded in place
- connected to cultural stories
- focused on relationships
- inquiry based
- requiring personal and collective agency.

Resourcing

Problems associated with accessing quality mathematics resources are not unique to remote areas. While it can be a problem in terms of funding and availability, there are many schools in regional and urban areas that do not have adequate funding allocated to the mathematics key learning area. It is very disappointing when one visits a school that has identified mathematics development as an area requiring attention to find that motivation does not extend to making it a funding priority.

> **PAUSE AND REFLECT**
>
> Consider a situation where you are teaching a Year 4 class at a school that has not provided an equitable funding strategy, and mathematics resources are at a critically low level. What would you consider to be an essential starter pack of resources? What key elements would you include in your justification to the school executive?

In terms of resources, there are many options beyond commercially produced material that can be sourced and organised. Organising the resources is the key to reducing teacher time required, by having the equipment ready to go for the appropriate teaching activity. Take, for example, resources that could be used for developing place value concepts. It is not necessary to hold back on using concrete materials if the commercially produced base-10 materials (see Figure 14.4) are not in the school.

Figure 14.4 Developing understanding of place value using base-10 materials

It is possible to:

- make your own 10-frames using recycled cardboard and objects such as stones and pebbles (see Figure 14.5)

Figure 14.5 Making your own 10-frames

- use paddle-pop sticks and rubber bands when investigating two-digit numbers (see Figure 14.6)

Figure 14.6. Exploring place value through paddle-pop stick bundling

- cut palm leaves into sections with 10 leaves per stem to make an interesting display for whole-class investigations
- use egg cartons to make individual 10-frames by cutting off the two egg compartments on one end of the carton. Use bottle caps for counters and ask the students to individually decorate their cartons in their own particular style (creative arts link).

Schools in remote areas, and particularly in small Pacific Island nations, have limited – if any – access to photocopiers. Even if innovative mathematics tasks can be accessed on the internet, there can be a problem with printing and copying. There are many activities designed for copying recording sheets that can still run smoothly and provide an innovative catalyst for student discussion without requiring photocopying.

CLASSROOM SNAPSHOT 14.2

Jessica found an interesting activity called Brainy Fish (New South Wales Department of Education and Training, Curriculum 2008, p. 28), targeting addition and subtraction strategies. The task required pairs of students to play on a game board (which was an outline of a fish with various one- and two-digit numbers). Each pair had a die and a spinner that pointed towards the following instructions:

- Double it.
- Double it plus 1.
- Double it, take away 1.
- How many more to make 10?

Each student was provided with counters to put over the numbers as they played the game. The students then took it in turns to roll the die, spin the spinner, follow the instruction and colour in the number on the fish that was the result of following the instruction – for example, if the student rolled a 2 then spun double it, they would need to colour in 4. The winner was the first student to place three counters in a row.

Jessica did not have access to a photocopier, but thought the students would find the activity stimulating and appropriate to their developmental stages. She decided to draw the fish on the board for the students to copy into their books, have one spinner and one die for the students to take turns to throw and spin, and complete the activity as a class activity with discussion along the way. At a later date, when learning chance concepts, the students made their own Brainy Fish spinners and they repeated the game in pairs.

In the Measurement and Geometry strand, a wide variety of everyday materials can be used as manipulatives. These include plastic straws and Blu-Tack, packaging of all kinds, paper, card and adhesive tape.

CLASSROOM SNAPSHOT 14.3

A Year 6 class in a small remote area school was completing a unit on the volume of a rectangular prism. Textbooks were in short supply, and the school did not have a photocopier due to a lack of access to servicing for the machine.

The teacher, David, began with some short revision questions and a worked example, then proceeded to draw some rectangular prisms on the board for the students to copy and work out the volume. The students began moving in their seats, chatting and disrupting one another.

David decided that he had to tackle this in a different way. He looked around the room and found five different rectangular prisms – packages that stored items or were ready to be thrown away.

The class was divided into five groups, and each group was provided with a rectangular prism package (two large cardboard boxes, a tissue container, a long-life milk carton, and a staples box), a ruler, pencils, calculator and a book in which to record their solutions.

Student discussions concerned which dimensions were important, measuring techniques, rounding measurements, perspectives when drawing, organising the recording of the solution, estimation and determining the appropriateness of calculator solutions. The rectangular prisms were rotated between groups, with most groups completing three tasks.

Figure 14.7 Practical measurement experiences

PAUSE AND REFLECT

What characteristics of this lesson changed the learning environment?

CLASSROOM SNAPSHOT 14.4

Kathy had been exploring length units in a Year 4 class on a small Pacific Island. The children were enjoying the practical activities she had facilitated throughout the teaching and learning sequence. It was time to assess the children on their understanding of the formal units – metres and centimetres.

The school had a culture of end-of-term tests and yearly exams driving teaching assessment, and many of the students had experienced failure and negative feedback as a result of the reliance on formal tests.

Kathy decided to continue with her student-centred teaching and learning practices, and assessed the students with an activity known as Towering Metres (New South Wales Department of Education and Training 2004), where the students needed to create a tower that was exactly 1 metre in height. Kathy was surprised that by observing, listening to and interacting with the students, she had a window through which to view understanding of a metre as a unit of length, measuring techniques with a tape measure and metre ruler, and use of language to compare lengths. She was particularly interested to observe one group's determination to add 1 centimetre of cardboard to the top of the tower when it was repeatedly measured to be 99 centimetres instead of 1 metre. Kathy noticed that the student-centred activity targeting the concept of metre as a unit of length provided valuable assessment evidence.

Figure 14.8 Towering metres

As with the exploration of measurement concepts indicated in the scenario in Classroom snapshot 14.4, the exploration of statistics concepts provides a wonderful opportunity to use freely available resources – in fact, in the primary setting, the investigations are enhanced through the collection of

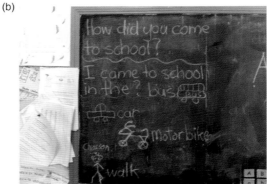

Figure 14.9 Real-life statistics experiences

data about everyday events. As illustrated in Chapter 8, collecting, displaying and analysing data concerning students' lives, the environment, body measurements and sporting competitions make use of readily available resources.

Simple spreadsheets should be introduced with a purpose. This could include organising a class or school celebration using local merchandise, setting up a market stall or organising a school café – thus providing opportunities for literacy in terms of sourcing recipes or creating promotional material. Some primary schools in remote areas are closely linked to secondary schools offering TAFE-delivered education and training (TVET) programs. Real-life experiences such as these enable links to be formed with the hospitality programs, which may assist in providing a smoother transition between primary and secondary school, with purposeful connections between feeder schools and local secondary schools.

It is essential in any mathematics classroom to avoid a situation where a textbook series is driving your lesson planning and teaching. Engaging activities often come in 'small packages' that are driven by student activity and experiences rather than glossy products. Students gain more from developing their own display materials in the form of models and posters than from purchased resources on display. This is evident in Figure 14.10, which shows students' images of cubes in the world around them and their own 3D constructions, beginning with student-designed nets of cubes.

Figure 14.10 Students' creations on display

Reflection

This chapter has focused on engaging students via open-ended practical tasks. While many areas have limited internet access, some remote areas have established connections that lend themselves to the techniques described in the content chapters of this book. The world has expanded for many students in remote areas through the use of connected classrooms (linking classrooms in real time through class video streaming, screen sharing and IWB sharing).

Through reading this chapter and completing the activity tasks, you will have explored issues related to teaching out of urban areas and further afield in remote and rural locations. While there are many challenges, such as remoteness, fewer professional development opportunities, fewer role models with regard to careers in mathematics and science, varied class structures and a smaller pool of experienced teachers, there are many positives to a teaching career in a remote setting. In regional, and particularly in remote, settings, the community is closely related to the school, forming an invaluable resource. With lateral thinking, resources can be sourced both online and using every-day materials. Flexibility in planning and open-ended tasks will assist in catering for individual developmental needs in multi-grade classrooms.

Websites for exploration

Apps in Education: <http://appsineducation.blogspot.com.au/p/maths-ipad-apps.html>

Australian Institute of Family Studies: <http://www.aifs.gov.au/institute/pubs/factssheets/2011/fs201103.html>

HOTmaths: <http://www.hotmaths.com.au>

iPads for Learning: <http://www.ipadsforeducation.vic.edu.au>

Rural and Distance Education – New South Wales Department of Education and Communities: <http://rde.nsw.edu.au>

CHAPTER 15

Planning and sustainability in the mathematics classroom

LEARNING OUTCOMES

By the end of this chapter, you will:

- be reacquainted with the TPACK framework and the different types of teacher knowledge required for effective teaching for numeracy
- understand how technology impacts on mathematics teaching
- develop strategies for evaluating and reflecting upon your own use of technology
- understand what a community of practice is and how to establish one within your school
- develop strategies for evaluating the use of online resources
- gain insight into the change process, and learn how school-wide approaches to teaching mathematics with ICT can be sustained.

KEY TERMS

- **Community of practice (CoP):** The social learning that occurs when practitioners who share a common interest collaborate over a period of time, sharing ideas, best practices and resources
- **Interactive whiteboard (IWB):** A board connected to a computer, capable of displaying a projected image, which allows the user to control the computer by touching the board or using the computer mouse

- **Technological pedagogical and content knowledge (TPACK):** A framework developed by Koehler and Mishra (2008a), which describes how teachers' understandings of technology and pedagogical content knowledge interact to produce effective teaching with technology

This chapter looks at how the use of technology in the mathematics classroom can be planned for and sustained in a school environment. We believe this is addressed best through developing a shared understanding of what it means to teach effectively with technology, which includes a consistent approach to planning and teaching. The chapter begins with reminding readers of the **technological pedagogical and content knowledge (TPACK)** framework, discussed in Chapter 1 and referred to throughout this book. TPACK considers how pedagogical content knowledge (PCK), content knowledge and technology knowledge intersect and interact with each other to enable effective teaching with technology to occur. This framework forms a useful basis from which teachers can develop a shared understanding of what knowledge is required for teaching with technology before then developing a consistent approach to integrating technology in their classrooms and lessons. The issue of sustaining a whole-school approach to teaching with technology is then examined, including ways to establish a **community of practice (CoP)** in order to foster accessibility and to promote a collaborative learning environment.

The TPACK framework

As discussed in Chapter 1, technological pedagogical content knowledge (TPACK) is a framework that builds on Shulman's (1987) formulation of PCK and describes how teachers' understanding of technology and pedagogical content knowledge interact with one another to produce effective teaching with technology (Koehler & Mishra 2008a). As Figure 15.1 shows, there are seven components to the framework, which were described fully in Chapter 1.

According to Koehler and Mishra (2008a, pp. 17–18), TPACK represents:

- an understanding of the representation of concepts using technologies
- pedagogical techniques that use technologies in constructive ways to teach content

- knowledge of what makes concepts difficult or easy to learn, and how technology can redress some of the problems faced by students
- knowledge of students' prior knowledge and theories of epistemology
- knowledge of how technologies can be used to build on existing knowledge and to develop new epistemologies or strengthen old ones.

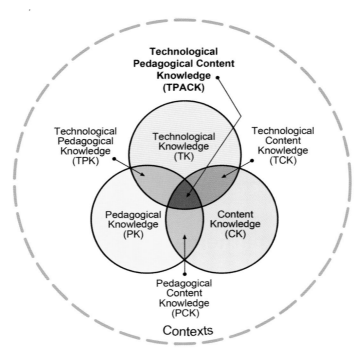

Figure 15.1 TPACK framework
Source: <http://www.tpack.org>.

PAUSE AND REFLECT

Chapter 1 provided an overview of the TPACK framework. For more information about the components of the framework, visit <http://www.tpack.org> and access Mishra's and Koehler's original article, 'Technological pedagogical content knowledge: A framework for teacher knowledge, <http://one2oneheights.pbworks.com/f/MISHRA_PUNYA.pdf>. Read through the descriptors for each component and then write your own definition of what you understand TPACK to mean.

ACTIVITY 15.1

1. Think about what the implications of TPACK are for teaching mathematics. Copy the diagram in Figure 15.1, and label each circle either 'content', 'technology' or 'pedagogy'.
2. Think about a particular area of mathematics you want to teach – for example, equivalent fractions. Write this in the content part of the circle, then identify two or three learning outcomes that you want students to achieve.
3. In the technology circle, write down what technology you could use to help students understand.
4. In the pedagogy circle, write down how you will use the technology and other teaching approaches.

TPACK in the mathematics classroom

Technology has had a considerable impact on the development and expansion of new and existing mathematical concepts and applications over the past few decades (Guerrero 2010). Technology has influenced content development and exploration in areas such as statistics, combinatorics, algebra, probability, geometry and matrices through providing increased access to, understanding of and applications for advanced mathematical concepts through concrete modelling, iterative applications and recursive functioning (Grandgenett 2008). Dynamic software programs such as Geometer's Sketchpad and TinkerPlots have made the exploration of core mathematical concepts tangible and interactive for students. Hand-held graphing devices have allowed students to develop a deeper understanding of mathematical concepts and their multiple representations (for example, their functions and their graphical, tabular and symbolic representations) (Guerrero 2010). Wireless network technologies, such as the TI Navigator, and virtual learning environments can also be used to actively involve students in interactive mathematics instruction.

Technology has also had an impact on how we think about teaching mathematics. The National Council of Teachers of Mathematics (NCTM) states in its Principles and Standards that electronic technologies are essential tools for teaching, learning and doing mathematics, and enable students to focus on decision-making, reflection, reasoning and problem-solving. In Australia, the Australian Association of Mathematics Teachers (AAMT) recommended in 1996 that:

- all students have ready access to appropriate technology as a means to both support and extend their mathematics learning experiences
- priority be given to the use of calculators and computers as natural media for mathematics learning within a technologically rich learning environment
- teachers at all levels be actively involved in exploring ways to take full advantage of the potential of technology for mathematics learning within the total curriculum
- students who use calculator and computer technology in their learning of mathematics have access to the same technological resources when their understanding of mathematics is being evaluated
- education authorities make available to teachers professional development opportunities to support the development of knowledge and skills necessary for the successful use of calculator and computer technologies in classrooms.

The Australian Curriculum: Mathematics (ACARA 2011) explicitly mentions the use of digital technologies in relation to the mathematics curriculum (e.g. ACMSP069: Collect data, organise into categories and create displays using lists, tables, picture graphs and simple column graphs, with and without the use of digital technologies) and through the Australian Curriculum: Technologies.

While some of the content in the New Zealand mathematics curriculum implies that technology could be used, it is not explicitly mentioned until Level 8: 'calculating probabilities, using such tools as two-way tables, tree diagrams, simulations, and technology', <http://nzcurriculum.tki.org.nz>.

The New Zealand Technology Curriculum contains three learning areas: Technological practice, Technological knowledge and Nature of technology. While mathematics is not explicitly mentioned, it does have links to the strands of technical knowledge and the nature of technology, particularly in terms of its impact on mathematics (e.g. graphic calculators).

One effect of the use of technology in the mathematics classroom has been to enable teachers to focus on developing students' conceptual understanding over rote procedural skills through incorporating devices such as calculators to carry out routine procedures.

Guerrero (2010) argues that TPACK in mathematics can be characterised by four central components: a teacher's conception and use of technology; technology-based classroom management; technology-based mathematics instruction; and depth and breadth of content. These components are manifested in the classroom through a teacher's practice. When making decisions about what technology to include in a lesson, it may be helpful to consider the following:

- Am I using the technology appropriately and authentically, rather than as a sideshow tool?
- Am I aware of the growing variety of tools available for mathematical instruction?
- Is it the most appropriate tool?
- What adaptations might I need to make in order to accommodate the technology?
- What management issues do I need to consider?
- Is my content knowledge rich enough to explore the depth and breadth of content through technology?

Guerrero (2010) used the four components in conjunction with the TPACK framework to examine a teacher's use of technology in a secondary mathematics classroom. She found that the teacher, 'Barbara', had comprehensive TPACK, which was demonstrated through her balanced use of technology as one tool within her instructional repertoire and her belief in the benefits of technology. The questions in Table 15.1 can be used as a guide to analysing your own use of technology within the context of a lesson, as a planning tool prior to the commencement of the lesson or as a reflective tool following the lesson – especially in collaboration with a colleague.

Table 15.1 Questions to guide the development and use of technology by analysing each component of TPACK

Component	Question
Instruction	Is technology the most appropriate instructional tool for teaching and learning this topic?
	How will technology affect the collaborative nature of my classroom?
	Will I be able to adapt my instruction using technology, based on student feedback, progress and/or inquiry?
	How will I adjust my instruction and the use of technology to meet individual student needs?
Management	How will I manage the physical logistics of technology? Where will we use the technology? How many students per computer/calculator?
	Can I troubleshoot technical and/or application problems?
	How do I manage student progress and behaviour?
	How do I encourage and maintain student engagement with technology-based lessons?

Table 15.1 (*cont.*)

Component	Question
Depth and breadth of content	Do I have the mathematical knowledge to handle student inquiries that may take us beyond the intent of this lesson?
	Am I willing to acknowledge my own content-related shortcomings and invest the time and energy to investigate student-generated 'content trails'?
Conception and use of technology	Is this topic best addressed through the use of technology? If so, how?
	What should students learn about this topic through the use of technology?
	How does technology improve teaching and learning of this topic?
	Is the use of technology in this lesson pedagogically appropriate?
	Do I have the skills to operate, navigate and apply the various features of mathematics-related technology tools?
	Is technology fully integrated into this lesson or is it an add-on?
	Which technology will best support teaching and learning of a specific topic?

Source: Adapted from Guerrero (2010).

PAUSE AND REFLECT

Think about the last lesson you took (or observed) that involved the use of technology. How many of the components in Table 15.1 did you consider when designing the lesson? In what ways do you think the questions in Table 15.1 could be used to help design future lessons?

ACTIVITY 15.2

The AITSL website, <http://www.teacherstandards.aitsl.edu.au/Illustrations>, contains a number of 'illustrations of practice' that are designed to help teachers situate their own practice within the context of the national professional standards for teachers. A number of illustrations are provided for each of the standards – graduate, proficient, highly

Figure 15.2 AITSL photo of practice

accomplished and lead – across a range of teaching areas. Although the standards are Australian, they would also be applicable to the New Zealand context.

Access the following illustration of practice, produced by the Mathematics Association of Victoria (MAV) in conjunction with the Mathematics Education Research Group of Australasia (MERGA), at <http://www.teacherstandards.aitsl.edu.au/Illustrations/Details/IOP00155>. It shows a graduate teacher demonstrating knowledge of teaching strategies that are responsive to the learning strengths and needs of students from diverse linguistic, cultural, religious and socio-economic backgrounds (standard 1, focus area 1.3). The descriptor is as follows:

> In her first year of teaching, a teacher designs and implements explicit learning experiences for her class, comprising students at a Year 3/4 level. She selects and uses content and resources for her numeracy lessons that are appropriate to the strengths and needs of individual students from diverse backgrounds. In the illustration, she develops teaching activities to assist students to understand the mathematical concepts of whole numbers, fractions and decimals. She supports differentiated learning through small-group work. She is also able to use a literacy approach to support these numeracy activities.

1 Look at the lesson excerpt through the lens of incorporating ICT. Is there any evidence of ICT incorporation in this lesson?
2 If you were conducting a similar lesson, how could you capitalise on the use of ICT to enhance the lesson? (Refer to the guiding questions in Table 15.1 to help you.)

Planning for teaching mathematics with technology

Table 15.1 provided some questions that would be useful when considering what type of technology to include when teaching mathematics topics. This section looks specifically at how to approach planning mathematics lessons, with the aim being to develop a consistent school-wide approach. While it is recognised that lesson planning is influenced by personal and individual preferences, we believe that there are some common elements that should be considered in order to maximise the learning opportunities for students.

PAUSE AND REFLECT

Before reading further, consider the following questions:

- Do you think planning for mathematics is different from planning for other curriculum areas? If so, how is it different?
- Are there aspects to consider that are more relevant to mathematics than to other curriculum areas?
- What do you think you need to know before teaching a mathematical topic?
- Where do you start when planning?

Take some time to record your responses to these questions. At the end of this section, revisit your answers and consider them again in the light of what you have read.

The first step in planning a lesson is to identify exactly what it is you want students to learn. This is typically guided by observations of students and by the relevant curriculum documents. Once you have decided on this, we recommend you then consider it in relation to your own knowledge – that is, what do you need to know to teach this topic?

CLASSROOM SNAPSHOT 15.1

A Year 1 teacher was preparing to teach a topic from the New Zealand Mathematics Curriculum, Geometry and Measurement strand, which is similar to the Australian Curriculum: Mathematics – Measurement and Geometry strand. She identified the following achievement objective as being relevant to her class:

> Order and compare objects or events by length, area, volume and capacity, weight (mass), turn (angle), temperature, and time by direct comparison and/or counting whole numbers of units (GM1-1). (The Australian Curriculum has a similar descriptor: Measure and compare the lengths and capacities of pairs of objects using uniform informal units – ACMMG019.)

She then brainstormed a list of understandings that she needed to consider before planning her actual lesson, which were as follows:

- Length refers to the measurement of something from end to end.
- Sequence for teaching measurement is needed.
- Children need to experience the usefulness of non-standard units.
- There are a number of principles to consider when asking students to measure with non-standard units:
  - The unit must not change – for example, we should select one type of informal unit, such as straws, to measure the length of the table, rather than a straw, a pencil and a rubber.
  - The units must be placed end to end (when measuring length), with no gaps or overlapping bits.
  - The units need to be used in a uniform manner – that is, if dominoes are being used to find the area of the top of a desk, then each domino needs to be placed in the same orientation in order to accurately represent the standard unit.
  - There is a direct relationship between the size of the unit and the number required – that is, the smaller the unit, the bigger the number and vice versa.

As Classroom snapshot 15.1 shows, a number of factors need to be taken into account when considering even a straightforward topic such as measuring length with informal units. Other questions that are useful to consider, and that are particularly helpful for mathematics planning, include the following:

- What are students likely to know about this topic already?
- What are the common or likely misconceptions associated with this topic?

- What questions could I ask or what activities could I use to uncover or address these misconceptions?
- What resources could I make use of (including digital technologies)?
- How can I cater for individual differences and student diversity?
- How will I know if students have understood the concept?

Once you have considered these elements, you are ready to frame your objectives, and to decide how the lesson will be introduced, what key questions to ask, what activities and tasks you will provide, how you will organise groupings in the lesson and how you will monitor and evaluate students' understanding.

In Australian and New Zealand classrooms, lessons typically follow a three-part approach. In the introductory part of the lesson, the teacher may use a hook or lesson starter to engage students and elicit prior knowledge. In the lesson discussed in Classroom snapshot 15.1, the teacher went on to use the storybook *How Big is a Foot?* by Rolf Myller (1990) to engage students in considering the potential difficulties caused by measuring with informal units. Other starters could include mental computation activities such as 'Today's Number is …', sharing of a YouTube clip or interactive website, or the presentation of a particular scenario. The introductory part of the lesson is usually followed by the main teaching activity, which can occur in small groups, individually or a combination of both. The third part of the lesson is known as the plenary, and focuses on the mathematics in which the students have been engaged. It is important to pay attention to how the lesson is brought to a conclusion and to plan for this.

Leading the teaching of ICT within your school

You may find yourself in a situation where you are called upon to take the lead in developing a whole-school approach to teaching with ICT. This next section looks at ways of supporting teachers and/or school leaders in developing and sustaining a whole-school approach to the effective integration of ICT into both their general teaching practice and mathematics teaching.

Developing a community of practice

The term 'community of practice' was used by Lave and Wenger (1991) to describe the social learning that occurs when practitioners who share a common interest collaborate over a period of time, sharing ideas, best practices and resources. CoPs enable practitioners to take collective responsibility for managing the knowledge they need, and can serve to create a direct link between learning and performance. While CoPs can evolve naturally, within this context a CoP can be created specifically with the goal of gaining knowledge in teaching with ICT. Through the process of sharing information and experiences with the group, members can learn from each other, and have the opportunity to develop themselves both personally and professionally (Lave & Wenger 1991).

According to Wenger (2006), there are three elements that distinguish a community of practice from other communities:

- *The domain.* A community of practice is something more than a club of friends or a network of connections between people. It has an identity defined by a shared domain of interest. Membership therefore implies a commitment to the domain, and therefore a shared competence that distinguishes members from other people.
- *The community.* In pursuing their interest in their domain, members engage in joint activities and discussions, help each other, and share information. They build relationships that enable them to learn from each other.
- *The practice.* Members of a community of practice are practitioners. They develop a shared repertoire of resources: experiences, stories, tools, ways of addressing recurring problems – in short, a shared practice. This takes time and sustained interaction.

Table 15.2 is adapted from Wenger (2006). It illustrates what a CoP might look like in a school community with a focus on developing a shared understanding of teaching mathematics effectively with ICT.

Conducting an ICT skills audit

The last activity in Table 15.2 refers to the idea of mapping knowledge and identifying gaps. One way to do this is to conduct an audit of the staff's skills and levels of confidence with using ICT in their mathematics teaching. This can serve two purposes: first, it can identify for the individual what their respective strengths and weaknesses are, leading to focused professional learning to improve practice; and second, the collective expertise of the staff can be capitalised upon, and areas of school need identified and addressed.

Chapter 15: Planning and sustainability in the mathematics classroom

Table 15.2 Example of a CoP in a school community

Focus	Catalyst
Problem-solving	How can I incorporate the use of technology in my classroom without access to an IWB?
Requests for information	How can I get access to Geometer's Sketchpad?
Seeking experience	Does anyone know how to use the hat feature in TinkerPlots?
Reusing assets	How can I use the save features in EasiTeach to save lessons?
Coordination and synergy	How do I know what other classes are doing? How can we find time to meet and plan collaboratively?
Discussing developments	What do others think of the latest version of EasiTeach? Are there other alternatives we should consider?
Documentation projects	Here is a lesson that I conducted with my Year 5 class – I would love to receive feedback from other teachers.
Visits	I would love to see how other Year 1 teachers are using IWBs with their classes – is it possible to organise some inter-school visits?
Mapping knowledge and identifying gaps	I'm not aware of what I don't know … could we conduct an audit of staff skills and identify areas in which we are strong and which need attention?

CLASSROOM SNAPSHOT 15.2

The following provides an explanation of an example resource that was set up by a Year 4 teacher, Miss Brown. The students were required to carry out a 'jump strategy' for subtraction using the **interactive whiteboard (IWB)**. Miss Brown pre-recorded instructions for those students who needed reminding of what the 'jump strategy' meant, along with instructions on how to save their file when they had completed the activity. The pre-recording of this enabled Miss Brown to work with the rest of the class, as any difficulties could be overcome through accessing the pre-recorded tutorial.

MISS BROWN: I've set this up really to be an independent activity, so I don't need to be here as the students can do this on their own. What I want them to focus on during this lesson is just to consolidate the use of the jump strategy for subtraction. So I have highlighted how we go about doing that and the students are well aware of this strategy and use it a lot in their daily mathematics. And here I've listed just the instructions of how to actually complete this activity.

So the first thing you notice it says is to think about the jump strategy for subtraction, just to remind yourself. So I made a quick recording in case students were unsure.

The next recording focuses on recalling the steps for how to make a recording of what they do on the Smartboard:

Miss Brown: You come down to Start, All Programs, go across to Science and Technology, and notebook software.

When you've done that, go to Setup box, just drag it down to the bottom and press record then you'll be able to integrate that on the whiteboard. So what you need to do is to remember to press Record when you find your file name and it will record your working out. I'd like you to focus on just using your jump strategy for today's lesson and if you'd like revision of the jump strategy, come back up to slide number two, where it will explain 67 take away 42 and you'll see, in outline, using the MAB blocks. The next line of working is where I've subtracted the tens and now we're left with two tens and seven units. So 27. And then the next line I've subtracted the units again, so here the units will be two. Taking away two units and left with five. So the answer of 67 take away 42 is 25.

Remember you also need to save your recording and save your filename, so I'll show you an example here. Here I've labelled the file as 'Miss Brown 47 take 33', and then I'll show you how to upload that on to your page.

Okay, that was just a quick demonstration to remind the students of the steps to go through and that's enough of a refresher really for them now to go ahead and work through the whole thing independently. We can then use it as an authentic assessment of their thinking throughout the term and compare that to initial assessments of how they might have approached subtraction at the start and at the end of the term.

So now I can just change that for anything. So you know, I've done the similar things for addition, similar things for multiplication and division. And the fact that you can tape yourself explaining it to the students means that those students who need a refresher feel that they can quickly go and listen to that or if they are doing it independently, take a set of headphones and feel confident in doing that.

Classroom snapshot 15.2 highlights an example of the type of expertise that individual staff members may have, which can then be accessed by other teachers. It may well be that, within a given school, little is known about the ICT skills possessed by other teachers or about resources that may have been developed and could be shared with other staff. In order to investigate and document the skills and knowledge held by staff members, we recommend that an ICT audit of staff skills be carried out. This may be one of the earliest activities undertaken through the community of practice.

ACTIVITY 15.3

Conduct a self-audit of your ICT skills using the following rating scale and questions. Put a tick in the box to indicate your level of confidence (VC – very confident; C – confident; SC – somewhat confident; U – unsure; NC – not confident).

Skill/knowledge	VC	C	SC	U	NC
Basic word processing skills					
Use of spreadsheets					
Use of presentation software (e.g. PowerPoint)					
Use of IWB for presenting/whole-class sharing					
Use of IWB for planning and structuring lessons					
Use of IWB facilities and features (e.g. recording, saving work)					
Creation of animations					
Use of specialist software (e.g. Geometer's Sketchpad)					
Knowledge of interactive websites for mathematics					
Knowledge of appropriate apps for mathematics					
Other (please name)					

The table in Activity 15.3 provides a starting point for thinking about your own skills and knowledge. The table can be extended to provide further information to assist ICT leaders to identify the staff expertise within a school and areas for future professional learning. The following questions may be useful to consider:

- In which areas of ICT use are you most confident?
- Do you have an IWB in your classroom? How confident are you with using the IWB? How would you typically use the IWB?
- Are there any particular websites that you use regularly in your teaching of mathematics? Please list some.
- Do you make use of blogs, wikis or online communities in your classroom?
- Do you have an iPhone or an iPad and, if so, do you regularly download apps?
- Have you attended any professional learning focused on the use of ICT within the last 12 months?
- With what areas of ICT use would you like more support (consider both technical knowledge and pedagogical knowledge)?

Professional reading and viewing

Depending upon the expertise identified, the audit may reveal that there are members of staff who would be suitable and willing to lead professional learning in particular areas. It may also be necessary to source professional learning from outside the school, such as by attending conferences or arranging school visits. One easily accessible and legitimate form of professional learning that can occur as a community of practice involves professional reading. As a school leader, or member of the CoP, you could suggest accessing particular articles and establish a reading group to respond to those articles on a regular basis. Following are some articles and readings that could be used as starting points for reading groups:

- Catherine Attard and Christina Curry, 'Exploring the use of iPads to engage young students with mathematics', <http://www.merga.net.au/documents/Attard_&_Curry_2012_MERGA_35.pdf>. In this research article, the authors describe how iPads were implemented with a Grade 3 class and what impact this had on teaching and learning practices and student engagement. Points for discussion could include the influence of ICT on engagement, the management issues associated with class use of devices such as iPads and the advantages and disadvantages identified.
- Gary Beauchamp and John Parkinson, 'Beyond the "wow" factor: Developing interactivity with the interactive whiteboard', <http://core.ac.uk/download/pdf/30965.pdf>. This article looks at whether the IWB is simply the next stage of development in methods of presenting information to pupils, or

can have a significant impact on the quality of pupils' learning. Although this notion is examined in a science context, the article contains many points that are applicable to all subject areas, such as the general agreed benefits of using IWBs, features of IWBs, development of skills and teaching methods. It would provide a good starting point for evaluating and discussing teachers' use of IWBs within the school.
- Issues of the *Australian Primary Mathematics Classroom* often include articles describing 'Teaching with technology'. For example, in vol. 19, no. 2, 2014, Lorraine Day talks about the changing landscape of ICT and Kevin Larkin identifies iPad apps that promote mathematical knowledge. The articles are limited to two or three pages, and are written specifically for primary teachers – see <http://www.aamt.edu.au>.

Policy and curriculum documents

For access to a range of Australian resources and documents, join the Connect with Maths Digital Learning Community through AAMT, <http://www.aamt.edu.au>. Access the resource, 'ICT for Everyday Learning: A teacher's toolkit', and use it as a basis for discussion at a staff meeting. The NZmaths website, <http://nzmaths.co.nz/numeracy-references>, has a number of numeracy references that can be accessed and used as a basis for professional conversations. Annual evaluation reports and compendium papers about the Numeracy Development Projects are available for download, along with PowerPoint presentations of the National Numeracy Facilitators' Conferences. Other sources for reading can be found in the Further Reading sections throughout this book. You may also consider other avenues for discussion, which could be based around collaborative viewing of particular websites, including YouTube, Teachers TV, <http://www.teacherstv.com.au> and TED Talks, <http://www.ted.com/talks>.

ACTIVITY 15.4

View the TED Talk by Sugata Mitra on building a school in the cloud at: <http://www.ted.com/talks/sugata_mitra_build_a_school_in_the_cloud.html>.

1. What implications does his message have for today's schooling practices?
2. How well do you think our schools are preparing our students for the future?
3. How important do you think it is that all students have equitable access to technology?

The following two YouTube clips provide a more light-hearted starting point for talking about the impact of technology on learning. The first one could be used as a stimulus for discussing the role of books in an online world, while the second shows a very engaging lecturer who certainly capitalised on technology to trick and amaze his students.

- <http://www.youtube.com/watch?v=pQHX-SjgQvQ>
- <http://www.youtube.com/watch?v=aP4fWMLofvo>.

Website evaluation

Another useful professional learning activity for a whole staff would be to consider using or adapting a framework to evaluate the use of interactive websites. Staff discussion could occur around what makes a good website and

Type title (e.g. print based) Source/publishing details
Criteria and rating *(1. N/A or not evident; 2. Evident; 3. Very evident)*
Relevance to Australian Curriculum: Mathematics *Content strand (include year level and content descriptors)* *Proficiency strands (relevant to year level if possible)*
Focus on substantive mathematics *(Cognitive demand, 'big ideas: not superficial, clear (mathematical) outcomes)*
Differentiation *(Flexible entry level, ease of adaptability, cater for a range of abilities and learning)*
Engagement/motivation *(Is the task intrinsically motivating, interesting?)*
Sense-making *(Is the task of an inquiry/investigative nature? Open-ended? Connects concepts? Focuses on conceptual understanding? Critical thinking?)*
Student relevance/context *(Is the task likely to be relevant to students' experiences, interests, context?)*
Communication *(Does the task encourage communication between peers? Verbal and written? Promote discussion?)*
Justification and reflection *(Are students expected to justify their approaches and reflect on their learning?)*
Use of different representations *(Does the task encourage use of different representations and approaches, such as manipulatives, ICT, group work, etc?)*
Ease of implementation *(Are teacher instructions clear? Are materials or additional resources readily available? Is preparation time minimal?)*
Teacher support materials *(Is the task accompanied by resources such as teacher advice, objectives, teaching notes, assessment suggestions or tools?)*
Any concerns/issues noted?

Figure 15.3 Framework for evaluating teaching resources (including digital resources)

particular criteria could be used to evaluate them. Figure 15.3 shows a framework that was developed by one of the authors to assess teaching resources. It contains a number of criteria to consider, including mathematical relevance, engagement, differentiation and use of different representations. Although some of the criteria may not be applicable to all digital resources, it would provide a useful starting point for collaborative discussions around the use of online resources.

Collating resources

Two teachers, Helen Edmonds and Pam Wright from the Concord School in Bundoora, Victoria, <http://www.concordsch.vic.edu.au/elearning/MAV/apps_numeracy.html>, have developed a database describing apps that support personalised learning in numeracy. Their website contains descriptions of almost 50 apps, which are all linked to content strands and the Victorian Essential Learning Standards (VELS) levels.

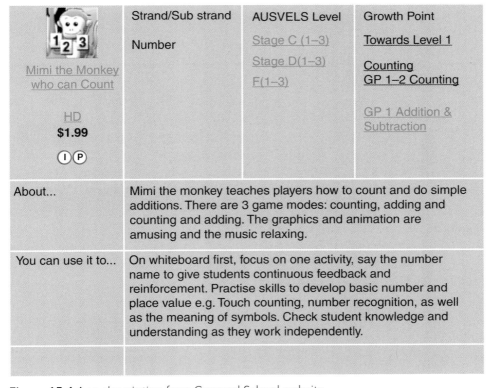

	Strand/Sub strand	AUSVELS Level	Growth Point
Mimi the Monkey who can Count HD $1.99 Ⓘ Ⓟ	Number	Stage C (1–3) Stage D(1–3) F(1–3)	Towards Level 1 Counting GP 1–2 Counting GP 1 Addition & Subtraction
About…	Mimi the monkey teaches players how to count and do simple additions. There are 3 game modes: counting, adding and counting and adding. The graphics and animation are amusing and the music relaxing.		
You can use it to…	On whiteboard first, focus on one activity, say the number name to give students continuous feedback and reinforcement. Practise skills to develop basic number and place value e.g. Touch counting, number recognition, as well as the meaning of symbols. Check student knowledge and understanding as they work independently.		

Figure 15.4 App description from Concord School website

Schools often use their intranets to house their own resources and provide customised links to commonly used resources. One role of the CoP could be to collate a school database on apps and websites, using the framework in Figure 15.4 as a guide for determining selection.

NZmaths, <http://www.nzmaths.co.nz>, hosts a repository of digital resources that support the New Zealand Mathematics Curriculum. The Digital Learning Objects that are housed in the Digistore require an Education Sector logon, but there are other learning objects which are accessible to all. Figure 15.5 shows a screenshot of a number line, which is one of the freely available Learning Objects (see <http://www2.nzmaths.co.nz/LearningObjects/NumberLine/index.html>).

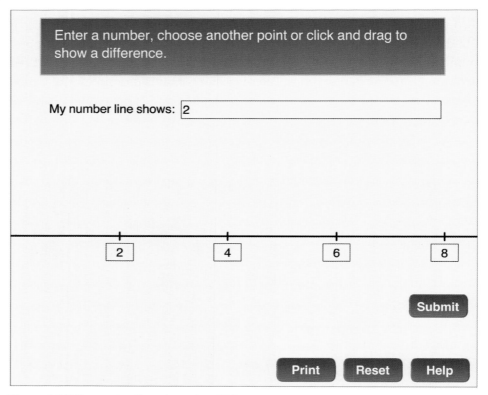

Figure 15.5 Screenshot from Learning Objects

Involving the community

Parental involvement in a child's education is positively correlated with academic achievement, and can even assist in improving levels of student health, reducing student dropout rates, fostering positive attitudes towards learning and school, and promoting productive student behaviours (Peressini 1998). Parents are an integral part of the school community, and therefore should be included and encouraged to actively participate in their children's education. An inclusive approach to school decision-making and parental involvement creates a sense of shared responsibility among parents, community members and teachers.

However, it can sometimes be challenging to encourage parents to participate in their children's mathematical education, as many often find themselves in unfamiliar territory when they attempt to work with their children on mathematics. One way to address this issue is to provide parents with information, workshops and take-home activities to help familiarise them with the mathematical content and pedagogical approaches that are part of today's classroom. It is also reasonable to expect that many parents would be unfamiliar and/or not confident with using the technology that is so readily available and accessible to their children. The following suggestions could be used as starting points for involving parents in school-based activities designed to inform them and make them more comfortable with both mathematics education and the role played by technology in this:

- *Host a maths night or maths afternoon:* Set up a number of games and activities in the school's hall or gymnasium, and invite parents to participate with their children. Set up a number of computers and an IWB, and have children instruct parents on their use.
- *Take-home numeracy activities:* A number of schools have weekly take-home numeracy bags, which encourage parents and children to complete mathematical activities at home. This experience can be especially valuable if the purpose of each activity is included and links to ICT can be made with suggestions for websites that reinforce the concept.
- *Publish newsletter links to websites:* A regular feature of the newsletter could be 'Website of the Week', which would encourage parents to visit websites with their children and build up their own bank of online resources.
- *Host 'Appy Hour':* Once a month or once a term, host an 'Appy Hour', where participants share afternoon tea along with their favourite apps. A database could be established where the apps are named and described, with selected ones being included in the weekly school newsletter.

It seems logical to expect that if any change or practice is to be sustained in a school community, parents will have an integral role to play in helping this to occur. The above suggestions are starting points for actively involving parents and making them part of the school's community of practice.

Leadership, ICT and change

As a school leader, and one who is committed to encouraging staff within your school to effectively utilise ICT tools in their teaching, it is useful to understand the change process and the key drivers for achieving effective and lasting change. *ICT in Everyday Learning: A Toolkit for Teachers*, <http://www.ictineverydaylearning.edu.au/leadership-and-ict/leadership-and-ict.html> (requires users to register), illustrates how pedagogy, content and technology can successfully and effectively be integrated to promote learning. Leaders in schools discuss theoretical and practical approaches to technological change and how technology has affected teaching and learning. They offer insights into understanding and implementing innovative teaching practices in education. The website contains many resources that would be useful for guiding the change process and for professional learning. Links are made with TPACK, and there is a video provided where Punya Mishra and Matthew J Koehler discuss how technology impacts on leadership to effect change within the classroom.

It is important to understand that people adopt change at different rates, and that participants' perception of a need to change is an influential factor in the success of any reform or innovation (Fullan 1992). Imposed professional learning or change often results in resistance or rejection, with participants tending to select suitable 'bits', or delaying until the innovation has been superseded (Hargreaves 1996). This is particularly applicable to technological change, which happens so fast, with many teachers experiencing trends that come and go, making them understandably wary of adopting new practices that may be superseded in the near future.

The 'diffusion of innovations' is a theory popularised by Rogers (2003) that is useful for understanding how, why and at what rate new ideas and technology spread through cultures. Rogers uses the terms 'innovators', 'early adopters', 'early majority', 'late majority' and 'laggards' to describe the levels of adoption for innovations. Innovators, for example, are the first individuals to adopt an innovation; they are willing to take risks and interact closely with other innovators. Laggards, on the other hand, are the last to adopt an innovation, and show little or no opinion leadership. According to Howell (2012), it would be expected that all categories would be present in most schools, and her book, *Teaching with ICT*, contains examples of case studies of teachers who fall into each of these categories. The issue of different rates of adoption is also raised by Beauchamp (2004), who uses the terms 'black/whiteboard substitute', 'apprentice', 'initiate', 'advanced' and 'synergistic' to describe teachers' levels of use with an IWB. In his research, Beauchamp found that teachers operating at the lower end of the framework were primarily using the IWB as they would a whiteboard, displaying presentations and demonstrations. At the other end of the framework, synergistic users used the IWB in an intuitive way, which created new learning scenarios.

PAUSE AND REFLECT

Robinson (2009) provides a useful overview of Rogers' (2003) 'Diffusion of innovations' at <http://www.enablingchange.com.au/Summary_Diffusion_Theory.pdf>. Beauchamp's (2004) article is available at <http:// marynabadenhorst.global2.vic.edu.au/files/2009/11/IWB-Transition- Framework-article.pdf>.

Read both articles and reflect on the following:

1 What are the implications for your school if you have 'laggards' on your staff?
2 What are the implications for students if their teacher is a 'laggard'?
3 How will you engage 'laggards' in the change process?

Reflection

This chapter has provided an overview of the teacher knowledge required for effective teaching with ICT, and some practical suggestions on how teachers can become aware of this knowledge and audit their own technological skills and expertise. After reading the chapter and participating in the activities, you should have a good awareness of the issues involved with planning and sustaining a consistent and effective approach to teaching ICT within a school.

Websites for exploration

AAMT: <http://www.aamt.edu.au>

Concord School: <http://www.concordsch.vic.edu.au/elearning/MAV/apps_numeracy.html>

NZmaths: <http://nzmaths.co.nz/numeracy-references>

Glossary

Algebraic thinking: Thinking that considers the general relationships between numbers, rather than manipulation of numbers

Algorithm: A step-by-step procedure for undertaking a computation

Angle: A measure of the amount of turn

Area: The amount of space contained within a closed shape

Array: An arrangement of rows and columns

Assessment: To identify in measurable terms the knowledge, skills and beliefs of an individual or group

Assessment as learning: The use of assessment information by teachers and students to guide curriculum planning that influences future student learning

Assessment for learning: The central purposes of assessment for learning are to provide information on student achievement and progress and set the direction for ongoing teaching and learning.

Assessment of learning: The process of gathering information about student achievement and communicating this information

Attribute: A property or characteristic of something

Bisect: Divide into two equal parts

Capacity: The measure of how much a three-dimensional object can hold

Column or bar graph: A way of displaying categorical data using vertical columns or horizontal bars and frequency counts

Community of practice (CoP): The social learning that occurs when practitioners who share a common interest collaborate over a period of time, sharing ideas, best practices and resources

Computation: All operations on numbers that are used to calculate a result or answer

Congruent figures: Figures that are exactly the same size and shape

Constructivism: A theory of learning whereby new knowledge is constructed on the basis of past experiences

Counting principles: The principles that govern and define counting: the one-to-one principle, the stable order principle, the cardinal principle, the abstraction principle and the order irrelevance principle

Data: Information collected in a systematic way

Digital technology: Any technology controlled using digital logic, including computer hardware and software, digital media and media devices, digital toys and accessories, and contemporary and emerging communication technologies

Dissect: Divide into two parts

Distribution: The look or shape of the data when displayed systematically on an axis

Distributive property: When multiplying one number by another number, the result is the same as multiplying its addends and summing the products. It can be expressed as $a(b + c) = ab + bc$ (e.g. 5 x 17 = 5 x 10 + 5 x 7 = 50 + 35 = 85).

Drill and practice: Repetition of a skill or procedure

Edge: The interval where two faces of a solid meet

Equation: A number sentence that states two quantities are equal

Equivalence: A state of being equal or equivalent; balanced

Estimation: An approximation or judgement about the attribute(s) of something

Face: A flat surface of a polyhedron

Factor: A whole number that divides exactly into another number or a whole number that multiplies with another whole number to make a third number (e.g. 3 and 4 are both factors of 12)

General capabilities: A key dimension of the Australian Curriculum: Mathematics; addressed explicitly in the content of the learning areas

Generalisation: The formalisation of general concepts; general statements that can be used to form conclusions

Hefting: Estimating the mass of objects by holding them in the hands

HOTmaths: An interactive online mathematics learning and teaching resource for students, teachers and parents

Hundreds square: An array of 10 columns and rows depicting the numbers 1 to 100

Inclusion: Policy of including all students in mainstream classrooms wherever possible

Independent events: Two or more events that have no dependence on each other – for example, the chance of getting heads when tossing a fair coin is always 50 per cent, regardless of any previous tosses

Interactive whiteboard (IWB): A board connected to a computer, capable of displaying a projected image, which allows the user to control the computer by touching the board or using the computer mouse

Interquartile range: Contains the 'middle half' of the data set

Isometric projection: A corner view of an object

Length: A measure of something from end to end

Mass: The amount of matter in an object

Mathematical content knowledge: Understanding the necessary mathematics beyond the immediate demands of the curriculum

Maths300: An online resource containing over 170 inquiry-based lesson ideas for teachers of students from Foundation to Year 12

Mean: A balancing point that provides a summary of the data

Median: The middle value of an ordered set of data

Mental computation: Performance of mathematical computations in the mind without the aid of pen and paper or calculator

Mode: The most frequently occurring value

Multiplicative thinking: Thinking and reasoning about more than one quantity or value at once (e.g. doubling the side length of a square means that the area quadruples)

Non-standard units/informal units: Everyday materials or objects that can be used to measure various attributes – for example, hands, feet, straws, tiles, marbles

Odds: A way of expressing probability as a ratio, usually expressed as the probability of an event not being likely to occur

'Out of field' teacher of mathematics: A teacher of mathematics who has not completed the minimum requirements to be registered to teach secondary mathematics. These teachers may be qualified to teach in the primary years or in secondary areas other than mathematics.

Outcomes: The set of possible predictable events in a given situation, such as heads and tails when tossing a single coin

Part-part-whole: The relationship between the parts of something and its whole

Part-whole numbers: Those representations of number that relate a part to the whole, such as fractions (e.g. $\frac{3}{5}$ is three parts taken from five parts of a whole), decimals (e.g. 0.6 is six-tenths of a whole) and percentages (e.g. 35% is 35-hundredths of a whole)

Pattern: A set of objects or numbers in which each is related to the others according to a particular regularity or rule

Pedagogical content knowledge: The complex blend of content and pedagogical knowledge that provides the basis for classroom decisions made on a day-to-day basis

Pictograph: A column graph that uses pictures or symbols to show the data

Pie chart: A sector graph based on a circle, with each sector in proportion to the percentage of the whole

Place value: The value of a digit depends on its place in the number. For example, in the number 361, the 3 has the value of 300 or three hundreds, 6 has a value of 60, or six tens, and there is one unit.

Population: The entire group of interest – for example, all Year 3 students in the school

Probability: A quantification of the chance of an event based on the possible outcomes

Proportional reasoning: Reasoning about relationships (e.g. 6 being two threes or three twos rather than one more than 5) or comparing quantities or values (e.g. 1 km is 1000 m)

Randomness: A situation where any particular outcome is uncertain

Range: A measure of the spread of the data

Rational number: Any number that can be represented as a fraction

Relative frequency: The number of times a given outcome occurs relative to other possible outcomes. This is the basis for classical probability theory.

Remote area: A location far away from an urban centre or large regional town

Sample: A sub-group of the population, often intended to be representative of the whole group – for example, Ms Pitt's Year 3 class is a sample of the population of Year 3 children in the school

Scootle: An online repository of digital resources for Australian teachers

Simulation: Any created activity that aims to represent a 'real-life' situation

Spatial awareness: The ability to mentally visualise objects and spatial relationships

Standard deviation: A measure of the variability or spread of a set of data

Stem-and-leaf plot: A display based on splitting each data value into a 'stem' and 'leaves' – particularly useful for comparing two groups on the same attribute

Subitise: Recognise how many are in a set without counting

Tangram: An ancient Chinese dissection puzzle comprising seven flat shapes

Technological pedagogical content knowledge (TPACK): A framework developed by Koehler and Mishra (2008a), which describes how teachers' understandings of technology and pedagogical content knowledge interact with one another to produce effective teaching with technology

Technology metaphor: A way of describing students' use of technology and technological tools as proposed by Galbraith et al. (2000)

Temperature: The measure of how hot or cold something is

10-frame: An array of two rows of five used to provide a visual representation of the base-10 number system

Three-dimensional (3D) object: Having three dimensions, requiring three coordinates to specify a point

Time: The duration of an event from its beginning to its end

TinkerPlots: Dynamic data software that aims to develop students' understanding of data, number, probability and graphs

Transformation: Shifting or modifying a shape, including reflecting, enlarging, translating and rotating

Two-dimensional (2D) figure: Having two dimensions, a flat surface with no depth, requiring two coordinates to specify a point

Value: The measure of a cost placed on something

Variable: A value that can change within a problem or set of operations

Variation: The concept that underpins statistics

Vertices: Plural of vertex, a point on a 3D shape where three or more straight edges meet to make a corner

Virtual manipulative: A virtual representation of a physical manipulative

Visualisation: A mental image that is similar to a visual perception

Volume: The amount of space occupied by a three-dimensional object

Whole number: A number that has fractional parts; an integer (e.g. 71 is a whole number but 71.5 is not)

References

Alexander, L & James, HT 1987, *The nation's report card*, Cambridge, MA: National Academy of Education.

Allen, E 2011, 'Scandal of the primary pupils who can get full marks in maths without even knowing their times tables', *Mail Online*, 7 September, <http://www.dailymail.co.uk/news/article-2034442/Pupils-passing-maths-exams-good-marks-dont-know-times-tables.html>, viewed 20 May 2013.

Allen, P 1988, *A Lion in the Night*. Ringwood: Penguin.

Askew, M 2008, 'Mathematical discipline knowledge requirements for prospective primary teachers, and the structure and teaching approaches of programs designed to develop that knowledge', in P Sullivan & T Wood (eds), *Knowledge and beliefs in mathematics teaching and teaching development (The international handbook of mathematics teacher education*, vol. 1, Rotterdam: Sense, pp. 13–36.

Askew, M, Brown, M, Rhodes, V, Johnson, D & Wiliam, D 1997, *Effective teachers of numeracy*, London: School of Education, King's College.

Aspect (Autism Spectrum Australia) 2013, 'Media release: World Autism Day', <http://www.autismspectrum.org.au/index.php?option=com_content&view=article&id=700:world-autism-awareness-day-tuesday-2-april&catid=94:media-releases>, viewed 20 May 2013.

Asplin, P, Frid, S & Sparrow, L 2006, 'Game playing to develop mental computation: A case study', in P Grootenboer, R Zevenbergen & M Chinnappan (eds), *Identities, cultures, and learning spaces* (proceedings of the 29th annual conference of the Mathematics Education Research Group of Australasia, Canberra), Adelaide: MERGA, <http://www.merga.net.au>, viewed 20 January 2013.

Assessment Reform Group 2002, *Assessment for Learning: 10 Principles – Research-based principles to guide classroom practices*. Bristol: Assessment

Reform Group, <http://methodenpool.uni-koeln.de/benotung/assessment_basis.pdf>.

Assessment Resource Banks 2011, Website, <http://arb.nzcer.org.nz>, viewed 20 January 2013.

Australian Association of Mathematics Teachers (AAMT) 1996, 'Statement on the use of calculators and computers in Australian schools', <http://www.aamt.edu.au/Publications-and-statements/Position-statements/Calculators-and-Computers>, viewed 20 January 2013.

—— 2008, *Position paper on the practice of assessing mathematics learning*, <http://www.aamt.edu.au/content/download/9892/126724/.../aamt-assess-bw.pdf>, viewed 20 January 2013.

—— 2013, *Make it count: Blueprint for supporting best teaching of mathematics for Indigenous learners*, viewed 20 May 2013, <http://www.aamt.edu.au/Media/Files/Make-It-Count-Blueprint>.

Australian Council of Deans of Science 2006, *The preparation of mathematics teachers in Australia: Meeting the demands for suitably qualified mathematics teachers in secondary schools*, report prepared by K Harris & F Jensz, Melbourne: Centre for the Study of Higher Education, University of Melbourne.

Australian Curriculum Assessment and Reporting Authority (ACARA) 2011, *The Australian Curriculum: Mathematics V3.0*, <http://www.australiancurriculum.edu.au/Mathematics/Curriculum/F-10>, viewed 20 February 2013.

Australian Institute of Family Studies (AIFS) 2011, Families in regional, rural and remote Australia, Fact Sheet, <http://www.aifs.gov.au/institute/pubs/factssheets/2011/fs201103.html>.

Banks, S 2011, 'A historical analysis of attitudes toward the use of calculators in junior high and high school math classrooms in the United States since 1975', MA thesis, Cedarville, OH: University of Cedarville, <http://digitalcommons.cedarville.edu/cgi/viewcontent.cgi?article=1030&context=education_theses>, viewed 20 February 2013.

Beauchamp, G 2004, 'Teacher use of the interactive whiteboard in primary schools: Towards an effective transition framework', *Technology, Pedagogy and Education*, vol. 13, no. 3, pp. 327–48.

Beauchamp, G & Parkinson, J 2005, 'Beyond the "wow" factor: Developing interactivity with the interactive whiteboard, *School Science Review*, vol. 86, no. 316, pp. 97–104.

Beswick, K 2007–08, 'Influencing teachers' beliefs about teaching mathematics for numeracy to students with mathematics learning difficulties', *Mathematics Teacher Education and Development*, no. 9, pp. 3–20.

Beswick, K & Jones, T 2011, 'Taking professional learning to isolated schools: Perceptions of providers and principals, and lessons for effective professional learning', *Mathematics Education Research Journal*, vol. 23, no. 2, pp. 83–105.

Biggs, JB & Collis, KF 1982, *Evaluating the quality of learning: The SOLO taxonomy*, New York: Academic Press.

Bills, T & Hunter, R 2015, The role of cultural capital in creating equity for Pāsifika learners in mathematics, in M. Marshman, V Geiger & A Bennison (eds), *Mathematics education in the margins* (Proceedings of the 38th annual conference of the Mathematics Education Research Group of Australasia), Mooloolaba, Qld: MERGA, pp. 109–116, <http://www.merga.net.au>, viewed 20 October 2015.

Blank, M 2002, 'Classroom discourse: A key to literacy', in K Butler & E Silliman (eds), *Speaking, reading and writing in children with learning disabilities: New paradigms in research and practice*, Mahwah, NJ: Lawrence Erlbaum, pp. 151–73.

Boaler, J 2008, 'Promoting "relational equity" and high mathematics achievement through an innovative mixed ability approach', *British Educational Research Journal*, vol. 34, no. 2, pp. 167–94.

Board of Studies, Teaching & Educational Standards NSW 2012, *Assessment for Learning Practices*, Sydney: NSW Government, <http://syllabus.bostes.nsw.edu.au/support-materials/assessment-for-as-and-of-learning>, viewed 20 October 2015.

Bobis, J 2009, *Count me in too: The learning framework in number and its impact on teacher knowledge and pedagogy*, Sydney: Department of Education and Training.

Booker, G, Bond, D, Sparrow, L & Swan, P 2010, *Teaching primary mathematics*, 4th edn, Sydney: Pearson.

Braden, JP 1994, *Deafness, deprivation and IQ*, New York: Plenum Press.

Brady, J, Clarke, B & Gervasoni, A 2008, 'Children with Down syndrome learning mathematics: Can they do it? Yes they can!' *Australian Primary Mathematics Classroom*, vol. 13, no. 4, pp. 10–15.

Bryant, P & Nunes, T 2012, *Children's understanding of probability*, London: Nuffield Foundation, <http://www.nuffieldfoundation.org/news/childrens-understanding-probability>. viewed 20 February 2013.

Callingham, R 1993, 'Cemetery maths', *Australian Mathematics Teacher*, vol. 49, no. 3, pp. 38–9.

—— 2008, 'Dialogue and feedback: Assessment in the primary mathematics classroom'. *Australian Primary Mathematics Classroom*, vol. 13, no. 3, pp. 18–21.

Carpenter, TP, Fennema, E, Franke, ML, Levi, L & Empson, S 1999, *Children's mathematics: Cognitive guided instruction*, Portsmouth, NH: Heinemann.

Chua, B & Wu, Y 2005, 'Designing technology-based mathematics lessons: A pedagogical framework', *Journal of Computers in Mathematics and Science Teaching*, vol. 24, no. 4, pp. 387–402.

Clements, DH 1999, 'Subitising: What is it? Why teach it?' *Teaching Children Mathematics*, vol. 5, no. 7, pp. 400–5.

Collis, KF & Romberg, TA 1991, 'Assessment of mathematical performance: An analysis of open-ended test items', in MC Wittrock & EL Baker (eds), *Testing and cognition*, Englewood Cliffs, NJ: Prentice Hall, pp. 82–130.

Commonwealth of Australia 2009, *Shape of the Australian Curriculum: Mathematics*, Canberra: Commonwealth Government, <http://www.acara.edu.au/verve/_resources/Australian_Curriculum_-_Maths.pdf>, viewed 2 December 2012.

Cornish, L 2006, 'What is multi-grade teaching?', in L Cornish (ed.), *Reading EFA through multi-grade teaching: Issues, contexts and practices*, Armidale, NSW: Kardoorair Press, pp. 9–26.

Cotton, T 2010, *Understanding and teaching primary mathematics*, London: Pearson.

Cowan, R 2011, *The development and importance of proficiency in basic calculation*, London: Institute of Education, viewed 20 January 2013, <http://www.ioe.ac.uk/Study_Departments/PHD_dev_basic_calculation.pdf>.

Dale, P 2007, *Ten in the bed*, London: Candlewick Press.

—— 2010, *Ten out of bed*, London: Walker Books.

Department of Education and the Arts, Tasmania 1994, *K–8 Guidelines: Measurement*, Hobart: Department of Education and the Arts, Tasmania.

Department of Education, Employment and Workplace Relations (DEEWR) 2009, *Belonging, being and becoming: The early Years Learning Framework for Australia*, <http://www.deewr.gov.au/Earlychildhood/Policy_Agenda/Quality/Documents/Final%20EYLF%20Framework%20Report%20-%20WEB.pdf>, viewed 20 December 2013.

Department of Education and Training, NSW 2003, *Fractions, pikelets and lamingtons*, Sydney: Department of Education and Training Professional Support and Curriculum Directorate, <http://www.schools.nsw.edu.au>, viewed 20 December 2013.

—— 2012, *Curriculum support material: Newman's prompts*, Sydney: NSW Government, <http://www.curriculumsupport.education.nsw.gov.au/primary/mathematics/numeracy/newman>, viewed 20 December 2013.

Department of Education and Training WA 2004, *First steps in mathematics: Number*, Melbourne: Rigby Harcourt Education.

DeSimone, JR & Parmar, RS 2006, 'Middle school mathematics teachers' beliefs about inclusion of students with learning disabilities', *Learning Disabilities Research and Practice*, vol. 21, no. 2, pp. 98–110.

Donlan, C, Cowan, R, Newton, EJ & Lloyd, D 2007, 'The role of language in mathematical development: Evidence from children with specific language impairments', *Cognition*, no. 103, pp. 23–33.

English, L 1991, 'Young children's combinatoric strategies', *Educational Studies in Mathematics*, no. 22, pp. 451–74.

Fischbein, E, Nello, MS & Merino, MS 1991, 'Factors affecting probabilistic judgements in children and adolescents', *Educational Studies in Mathematics*, no. 22, pp. 523–49.

Fullan, M 1992, *Meaning of educational change*, New York: Teachers College Press.

Galbraith, P, Goos, M, Renshaw, P & Geiger, V 2000, 'Emergent properties of teaching-learning in technology-enriched classrooms', in J Bana & A Chapman (eds), *Mathematics education beyond 2000*, Sydney: MERGA, p. 690.

Gelman, R & Gallistel, C 1978, *The child's understanding of number*, Cambridge, MA: Harvard University Press.

Gigerenzer, G 2002, *Reckoning with risk*, Harmondsworth: Penguin.

Glenday, C (ed.) 2007, *Guinness World Records 2008*, London: Jim Paterson Group.

Goos, M 2012, 'Digital technologies in the Australian Curriculum: Mathematics – a lost opportunity?' in B Atweh, M Goos, R Jorgensen & D Siemon (eds), *Engaging the Australian Curriculum: Mathematics – perspectives from the field*, Sydney: Mathematics Education Research Group of Australasia, pp. 135–52.

Goos, M, Stillman, G & Vale, C 2007, *Teaching secondary school mathematics*, Sydney: Allen & Unwin.

Gould, P, Outhred, L & Mitchelmore, M (2006). One-third is three-quarters of one-half. In P Grootenboer, R Zevenbergen & M Chinnappan (eds), *Identities, cultures, and learning spaces* (Proceedings of the 29th annual conference of the Mathematics Education Research Group of Australasia, Canberra, Adelaide: MERGA, pp. 262–70, <http://www.merga.net.au>, viewed 20 December 2013.

Grandgenett, NF 2008, 'Perhaps a matter of imagination: TPCK in mathematics education', in AACT Committee on Innovation and Technology (ed.), *Handbook of technological pedagogical content knowledge (TPCK) for educators*, New York: Routledge, pp. 145–64.

Gregory, S 1998, 'Mathematics and deaf children', in S Gregory, P Knight, W McCracken, S Powers & L Watson (eds), *Issues in deaf education*, London: David Fulton, pp. 121–7.

Guerrero, S 2010, 'Technological pedagogical content knowledge in the mathematics classroom', *Journal of Digital Learning in Teacher Education*, vol. 26, no. 4, pp. 132–9.

Guldberg, K, Porayska-Pomsta, K, Good, J & Keay-Bright, W 2010, 'ECHOES II: the creation of a technology enhanced learning environment for typically developing children and children on the autism spectrum', *Journal of Assistive Technologies*, vol. 4, no. 1, pp. 49–53.

Hargreaves, A 1996, 'Development and desire: A postmodern perspective', in TR Guskey & M Huberman (eds), *Professional development in education*, New York: Teachers College Press, pp. 9–34.

Haylock, D & Cockburn, A 2008, *Understanding mathematics for young children*, London: Sage.

Higgins, J 1990, 'Calculators and common sense', *The Arithmetic Teacher*, vol. 37, no. 7, pp. 4–5.

Highfield, K 2010, 'Robotic toys as a catalyst for mathematical problem solving', *Australian Primary Mathematics Classroom*, vol. 15, no. 2, pp. 22–7.

Hodgson, T, Simonsen, L, Lubek, J & Anderson, L 2003, 'Measuring Montana: An episode in estimation', in DH Clements & G Bright (eds), *Learning and teaching measurement*, Reston, VA: National Council of Teachers of Mathematics, pp. 221–30.

Howell, J 2012, *Teaching with ICT: Digital pedagogies for collaboration and creativity*, Melbourne: Oxford University Press.

Jacob, L & Willis, S 2001, 'Recognising the difference between additive and multiplicative thinking in young children', in J Bobis, B Perry & M Mitchelmore (eds), *Numeracy and beyond* (Proceedings of the 24th annual conference of the Mathematics Education Research Group of Australasia), Sydney: MERGA, <http://www.merga.net.au>, viewed 20 December 2012.

Jenkins, K, Reitano, P & Taylor, N 2011, 'Teachers in the bush: Supports, challenges and professional learning', *Education in Rural Australia*, vol. 21, no. 2, pp. 71–85.

Jorgensen, R 2010, 'Issues of social equity in access and success in mathematics learning for Indigenous students', in C Glascodine and KA Hoad (eds), *Teaching mathematics? Make it count. What research tells us about effective mathematics teaching and learning* (conference proceedings), Melbourne: ACER, pp. 27–30.

Koehler, MJ & Mishra, P 2008a, TPACK Framework diagram, <http://www.tpack.org>, viewed 20 December 2013.

—— 2008b, 'Introducing technological pedagogical knowledge', in AACTE Committee on Innovation and Technology (ed.), *Handbook of technological pedagogical content knowledge (TPCK) for educators*, New York: Routledge, pp. 3–29.

Konold, C & Miller, CD 2005, *TinkerPlotsTM: Dynamic data exploration* [computer software], Emeryville, CA: Key Curriculum Press.

Lave, J & Wenger, E 1991, *Situated learning: Legitimate peripheral participation*, Cambridge: Cambridge University Press.

Leach, C 2010, 'The use of Smartboards and bespoke software to develop and deliver an inclusive, individual and interactive learning curriculum for students with ASD', *Journal of Assistive Technologies*, vol. 4, no. 1, pp. 54–7.

Livy, S, Muir, T & Maher, N 2012, 'How do they measure up? Primary pre-service teachers' mathematical knowledge of area and perimeter', *Mathematics Teacher Education and Development*, vol. 14, no. 2, pp. 91–112.

Lucangeli, D & Cabrele, S 2006, 'Mathematical difficulties and ADHD', *Exceptionality*, vol. 14, no. 1, pp. 53–62.

Ma, L 1999, *Knowing and teaching elementary mathematics: Teachers' understanding of fundamental mathematics in China and the United States*, Mahwah, NJ: Lawrence Erlbaum.

MacDonald, A, Davies, N, Dockett, S & Perry, B 2012, 'Early childhood mathematics education', in B Perry, T Lowrie, T Logan, A MacDonald &

J Greenlees (eds), *Research in mathematics education in Australasia 2008–2011*, Rotterdam: Sense, pp. 169–92.

MacGregor, M & Stacey, K 1999, 'A flying start to algebra', *Teaching Children Mathematics*, vol. 6, no. 2, pp. 78–85.

McIntosh, A 2004, 'Developing computation', *Australian Primary Mathematics Classroom*, vol. 9, no. 4, pp. 47–9.

McIntosh, A & Dole, S 2005, *Developing computation*, Hobart: Department of Education, Tasmania.

Mewborn, D 2001, 'Teachers' content knowledge, teacher education, and their effects on the preparation of elementary teachers in the United States', *Mathematics Teacher Education and Development*, no. 3, pp. 28–36.

Ministry of Education 1996, *Te Whāriki: He Whāriki Mātauranga mō ngā Mokopuna o Aotearoa: Early Childhood Education*. Wellington: Learning Media.

—— 2007, *The New Zealand Curriculum*. Wellington: Learning Media Limited.

—— 2008, *New Zealand Number Framework*, <http://nzmaths.co.nz/new-zealand-number-framework>, Wellington: Ministry of Education

Morris, C & Matthews, C 2011, 'Numeracy, mathematics and Indigenous learners: Not the same old thing', in *Proceedings of the ACER Research Conference, 2011*, Melbourne: ACER, viewed 20 January 2013, <http://www.acer.edu.au/documents/RC2011_-_Numeracy_mathematics_and_Indigenous_learners-_Not_the_same_old_thing.pdf>.

Moyer, MB & Moyer, JC 1985, 'Ensuring that practice makes perfect: Implications for children with learning disabilities', *The Arithmetic Teacher*, vol. 33, no. 1, pp. 40–2.

Mozelle, S 1989, *Zack's alligator*, New York: HarperCollins.

Muir, T 2005, 'When near enough is good enough: Eight principles for enhancing the value of measurement estimation for students', *Australian Primary Mathematics Classroom*, vol. 10, no. 2, pp. 9–14.

—— 2012, 'What is a reasonable answer? Ways for students to investigate and develop their number sense', *Australian Primary Mathematics Classroom*, vol. 17, no. 1, pp. 21–8.

Mulligan, J & Mitchelmore, M 2009, 'Awareness of pattern and structure in early mathematical development', *Mathematics Education Research Journal*, vol. 21, no. 2, pp. 33–49.

Myller, R 1990, *How big is a foot?* New York: Random House.

New South Wales Department of Education and Training 2004, *Teaching measurement: Stage 2 and 3*, Sydney: NSW Department of Education and Training, Curriculum K–12 Directorate.

—— 2008, *Developing efficient numeracy strategies: Stage 2*, Sydney: NSW Department of Education and Training, Curriculum K–12 Directorate.

Newman, MA 1977, 'An analysis of sixth-grade pupils' errors on written mathematical tasks', *Victorian Institute for Educational Research Bulletin*, no. 39, pp. 31–43.

NZmaths 2010, *Geometry information*, <http://nzmaths.co.nz/geometry-information>, viewed 20 October 2013.

Nicol, C, Archibald, J & Baker, J 2010, *Investigating culturally responsive mathematics education*, Toronto: Canadian Council on Learning, viewed 20 May 2013, <http://www.ccl-cca.ca/pdfs/FundedResearch/201009NicolArchibaldBakerFullReport.pdf>, viewed 20 October 2013.

Pegg, J & Baker, P 1999, 'An exploration of the interface between van Hiele's levels 1 and 2: Initial findings', in *Proceedings of the 23rd International Group for the Psychology of Mathematics Education*, vol. 4, Haifa: Israel Institute of Technology, pp. 25–32.

Pegg, J & Davey, G 1998, 'A synthesis of two models: Interpreting student understanding in geometry', in R Lehrer & C Chazan (eds), *New directions for teaching and learning geometry*, Mahwah, NJ: Lawrence Erlbaum, pp. 109–35.

Pegg, J, Graham, L & Bellert, A 2005, 'The effect of improved automaticity of basic number skills on persistently low-achieving pupils', in HL Chick & JL Vincent (eds), *Proceedings of the 29th Conference of the International Group for the Psychology of Mathematics Education*, vol. 4, Melbourne: PME, pp. 49–56, <http://www.emis.de/proceedings/PME29/PME29RRPapers/PME29Vol4PeggEtAl.pdf>, viewed 20 December 2013.

Peressini, D 1998, 'What's all the fuss about involving parents in mathematics education?', in D Edge (ed.), *Involving families in school mathematics*, Reston, VA: NCTM, pp. 5–10.

Perry, B, Dockett, S & Harley, E 2012, 'The Early Years Learning Framework for Australia and the Australian Curriculum: Mathematics: Linking educators' practice through pedagogical inquiry questions', in B Atweh, M Goos, R Jorgensen & D Siemon (eds), *Engaging the Australian Curriculum: Mathematics – perspectives from the field* (conference proceedings), Sydney: Mathematics Education Research Group of Australasia, pp. 153–74.

Pfannkuch, M, Regan, M, Wild, C & Horton, NJ 2010, 'Telling data stories: Essential dialogues for comparative reasoning', *Journal of Statistics Education*, vol. 18, no. 1, pp. 1–38, <http://www.amstat.org/publications/jse/v18n1/pfannkuch.pdf>, viewed 20 January 2013.

Pierce, R & Stacey, K 2010, 'Mapping pedagogical opportunities provided by mathematics analysis software', *International Journal of Computers for Mathematical Learning*, vol. 15, no. 1, pp. 1–20.

Powell, D. 2012, 'A review of inclusive education in New Zealand', *Electronic Journal for Inclusive Education*, vol. 2, no. 10, Article 4, <http://corescholar.libraries.wright.edu/cgi/viewcontent.cgi?article=1147&context=ejie>, viewed 20 July 2013.

Robicheaux, R 1993, 'How can we design paper money for the visually impaired?' *The Arithmetic Teacher*, vol. 40, no. 8, pp. 479–81.

Robinson, L 2009, 'A summary of diffusion of innovations', <http://www.enablingchange.com.au/Summary_Diffusion_Theory.pdf>, viewed 20 December 2012.

Rodda, E 1986, *Pigs might fly*, Sydney: Angus & Robertson.

Rogers, EM 2003, *Diffusion of innovations*, 5th edn, New York: The Free Press.

Rowland, T, Turner, F, Thwaites, A & Huckstep, P 2010, *Developing primary mathematics teaching*, London: Sage.

Ruthven, K 1998, 'The use of mental, written and calculator strategies of numerical computation by upper primary pupils within a "calculator-aware" number curriculum', *British Educational Research Journal*, vol. 24, no. 1, pp. 21–42.

Ryan, J & Williams, J 2007, *Children's mathematics 4–15: Learning from errors and misconceptions*, Maidenhead: Open University Press.

Sayre A & Sayre, J 2003, *One is a snail, ten is a crab*, London: Walker Books.

Serow, P 2007a, 'Incorporating dynamic geometry software within a teaching framework', in K Milton, H Reeves & T Spencer (eds), *Mathematics: Essential for learning, essential for life* (Proceedings of the 21st biennial conference of the Australian Association of Mathematics Teachers), Adelaide: Australian Association of Mathematics Teachers, pp. 382–97.

—— 2007b, 'Utilising the Rasch model to gain insight into students' understandings of class inclusion concepts in geometry', in J Watson & K Beswick (eds), *Mathematics: Essential research, essential practice* (Proceedings of the 30th annual conference of the Mathematics Education Research Group of Australasia), Adelaide: Mathematics Education Research Group of Australasia, pp. 651–60.

—— 2015 'Education for sustainability in primary mathematics education', in N Taylor, F Quinn & C Eames (eds), *Educating for sustainability in primary schools: Teaching for the future*, Rotterdam: Sense, pp. 177–94.

Serow, P & Callingham, R 2011, 'Levels of use of interactive board technology in the primary mathematics classroom', *Technology, Pedagogy and Education*, vol. 20, no. 2, pp. 161–73.

Serow, P & Inglis, M 2010, 'Templates in action', *Australian Mathematics Teacher*, vol. 66, no. 4, pp. 10–16.

Shulman, LS 1987, 'Knowledge and teaching: Foundations of the new reform', *Harvard Educational Review*, vol. 57, no. 1, pp. 1–22.

Shumway, J 2011, *Number sense routines*, Portland, ME: Stenhouse.

Siemon, D 2007, *Developing the 'big' ideas in number*, <https://www.edu.vic.gov.au/edulibrary/public/teachlearn/student/devbigideas.pdf>, viewed 20 February 2013.

Siemon, D, Bleckly, J & Neal, D 2012, 'Working with the big ideas in number and the Australian Curriculum: Mathematics', in B Atweh, M Goos, R Jorgensen & D Siemon (eds), *Engaging the Australian National Curriculum: Mathematics – perspectives from the field*, Sydney: Mathematics Education Research Group of Australasia, pp. 19–45, <http://www.merga.net.au/node/223>, viewed 20 March 2013.

Simon, M 2000, 'Constructivism, mathematics teacher education, and research in mathematics teacher development', in LP Steffe & PW Thompson (eds), *Radical constructivism in action: Building on the pioneering work of Ernst von Glasersfeld*, London: Routledge Falmer, pp. 213–30.

Sowder, J 1990, 'Mental computation and number sense', *The Arithmetic Teacher*, vol. 37, no. 7, pp. 18–20.

Steen, LA 1999, 'Numeracy: The new literacy for a data-drenched society', *Educational Leadership*, vol. 57, no. 2, pp. 8–13, <http://www.ascd.org/publications/educational_leadership/oct99/vol57/num02/Numeracy@_The_New_Literacy_for_a_Data-Drenched_Society.aspx>, viewed 20 September 2012.

Steinle, V & Stacey, K 2001, 'Visible and invisible zeros: Sources of confusion in decimal notation', in M Mitchelmore, B Perry & J Bobis (eds), *Numeracy and beyond* (Proceedings of the 24th Annual Conference of the Mathematics Education Research Group of Australasia), Sydney: MERGA, pp. 434–41.

Sullivan, P 2011, 'Teaching mathematics using research-informed strategies', *Australian Education Review*, no. 39, <http://research.acer.edu.au/aer/13>, viewed 20 February 2013.

Sullivan, P & Lilburn, P 2004, *Open-ended maths activities: Using 'good' questions to enhance learning in mathematics*, Melbourne: Oxford University Press.

Swan, P & Marshall, L 2009, 'Mathematics games as a pedagogical tool', in *Proceedings: CoSMEd 2009 3rd International Conference on Science and Mathematics Education*, Penang, pp. 402–6, <http://www.recsam.edu.my/cosmed/cosmed09/AbstractsFullPapers2009/Abstract/Mathematics%20Parallel%20PDF/Full%20Paper/M26.pdf>, viewed 20 December 2013.

Thomas, N & Mulligan, J 1994, *Dynamic imagery in children's representations of number*, <http://www.merga.net.au/documents/RP_Thomas_Mulligan_1994.pdf>, viewed 20 February 2013.

Thompson, Z & Hunter, J (2015). Developing adaptive expertise with Pasifika learners in an inquiry classroom, in M Marshman, V Geiger & A Bennison (eds), *Mathematics education in the margins* (Proceedings of the 38th annual conference of the Mathematics Education Research Group of Australasia, pp. 611–618), Mooloolaba: MERGA, <http://www.merga.net.au>, viewed 20 February 2013.

Truran, KM 1995, 'Animism: A view of probability behaviour', in B Atweh & S Flavel (eds), *Proceedings of the Eighteenth Annual Conference of the Mathematics Education Group of Australasia*, Darwin: Mathematics Education Group of Australasia, pp. 537–41.

Vale, C 2010, 'Supporting out-of-field teachers of secondary mathematics', *Australian Mathematics Teacher*, vol. 66, no. 1, pp. 17–24.

Van de Walle, JA, Karp, KS & Bay-Williams, JM 2013, *Elementary and middle school mathematics*, 8th edn, Boston, MA: Pearson Education.

van Hiele, PM 1986, *Structure and insight: A theory of mathematics education*, New York: Academic Press.

von Glasersfeld, E 1996, 'Introduction: Aspects of constructivism', in CT Fosnot (ed.), *Constructivism: Theory, perspectives, and practice*, New York: Teachers College Press, pp. 3–7.

Watson, JM 1993, 'Pigs might fly!!', *Australian Mathematics Teacher*, vol. 49, no. 2, pp. 32–3.

—— 2006, *Statistical literacy at school: Growth and goals*, Mahwah, NJ: Lawrence Erlbaum.

Watson, JM & Callingham, R 1997, 'Data handling: An introduction to higher order processes', *Teaching Statistics*, vol. 19, no. 1, pp. 12–17.

Watson, JM & Caney, A 2005, 'Development of reasoning about random events', *Focus on Learning Problems in Mathematics*, vol. 27, no. 4, pp. 1–42.

Watson, JM, Collis, KF, Callingham, RA & Moritz, J 1995, 'A model for assessing higher order thinking in statistics', *Educational Research and Evaluation*, vol. 1, no. 3, pp. 245–75.

Watson, JM, Fitzallen, NE, Wilson, K & Creed, JF 2008, 'The representational value of hats', *Mathematics Teaching in the Middle School*, vol. 14, no. 1, pp. 4–10, <http://www.keycurriculum.com/docs/PDF/TinkerPlots/MTMS_Representational-Value-of-Hats.pdf>, viewed 20 December 2012.

Wenger, E 2006, 'Communities of practice: A brief introduction', <http://www.ewenger.com/theory>, viewed 20 January 2013.

White, A 2005, 'Active mathematics in classrooms: Finding out why children make mistakes – and then doing something to help them', *Square One*, vol. 5, no. 4, pp. 15–19.

Wild, CJ & Pfannkuch, M 1999, 'Statistical thinking in empirical enquiry', *International Statistical Review*, vol. 67, no. 3, pp. 223–65.

Wright, RJ 1994, 'A study of the numerical development of 5-year-olds and 6-year olds', *Educational Studies in Mathematics*, no. 26, pp. 25–44.

Yelland, N & Kilderry, A 2010, 'Becoming numerate with information and communication technologies in the twenty-first century', *International Journal of Early Years Education*, vol. 18, no. 2, pp. 91–106.

Yunkaporta, T 2009, 'Aboriginal pedagogies at the pedagogical interface', PhD thesis, Townsville: James Cook University.

Index

10-frames, 29–32, 351.
    See also hundreds square
Aboriginal Action Plans, 342
absenteeism, 346
abstraction principle, 16
accommodation, 15
activities
    extra-curricular, 340
    open-ended, 253, 258
addition, 33–34, 123–29
    commutative law, 125
    interactive resources, 36, 124–25, 126, 127, 128, 129, 306–08
    mental computation strategies, 35
    merge problems, 123
    and multiplication, 36–37, 39, 130
    parts-of-a-whole problems, 124
algebra
    in curriculum, 160–61, 173, 179
    interactive resources, 323–26
    issues to think about, 330
    in primary classroom, 160
    and proportional reasoning, 153–54
algebraic thinking, 160
algorithms, 125
American Council on Education (ACE), 5
angle turners, 56
angles, 55–57
    in curriculum, 107–08
    interactive resources, 56–57, 78
    and triangles, 78–79
Apples for the Teacher, 24
apps, 22, 117–19, 164, 377
area, 53, 72–73
    activities, 53–54
    circles, 79
    formulae, 75–78, 180–87
    interactive resources, 71, 76
    and perimeter, 71
arithmetic, 122
arithmetic blocks, 148
arrays, 40, 130–31
assessment
    as learning, 240, 265
    Blank's questioning framework, 249–50
    and curriculum, 241–42
    defined, 240–41
    developmental models, 248
    for learning, 240, 241–46
    national testing, 254–62
    Newman's Error Analysis, 248–49
    of learning, 240
    quality of student responses, 246–54
assessment items banks, 262–65
Assessment Reform Group, 241
Assessment Resource Centre, 258
assessment tasks
    adaptation of, 258–62
    construction of, 253–54
assimilation, 15
Attard, Catherine, 374

Index

Attention Deficit Hyperactivity Disorder (ADHD), 310
attributes of objects, 46, 47
Australia, 339
Australian Association of Mathematics Teachers (AAMT), 241, 255, 319, 342, 345, 363
   Top Drawer website, 129, 147
Australian Curriculum:
   Mathematics, 18, 20
   and assessment, 241, 242
   computation, 123, 129, 130, 135
   digital technologies, 363
   Foundation Year, 20
   geometry, 83, 93, 100, 108, 109
   inclusion, 296
   measurement, 51–52, 75–76
   multiplication and division, 36
   numbers, 33
   patterns and algebra, 160, 173, 179
   percentages, 152
   probability, 219
   statistics, 190, 191, 201
   strands, 8
   technology, 268, 269
Australian Institute for Teaching and School Leadership (AITSL), 365
Australian Institute of Family Studies (AIFS), 339
Australian Mathematics Sciences Institute, 170
Australian Primary Mathematics Classroom, 375

balance scales, 58, 172
bar graphs, 198
Bay-Williams, JM, 37, 53, 56, 59
Beauchamp, Gary, 374, 380
behavioural problems, 310
beliefs of teachers, 298–99
Beswick, K, 298, 350
Biggs, JB, 251
bisect, 100
Blank, Maureen, 249

Blank's questioning framework, 249–50
Bleckly, J, 142
booklets, 336
books, 21
Broken Rulers: Measure Me website, 70
Bryant, P, 217
Bureau of Meteorology, 236
bushfires, 236

Cabri (software), 105
cake icing, 53
calculators
   broken multiplication key, 37
   decimals, 151
   graphic, 6
   use of, 133–34, 300
Caney, A, 218
capacity, 54–55
cardinal principle, 16
Carpenter, TP, 34
Census at School New Zealand, 283–93
children
   who are gifted and talented, 313–14
   with English as additional language or dialect (EAL/D), 311–13
   with intellectual disabilities, 305–08
   with physical disabilities, 299–305
   with social, emotional and behavioural disabilities, 309–10
   with special needs, 297–99
circles, 79, 305
circumference, 79, 299
Classroom snapshots
   10-frames, 29–32
   addition, 125, 126, 306–08
   area, 72–73, 180–87
   assessment, 242–44, 245, 246–47, 256, 261
   circles, 305
   decimals, 150
   dynamic geometry software (DGS), 320–21
   equivalence, 171–72

estimation, 62–63, 74–75
games, 353
Indigenous students, 342–45
isometric projections, 95–96
length, 355
measurement, 368–69
multiplication, 135–37
numbers, 301–02
patterns, 169, 174–75, 176–78, 180–82, 313–14
place value, 272
probability, 220–21, 224–25, 230–33
proportional reasoning, 153–54, 155–56
quadrilaterals, 89–90, 101
statistics, 195–96, 198–99, 209–10
subtraction, 371–73
symmetry, 118–19
teachers, 317–18, 320
volume, 354
classrooms
  multicultural, 311–13
  multi-grade, 346–50
Clements, DH, 27
clocks, 59
clothes line activities, 22–23
cognitive guided instruction (CGI), 33–34
coin tosses, 229–30
Collis, KF, 251, 264
column graphs, 198
communities
  and Indigenous students, 345
  involving, 378–79
community of practice (CoP), 360, 370–71
commutative law, 125
comparing and ordering, 48
computation, 122
  in curriculum, 123, 129, 130, 135
Computer Algebra Systems (CAS), 105
concept maps, 334
congruent figures, 100
conservation, 67–68
constructivism, 15
content knowledge (CK), 4, 271, 292
Cornish, L, 346, 348

correlations, 218
Count Me in Too (CMIT) website, 17, 23, 24, 32, 65
counting
  apps, 22
  beyond 10, 23–26
  in curriculum, 20
  interactive resources, 17, 23, 154
  in other languages, 311
  rhymes and songs, 21–22
  sequencing, 22–23
counting principles, 16
Cowan, R, 135
crisis of thinking, 91
cross-sections, 96
cubes, 94, 95
Cuisenaire Rods, 124
cultural responsibility, 350
culture
  and geometry, 311–12
currency converters, 61
curriculum frameworks, 17–20
  and assessment, 241–42
  computation, 123, 129, 130, 135
  geometry, 83, 93, 99, 100, 107–08, 109–10
  inclusion, 295–97
  measurement, 51–52, 75–76
  numbers, 20, 33
  patterns and algebra, 160–61, 173, 179
  percentages, 152
  policy documents, 375–76
  probability, 217, 219
  statistics, 190, 191, 201
  technology, 268–69
Curry, Christina, 374

data, 190
  analysing, 200–08
  categorical, 196
  collection, 191, 194–99
  distribution, 208
  quantitative, 212
  reading, 232

data (*cont.*)
   recording, 196, 201
   representing, 197, 200–08, 211
   in social contexts, 212–13
   summarising. 339–41
   telling a story from, 209–12
data cards, 199
decimal number expanders, 148
decimals
   and fractions, 148
   and money, 151
   multiplying by 10, 150
   and probability, 234, 235
   understanding, 148–51
DeSimone, JR, 298
diagonals, 100
diamonds, 101
dice, 218–19
dictionaries, 299, 328
Digistore website, 132, 133, 212, 378
digital technologies, 268, 363
discontinuity, 92
dissect, 100
distribution, 208
distributive property, 135
division, 36, 37, 132–33
   grouping meaning, 132
   interactive resources, 133
   sharing meaning, 132
Dole, S, 35
doubling strategies, 29–30
drawing programs, 26
drill, 37
   and practice, 37–38, 135–37
dynamic geometry software (DGS), 105–07, 118
   lesson preparation, 320–21, 331–36

early number activities, 21–23
Early Years Learning Framework for Australia (EYLF), 18–19
edges, 93
education for sustainability (EfS), 341

Eight Ways approach, 342–43
English as additional language or dialect (EAL/D) students, 311–13
equality, 170–78
equations, 172, 179
equivalence, 146–47
   and equality, 170–78
estimation
   interactive resources, 65
   measurement, 46, 61–65, 74–75
Excel (spreadsheets), 197

faces, 93
factors, 131–32
First Steps Framework, 17
flip-books, 228
floor plans, 110
flowcharts, 334
formative assessment, 240
formulae, 51
   application of, 49
   area, 75–78, 180–87
   volume, 77
Fraction Estimation software, 277
fraction walls, 146–47
fractions
   as decimals and percentages, 152
   as operators, 148
   equivalent, 146–47
   interactive resources, 143–44, 146, 147, 277–83
   meanings of, 144–48
   and probability, 228, 234, 235
   understanding, 143–44
frequencies, 233
function machines, 169–70
functioning
   modes of, 251–53

Galbraith, P, 270
Gallistel, C, 16
games
   fractions, 146
   geometry, 104–05

percentages, 152
place value, 273–75
probability, 224
resourcing, 353
volume, 77
garden projects
edible, 341
Gelman, R, 16
general capability, 268
generalisations, 160, 178–87
geoboards, 71
GeoGebra (software), 105–07, 181
Geometer's Sketchpad, 105, 106
geometry, 83
apps, 117–19
concepts, 83–85
in classroom, 92–110
and culture, 311–12
in curriculum, 83, 93, 99, 100, 107–08, 109–10
interactive resources, 94, 96, 97, 102, 104–05, 114–17
issues to think about, 329–30
outside classroom, 85
van Hiele teaching phases, 112–17
van Hiele theory, 86–92, 331
See also dynamic geometry software (DGS)
GeoNet website, 236
gifted students, 313–14
Gigerenzer, G, 218
glossary, 382–87
Goos, M, 44, 270
Gould, P, 144
Graphing Stories website, 211
graphs, 74, 197, 202–03, 205, 208, 211, 232
group interactions, 310
Guerrero, S, 363, 364
Guinness Book of Records website, 62, 63

hand prints, 53
Healthy Life Survey, 194
hearing impairment, 302–03
hefting, 66

homework, 327
HOTmaths tools, 4
addition, 36, 124–25, 126, 127, 128, 129, 306–08
algebra, 323–26
angles, 78
area, 76
assessment items banks, 262–65
counting, 154
described, 7–8
dictionaries, 328
division, 133
fractions, 143–44, 278–83
functions, 155
geometry, 94, 96, 97, 102, 104–05, 114–17
hundreds square, 24, 41, 167, 174, 273
measurement, 66, 68, 69, 348
multiplication, 39, 40, 131, 132
numbers, 26, 29, 43, 301, 330
patterns, 155, 175, 178, 182–83, 186, 187
percentages, 152, 326
place value, 41
probability, 222, 226, 227, 229, 232
questions, 323
Roman numerals, 311
and sound, 309
statistics, 195–96, 202, 204–06, 211
time, 60, 80
volume, 77, 331
Howell, J, 380
hundreds square, 23–26, 165–68, 272
interactive resources, 24, 25–26, 41, 167, 174, 273

ICT Games website, 68
Illuminations website, 152, 172
area, 76
factors, 132
numbers, 26
probability, 224, 229, 236
and sound, 309
statistics, 206
inclusion
in curriculum, 295–97

independent events, 219–20
Indigenous students, 342–45
inferences, 209
informal units, 65
information and communication technologies (ICT).
    case study, 283–93
    in classroom, 276–78
    collating resources, 377–78
    community of practice (CoP), 370–71
    involving community, 378–79
    leading teaching of, 369, 374, 380–81
    policy documents, 375–76
    professional development, 374–75
    skills audits, 370–74
    technology metaphors, 270
    use of, 18
    website evaluation, 376
    See also technology use
innovation, 380
intellectual disabilities, 305–08
interactive resources
    addition, 36, 124–25, 126, 127, 128, 129, 306–08
    algebra, 323–26
    angles, 56–57, 78
    area, 71, 76
    collating, 377–78
    counting, 17, 23, 154
    division, 133
    estimation, 65
    fractions, 143–44, 146, 147, 277–83
    geometry, 94, 96, 97, 102, 104–05, 114–17
    hundreds square, 24, 25–26, 41, 167, 174, 273
    measurement, 66, 68–70, 348
    multiplication, 39, 40, 131–32
    number charts, 25–26
    numbers, 26, 29, 32, 38, 43, 259, 260, 301–02, 330
    patterns, 155, 162–64, 175–76, 178, 182–83, 186–87, 260
    percentages, 152–53, 326
    place value, 41
    probability, 222, 226, 227, 228–29, 232, 236
    proportional reasoning, 155–57
    statistics, 195–96, 202, 204–06, 211, 212, 312
    subtraction, 371–73
    time, 59–60, 80
    volume, 77, 331
interactive whiteboards (IWB), 7
    10-frames, 29–32
    geometry, 97–99, 118–19
    patterns, 176–78
    sharing/discussion, 127
    skills audits, 371–73
    See also Classroom snapshots
interquartile range, 191
isometric projections, 95–96

Jones, T, 350

Kahootit, 155–56
Karp, KS, 37, 53, 56, 59
Khan Academy website, 79
    decimals, 148
Kid Pix (drawing program), 26
Koehler, Matthew J, 7, 360, 380
Krieger Science website, 81

laggards, 380
language
    appropriate, 298
    and levels of thinking, 90, 92
    of mathematics, 101, 144, 191, 299, 303
    other than English, 311
    of probability, 220–23, 227
Lave, J, 369
leadership, 369, 374, 380–81
learning disabilities, 305–08
Learning Federation, 162–64
learning frameworks
    geometry, 86–92
    numbers, 16–17
Learning Today website, 172
length, 52, 355

Length Strength: Centimeters website, 70
lesson preparation, 367–69
  categories to assist, 322
  conclusion to lesson, 327
  dynamic geometry software (DGS), 320–21, 331–36
  homework setting/correcting, 327
  introduction to lesson, 323
  sharing/discussion, 327
  start of lesson, 323
  student tasks, 323–26
  teaching sequence, 331–36
levels of thinking
  crisis of thinking, 91
  hierarchical nature, 90
  and language, 90, 92
  level reduction, 91
  progression, 92
  van Hiele theory, 86
likelihood, 220
Lilburn, P, 258
Linear Arithmetic Blocks, 149
literature
  use of, 109
Livy, S, 71
location, 108–10

Ma, L, 71
MacGregor, M, 165, 170
Maher, N, 71
maps, 213
mass, 58, 69
matchstick activity, 180–82
mathematical content knowledge, 318
Mathematics Association of New South Wales (MANSW), 319
Maths300 website, 95, 96, 228, 277, 331
Matthews, Chris, 342
McIntosh, A, 35, 122
mean, 191, 204
measurement, 46
  concepts, 52–61
  concepts into older years, 74–81
  in curriculum, 51–52, 75–76
  estimation, 46, 61–65, 74–75
  interactive resources, 66, 68–70, 348
  issues to think about, 331
  learning sequence, 47–50
  misconceptions and difficulties, 67–73
  with non-standard units, 54, 56, 58, 61, 65–67
  teaching, 368–69
  media, 212
median, 191, 204
mental computation strategies, 35
mentors, 319
merge problems, 123
metric units, 51
Mishra, Punya, 7, 360, 380
Mitchelmore, M, 144
Mitra, Sugata, 375
mode, 191, 204
money, 51, 61, 300
  and decimals, 151
Muir, T, 71
Mulligan, J, 166
multi-age classes, 348
multicultural classrooms, 311–13
multi-grade classes, 346–50
multiplication, 36, 129–33
  and addition, 36–37, 39, 130
  as arrays, 40, 49, 130–31
  interactive resources, 39, 40, 131–32
  online resources, 132
  repeated equivalent groups, 131
  as scaled up, 129–30
  times tables, 130
  with larger numbers, 135–37
multiplicative thinking, 122, 141–42, 155
Murphy's Law, 229

National Assessment Program – Literacy and Numeracy (NAPLAN), 254–58
National Council of Teachers of Mathematics (NCTM), 276, 362
National Geographic website, 213
National Library of Virtual Manipulatives, 71, 152

National Rural Fire Authority, 236
national testing, 254–62, 350
Neal, D, 142
nets, 94
New Zealand, 340
New Zealand Association of Mathematics
    Teachers (NZAMT), 319
New Zealand Mathematics Curriculum
    (NZmaths), 18, 19
  and assessment, 241, 242
  computation, 123, 129, 130, 135
  counting skills, 20
  geometry, 83, 93, 99, 108, 109
  inclusion, 296
  measurement, 51
  patterns and algebra, 161, 173
  probability, 217
  statistics, 190, 191, 201
  strands, 8
  technology, 269
  technology use, 363
  van Hiele theory, 86
New Zealand Number Framework, 17
Newman, Anne, 248
Newman's Error Analysis, 248–49
non-standard units, 48
  use of, 54, 56, 58, 61, 65–67
nrich website, 124, 167, 310
  statistics, 208, 211
number bars, 124, 126
number charts, 23–26, 165–68, 272
  interactive resources, 25–26
number frameworks, 16–17
number lines, 127, 149
numbers.
  in curriculum, 20, 33
  early concepts, 15–16
  early operations with, 33–44
  interactive resources, 26, 29, 32, 38, 43,
    259, 260, 301–02, 330
  issues to think about, 330
  linking with quantities, 26
  operations with whole, 123–37
  part-whole, 143–57
  patterns with, 165–68, 169, 313–14
  place value, 41–44, 122, 127, 148, 272
  recall of number facts, 135
  See also hundreds square, counting
numerals, 23
Nunes, T, 217
nursery rhymes, 21–22

odds, 217, 234–35
one-to-one principle, 16, 21
open-ended activities, 253, 258
order irrelevance principle, 16
ordering, 48
ordinal numbers, 23
outcomes, 217, 218, 223–25, 234–36
Outhred, L, 144
outliers, 204

parallel lines, 91
parallelograms, 100
parents
  involving, 378–79
Parkinson, John, 374
Parmar, RS, 298
part-part-whole, 16, 27, 32
  10-frames, 29–32
part-whole numbers, 143–57
parts-of-a-whole problems, 124
pattern-recognition activities, 27, 155
patterns
  in curriculum, 160–61, 173, 179
  in early years, 162–64
  growing, 164–65
  interactive resources, 155, 162–64,
    175–76, 178, 182–83, 186–87, 260
  issues to think about, 330
  recognising, 173–75
  and rules, 175–78
  and structure, 161–62
  and transformations, 172
  with numbers, 165–68, 169, 313–14
pedagogical content knowledge (PCK), 6,
    271, 292, 318
pedagogical knowledge (PK), 6, 271, 292

percentages, 151–53
  in curriculum, 152
  as fractions, 152
  interactive resources, 152–53, 326
  and probability, 234, 235
perimeter, 71, 299
physical disabilities, 299–305
pi, 79
pictograms, 201
pie charts, 198, 202
Pierce, R, 268
place value, 41, 122, 127
  games, 273–75
  interactive resources, 41
  numbers less than 1, 148
  and technology use, 41–44
  understanding, 271–75
play dough, 58
playing cards, 49
policy documents, 375–76
populations, 191
PowerPoint, 27
PPDAC cycle, 191–92
  analyse, 200–08
  conclusions, 209–12
  plan, data, 194–99
  problems, 193–94
practice, 37
  illustrations of, 365–66
  See also drill and practice
prisms, 96
probability
  conditional, 236
  in curriculum, 217, 219
  early primary years, 220–22
  and fractions, 228, 234, 235
  games, 224
  importance of, 216–17
  interactive resources, 222, 226, 227, 228–29, 232, 236
  issues to think about, 328–29
  language of, 220–23, 227
  middle primary years, 222–26
  quantifying, 217

representing, 234–36
  social aspects, 216–17
  understanding, 217–20
  upper primary years, 226–34
Probability Explorer (software), 227
problem-solving, 191
  asking questions, 193–94
  understanding questions, 248–50
professional associations, 319
professional development, 374–75
  dynamic geometry software
    (DGS), 320–21
  mentors, 319
  remote areas, 350
Progressive Achievement Tests (PAT), 258
proportional reasoning, 141, 153–57
  interactive resources, 155–57
protractors, 56
pyramids, 96, 97–99

quadrilaterals, 89–90, 100, 101–04
quantitative data, 212
questions
  asking, 193–94
  Blank's questioning framework,
    249–50
  and children from other cultures, 312
  Newman's Error Analysis, 248–49
  start of lesson, 323

random generators, 229
randomness, 217, 218, 226
range, 191, 204
ratios, 145
relationships
  understanding, 168–70
relative frequencies, 233
remote areas, 339–40
  multi-grade classes, 346–50
  professional development, 350
  resourcing, 351–57
report writing, 211
resources, 3, 351–57
  See also interactive resources

responses
    quality of student, 246–54
rhymes, 21–22
risk, 235
Rogers, EM, 380
Roman numerals, 311
Romberg, TA, 264
routines, 309
rulers
    broken, 70–71
rural areas, 339–40
    See also remote areas
Ryan, J, 68

sample space, 217
samples, 206, 233
scales, 58, 60, 172
    making, 80
    reading, 68
school communities
    involving, 378–79
Scootle website, 132, 133, 193, 212
sequencing activities, 22–23
Serow, P, 113, 331
Shulman, LS, 4, 6, 360
Siemon, D, 15, 33, 41, 142
simulation, 219
    activities, 228–34
skills audits, 370–74
SOLO model, 251–53
songs, 21–22
sound, 309
spatial awareness, 46, 108
spinners, 229, 230–34
spreadsheets, 190, 197
stable order principle, 16, 21
Stacey, K, 165, 170, 268
standard deviation, 191
standard units, 49
standards, 365–66
statistics.
    Census at School New Zealand, 283–93
    in curriculum, 190, 191, 201
    importance of, 190–92
    interactive resources, 195–96, 202, 204–06, 211, 212, 312
    issues to think about, 328–29
    and media, 212
    resourcing, 355–57
    See also PPDAC cycle, data
stem-and-leaf plots, 208
Stillman, G, 44
students.
    Indigenous, 342–45
    quality of responses, 246–54
    talented, 313–14
    See also children
subitising, 26–29
subtraction, 33–34, 126–29
    interactive resources, 371–73
    mental computation strategies, 35
Sullivan, P, 258, 318
surveys, 234
symmetry, 118–19

tactile materials, 303–04
talented students, 313–14
tally marks, 196
tangrams, 53–54, 276
Te Whāriki, 18, 19
teacher aides, 297
Teacher Led website, 56
teachers
    absenteeism, 346
    being flexible, 340–45
    beliefs of, 298–99
    information and communication technologies (ICT) skills, 370–74, 380–81
    professional development, 374–75
    shortage of secondary mathematics, 317–21
    turnover, 341
teachers of mathematics
    and mentors, 319
    out of field, 317–18, 319, 320

teaching mathematics.
  culturally responsible, 350
  engaging students, 319
  issues to think about, 328–31
  lesson preparation, 322–28, 367–69
  multi-grade classes, 346–50
  and pedagogical content knowledge (PCK), 318
  resourcing, 351–57
  teaching sequence, 331–36
  with technology, 367–69
  See also Classroom snapshots
teaching phases
  geometry, 112–17
technological content knowledge (TCK), 6, 271, 291
technological knowledge (TK), 4, 291
technological pedagogical content knowledge (TPACK) framework
  case study, 291–93
  components, 4–9, 271
  described, 360–62
  development and use of technology, 364
  in mathematics classroom, 362–66
technological pedagogical knowledge (TPK), 6, 271, 291
technology
  in curriculum, 268–69
  development and use of, 364
  impact of, 363
  influence of, 362
  metaphors for, 270
  teaching mathematics with, 367–69
technology use.
  benefits of, 268
  and children with physical disabilities, 300
  and children with social difficulties, 309–10
  in classroom, 12
  and drill and practice, 38
  multiplication, 135–37
  and place value, 41–44
  proportional reasoning, 155–56

  and sound, 309
  statistics, 190
  See also websites, calculators, interactive resources
temperature, 60
tessellations, 110
Thomas, N, 166
three-dimensional objects, 84, 92, 93–99
time, 51, 58–59
  interactive resources, 59–60, 80
time zones, 80
times tables, 130
TinkerPlots (software), 6
  measurement, 74–75
  statistics, 209–10
  use of, 284
transformations, 84, 108–10, 172
tree diagrams, 229
triangles, 78–79
two-dimensional figures, 84, 100–07

uncertainty, 220–23
understanding, 92
units
  metric, 51
  non-standard, 48, 54, 56, 58, 61, 65–67
  standard, 49

Vale, C, 44
value, 61
Van de Walle, JA, 37, 53, 56, 59
van Hiele, PM, 92
van Hiele teaching phases, 112–17, 331
van Hiele theory, 86–92, 331
van Hiele-Geldof, Dina, 112
variables, 169
variation, 190, 218
vertices, 93
videos, 211
virtual characters, 309
virtual manipulatives, 43
vision impairment, 300, 303–05
visualisation, 84
volcanoes, 236

volume (*cont.*)
volume, 54–55
    formulae, 77
    interactive resources, 77, 331
    resourcing, 354

Watson, JM, 218, 328
websites
    and drill and practice, 38

evaluation, 376
weight, 58
Wenger, E, 369
whole number operations, 123–37
Williams, J, 68
Wright, RJ, 17

Yunkaporta, Tyson, 342